HOW TO ESTIMATE
BUILDING LOSSES
AND CONSTRUCTION COSTS

Fourth Edition

HOW TO ESTIMATE
BUILDING LOSSES
AND CONSTRUCTION COSTS

Fourth Edition

Paul I. Thomas

Prentice-Hall, Inc.

Englewood Cliffs, New Jersey

Prentice-Hall International, Inc., *London*
Prentice-Hall of Australia, Pty. Ltd., *Sydney*
Prentice-Hall of Canada, Inc., *Toronto*
Prentice-Hall of India Private Ltd., *New Delhi*
Prentice-Hall of Japan, Inc., *Tokyo*
Prentice-Hall of Southeast Asia Pte. Ltd., *Singapore*
Whitehall Books, Ltd., *Wellington, New Zealand*
Editora Prentice-Hall do Brazil Ltda., *Rio de Janeiro*

© 1983, 1976, 1971, 1960 *by*

Prentice-Hall, Inc.
Englewood Cliffs, N.J.

Fifth Printing . . . . . March 1987

Library of Congress Cataloging in Publication Data
Thomas, Paul I.
  How to estimate building losses and construction costs.

  Bibliography: p.
  Includes index.
  1. Building—Estimates.   2. Buildings—Repair and
reconstruction.   I. Title.
TH 435.T45   1983   692'.5   83–3243
ISBN 0-13-405902-6

Printed in the United States of America

In Memory of
My Mother and Father

Preface to the Fourth Edition

Since the Third Edition in 1976, there have been considerable changes in building construction materials and their costs—in estimating methods such as using computerized estimates—and in claim handling procedures and requirements. To keep abreast of these events, this text has been carefully updated. A glance through the Table of Contents will disclose the many revisions and additions.

Two new chapters have been added at the urging of a great many who use the book. (1) Fires of incendiary origin, in particular arson-for-profit, have increased more than 700 percent since 1964, according to the National Fire Protection Association. The importance of this increase is evidenced by the growing number of municipal, state and Federal agencies which, in cooperation with the insurance Industry, are now working to curtail this crime. Estimators of building losses are in an excellent position, when making a careful and thorough inspection of the premises, to observe and note "arson indicators" that may have been overlooked by others. Chapter 6, Estimating Losses of Suspicious Origin, outlines the estimator's role in these situations. Adequate information is provided as guidelines on *what to look for.*

(2) Condominiums are rapidly becoming major residential and commercial, multiple-occupancy type buildings, both in the initial construction and in the conversion of existing hotels, apartments and office buildings. Chapter 22, Estimating Condominium Losses, has been added to acquaint the estimator with the unique concept of the property covered under a unit owner's insurance policy and property covered under the insurance on common facilities, areas and services. The different types of condominiums are explained and also the importance of the principal legal documents as they relate to interests, value and loss.

Today it is essential to recognize that women are an important part of the business world, and this includes engineering and the insurance claim function. This edition has been updated by deleting all reference to sexism.

An effort has been made to delete unnecessary repetition where cross-reference accomplishes the purpose. A few subjects have been eliminated entirely as they are considered not essential to the objectives of the book. This includes Lumber Industry Abbreviations and Extracting Square Roots by formula. Square roots are readily obtainable from the Table or, as has been suggested, by pocket computer.

Twenty-two new Tables have been added so that quantities of materials may be read directly. These include:

(1) FBM of framing in gable and hip roofs with pitches 1/8 to 1/2; 2" × 4" and 2" × 6" rafters 16", 20" and 24" on center, for *ground-floor* areas of 200 to 3,000 square feet.

(2) FBM of joists $2'' \times 4''$ to $2'' \times 12''$, with centers $16''$, $20''$ and $24''$, for *ground-floor* areas 200 to 3,000 square feet.

(3) FBM and square feet of rough and finish floor materials for floor areas that are 200 to 3,000 square feet, including milling and cutting waste.

(4) Roof decking for gable and hip roofs with pitches 1/8 to 1/2, for *ground-floor* areas of 200 to 3,000 square feet.

(5) FBM in trusses with pitches 1/8 to 1/2, and $2'' \times 4''$ and $2'' \times 6''$ rafters, for *ground-floor* areas from 200 to 3,000 square feet.

(6) FBM and square feet of roof decking of various kinds, where roof pitches are 1/8 to 1/2, and *ground-floor* areas are 200 to 3,000 square feet.

(7) Number of cubic yards of concrete in walls 6 to 12 inches thick and 1 to 12 feet high.

(8) Number of cubic yards of concrete, base, and topping in slabs, walks and patios 200 to 3,000 square feet. A companion Table provides similar data for slabs, walks, and patios of small areas from 20 to 500 square feet.

(9) Number of concrete blocks, $8'' \times 8'' \times 16''$, per lineal foot of wall 8 to 120 inches high.

(10) Number of "standard" brick per square foot in walls 4 to 24 inches thick, with mortar joints 1/4 to 3/4 inches.

Several new formulas are included for quickly estimating quantities of materials. All illustrative examples showing material and labor have been updated in prices and wages to put them more in line with current conditions.

The pros and cons of "computerized estimating" are described and commented on. Also, the fine points of newer building materials are covered, including vinyl, hardboard, and aluminum siding; aluminum shakes and fiberglass shingles. The principle of Modular construction is explained. The National Flood Insurance Program has been brought up to date with the concurrence of General Adjuster, John E. McCormick.

The Table of Union Hourly Wage Scales for Major Building Trades has been deleted in this Fourth Edition because the average annual increases of 8 to 10 percent make it less than reliable as a basis for estimating current construction costs. The best source, as noted in Chapter 4, is the local prevailing wage rate.

45 WAYS THIS BOOK CAN HELP YOU

1. How do fire, heat and smoke affect building materials? What is *cold smoke* as distinguished from *hot smoke*? What effect does water have on building equipment and construction materials; water from sprinklers, fire hoses and from flood waters? What is the cause of mudslides? What steps can be taken to minimize damage in the early stages? What are the characteristics of tornadoes; of hurricanes? These and other causes of loss are discussed in Chapter 1.

2. There are 16 basic factors to be considered in estimating hours of labor. See Chapter 2.

3. There are 30 overhead items that are chargeable to specific job operations. There are 15 overhead items that are not chargeable to a specific operation. See Chapter 2.

4. What are the latest requirements of OSHA? What is Modular house construction? See Chapter 2.

5. When estimating the cost of demolition and removal of debris, there are 30 basic considerations. See Chapter 2.

6. Protecting property from further loss after a building has been damaged usually requires at least 20 considerations. See Chapter 3.

7. Previously existing damage may be discovered when taking off the details of a building loss. There are at least 10 common types of deterioration in buildings and 15 causes that contribute to such deterioration. See Chapter 3.

8. See how five outstanding and experienced building contractors set up their final estimate of a building loss. What is computerized estimating? How good is it? See Chapter 4.

9. How does estimating building losses differ when working in major catastrophes? What conditions can you expect to encounter in major catastrophes? What is the National Flood Insurance Program (NFIP); how does it operate? See Chapter 5.

10. There are seven Insurance Industry activities with long-range programs for curtailing and controlling the crime of Arson. What is the Arson Reporting-Immunity Bill? What is the Arson Resources Directory? See Chapter 6.

11. Building estimators can play a special role in building losses of suspicious origin. There are fundamental guidelines for observing and identifying evidence when the crime of arson is suspected. See Chapter 6.

12. What is the definition of Arson? What are the three essential elements needed to defend an action on an insurance policy when arson is suspected? What is *direct* evidence; what is *indirect* or circumstantial evidence? See Chapter 6.

13. This book lists many accidental causes of fire that must be eliminated before concluding that a fire is of incendiary origin. What are some of the arson-indicators to watch for? The book lists over 40 *motives* for incendiary fires. See Chapter 6.

14. Removing and preserving evidence of arson must be left to the proper authorities. See Chapter 6.

15. Is your arithmetic a little rusty? Do you need to refer to Metric Conversion Tables now and then? See Chapter 7.

16. Want to know how to estimate materials and labor to build, erect and strip forms for various concrete shapes such as walks, slabs, columns, and walls? There are concrete mixes for specific purposes. See Chapter 8.

17. The cubic yards of concrete in footings, foundations, columns, slabs and walls can be estimated simply by multiplying their cross-section by a factor such as .0031 for slabs and walls. Or, use one of the three Tables provided and read the number of cubic yards *directly*. See Chapter 8.

18. The labor to mix and place concrete, put in the reinforcing and finish the topping is shown in Chapter 8.

19. The damageability of concrete and methods of repair are discussed in Chapter 8.

20. Mixing mortar for various masonry units; how to estimate the quantity of mortar needed, and how to figure the labor to lay up masonry units are explained in Chapter 9.

21. How, and to what extent, masonry units, brick cement block, etc. can be injured in a fire is discussed in Chapter 9.

22. Studs and framing lumber in small quantities is sold in many areas by the piece or stick. But in quantity it is still sold at a price-per-thousand board feet (FBM). Learn to figure board feet; it's as easy as rolling off a log. Chapter 10 shows a Table that gives the number of board feet in framing lumber of various dimensions and lengths. Or you can use a companion table that provides a *multiplier* (factor) by which the length of any piece of framing lumber can be multiplied to obtain the number of board feet.

23. Do you know the Standard Sizes of Lumber, i.e. the *nominal* size as a 2″ × 4″, and the *actual* delivered size, which is 1.5″ × 3.5″? What do we mean by milling waste? Want to estimate the *number* of studs, joists, and rafters on various centers, in walls, floors, and roofs 8 to 36 feet long? See Chapter 10.

24. By a simple formula you can estimate the number of board feet of 2″ × 4″ and 2″ × 6″ studding on various centers in walls and partitions of any length and height. See Chapter 10.

25. To obtain the lineal feet of cross bridging cut on the job, multiply the length of the room by a factor shown in Chapter 10.

26. Chapter 10 also shows how to estimate the number of board feet of joists 2″ × 4″ to 2″ × 12″ on various centers, simply by multiplying the area of the floor by a factor such as .85 for 2″ × 6″, 16″ on center. End joists and cutting and fitting are included in the factor. Or, you can read directly from a Table for floor areas of 200 to 3,000 square feet.

27. The FBM of framing for gable and hip roofs may be read directly from Tables in Chapter 10, where 2″ × 4″ and 2″ × 6″ rafters are used, 16″ and 24″ on center, and roof pitches are 1/8 to 1/2. Or use the *ground-floor* area multipliers shown in another Table, same Chapter.

28. The material for roof trusses can be read directly from a Table in Chapter 10. All you need to know is whether the rafters are 2″ × 4″ or 2″ × 6″, and what the pitch of the roof is. Or, use the Table that gives the *ground-floor* area multipliers.

29. Average number of hours to install rough carpentry and how to estimate costs are discussed in Chapter 10.

30. Quantity of various kinds of roof decking can be read *directly* from Tables in Chapter 10, when the pitch and *ground-floor* area are known. Quantities of rough and finish floor may also be read directly from a Table for areas of 200 to 3,000 square feet.

31. How to lay strip oak flooring on a concrete base. All you want to know about finish flooring. See Chapter 11.

32. Want to know how to estimate windows, doors, storm sash and screens? How about interior trim such as baseboards, chair rails, ceiling moldings, etc.? See Chapter 11.

33. How is rough and finish carpentry damaged by fire and water? What are the choices for repair? See Chapter 11.

34. Drywall construction is the most common kind of interior wall finish. Learn all about its application, and the methods of estimating labor and material. See Chapter 12.

35. Learn to estimate the labor and material to install composition wallboard, plywood and ceiling tile in Chapter 12.

36. Want to know more about wood siding, composition siding, aluminum siding, vinyl siding and hardboard siding? See Chapter 13.

37. There are seven important considerations when estimating labor to apply roofing. Each is discussed in Chapter 14. If you cannot get on a roof to measure, or it has been destroyed, see Chapter 14 for a few suggestions.

38. Learn how to compute the area of any shape roof, and how to estimate the materials and labor to apply various kinds of roofing. See Chapter 14.

39. Just what are fiberglass shingles? What is corrugated roofing and siding? See Chapter 14.

40. You can read quantities of roofing for gable and hip roofs directly from Tables when the pitch is known and the *ground-floor* area can be determined. See Chapter 14.

41. What effect have fire, hail and wind on roofing? How are repairs made? See Chapter 14.

42. Want to know the covering capacity of paint materials; the methods of estimating painting; how to estimate the amount of paint for stairways, a lattice or cornice? See Chapter 15.

43. Use the Tables in Chapter 15 to read directly the wall and ceiling areas of rooms measuring 3' to 24' by 3' to 22' and floor heights 8' to 10'. In the same chapter a Table shows the number of rolls of paper for walls and ceilings in rooms 4' × 8' to 20' × 24' and room heights 8', 9' and 10'.

44. There are six methods of electric wiring. Determining the number of branch circuits is not difficult. See Chapter 19.

45. When estimating losses in condominiums it is important to understand the meaning of Master Deed or Declaration, Unit Deed, Bylaws, and Regulations. Also, what is meant by the *cube of space concept*. See Chapter 22.

ACKNOWLEDGMENTS

The author gratefully acknowledges the assistance and the contributions of many who helped to update and round-out this Fourth Edition. Special thanks go to the contractors submitting exibits of building loss estimates in Chapter 4: John Gunn, of Gunn and Gunn Construction Company in Miami, Florida; Robert B. Rogers III, of N. Frank Vought, Inc., New York City; Dewey C. White, of White Repair and Contracting Company, Atlanta, Georgia; George A. Pasqualucci, A.S.A., Boston, Mass.; and Robert L. Driver, of Driver-Eddy Construction Co., Santa Ana, California.

Also, special thanks to David T. Clayton, President, and staff, of Insurance Technical Training Institute (ITTI), Dallas, Texas; particularly to Joseph Blazek, Director of Property Programs, and George Epps, Property Instructor, both of the Insurance Technical Training Institute; Richard H. Greene, former Second Vice President of the Maryland, Baltimore, Md. currently with AFIA, London; P. Ken Brewer, Manager of Training and Education of Aetna Life & Casualty, Hartford, Conn.; and Herbert Larzelere, General Adjuster, Kemper Insurance, Summit, N.J., who took time to write suggestions and comments which were most helpful. Others who responded to a questionnaire and offered constructive suggestions include Ralph J. Deucker, Secretary of Claims, Atlantic Companies, New York City; George Guinane, Consultant, Bloomington, Ill; Harry R. Hanson, Property Claims Officer, Kemper Insurance, Long Grove, Ill.; John Hardman, General Adjuster S.E. Division, Kemper Insurance, Atlanta, Ga.; E.L. Lecomte, President, National Committee on Property Insurance, Boston, Mass.; John E. McCormick, General Adjuster, National Flood Insurance Program, Bethesda, Md.; Christopher Pitt, CPCU, Assistant Director, Legal and Claims Education, Insurance Institute of America, Malvern, Penn.; R.E. Pratt, R.E. Pratt & Company, Winston-Salem, N.C.; Phil G. Simkins, Firemans Fund Insurance Companies, San Francisco, Calif.; John Snyder, President of Marvin F. Bank & Company, Minneapolis, Minn.; and Robert Stanhope, Director of Vale Labs, Vale National, Chambersburg, Penn.

The following Industry Organizations were kind enough to furnish data or exibits used throughout the book: Andersen Corporation, Bayport, Minn.; Goldblatt Tool Company, Kansas City, Kansas; Independent Nail Inc., Bridgewater, Mass.; Patent Scaffolding Company, Fort Lee, N.J.; Portland Cement Association, Skokie, Ill., Porta Tools, Inc., Wilmington, N.C. 28401; Senco Products, Inc., Cincinnati, Ohio; Sumner Rider and Associates, New York, N.Y.; Underwriters Laboratories, Inc., Chicago, Ill.; Certain Teed Corporation, Valley Forge, Pa., and Hunter Douglas Inc., Technical Assistance Department, Durham, N.C.

Paul I. Thomas

Contents

HOW TO ESTIMATE
BUILDING LOSSES
AND CONSTRUCTION COSTS

Fourth Edition

CHAPTER 1

Causes of Damage to Buildings

Persons who wish to learn how to estimate the cost of restoring damaged buildings or other structures should familiarize themselves with the common causes of damage. They should understand how these causes, or perils, affect different materials of construction.

Policies of insurance that cover property damage may be divided into two broad types. The first group insures against named perils, while the second group insures against all risks of physical loss, with specifically named exclusions. The most frequently encountered causes of damage under both groups are: (1) Actual burning during a fire; (2) Heat; (3) Smoke; (4) Water; (5) Activity of Firefighters; (6) Chemicals; (7) Falling Debris; (8) Explosion; (9) Lightning; (10) Windstorm; (11) Hailstones; (12) Aircraft and Vehicle; (13) Sprinkler Leakage; (14) Flood, Backing of sewers, Seepage, etc.; (15) Sonic Boom; (16) Earthquake; (17) Miscellaneous Causes.

ACTUAL BURNING DURING FIRE

Actual ignition and burning of building materials, as distinguished from exposure to heat, is confined to the combustible materials. These are mainly wood and wood products, paints, asphaltic types of roofing and siding, electrical insulation and some types of composition wall board and insulation. Masonry, plaster, steel and other metals do not support combustion, but are subject to damage or complete destruction by heat from the burning of the building or its contents. The burning of paint, roofing materials, and insulation usually will require complete replacement of the part or area involved.

Wood is combustible, and its rate of burning varies with the density of the species, the moisture content, the bulk or size of the piece, the rate at which heat is supplied, and the amount of air available. Cedar ignites at about 375° F. while long leaf yellow pine ignites at approximately 425° F. Most woods will ignite within this range.

When framing lumber or large timbers have been subjected to burning, it is not always necessary to replace them. Much will depend on whether they have been structurally damaged, and on the importance of appearance. There are many instances when slightly charred joists, rafters, or studding, have been left in place. They may be scraped, whitewashed or painted to seal any odor, and then covered with sheetrock or some other type of wall board. Heavy girders in a dwelling or large timbers in mill-type buildings may be scraped and painted or boxed in, where inspection reveals no important structural damage. Surfaces of millwork, however, such as stairways, cabinets, built-in features, trim, or finished floors which have any degree of charring, in almost every case must be removed and replaced.

HEAT FROM FIRE

The extent of damage to building materials from the heat that is generated by the combustible parts of the building, but more significantly by the burning of the contents, depends on the temperatures reached and on the length of time the materials are subjected to these temperatures. Brick, stonework, structural steel, sheet metal and iron expand in the presence of heat. Lateral expansion of steel girders during a fire has many times caused them to push out exterior masonry walls. Light, unprotected steel trusses, columns or beams will soften and lose their tensile strength. They may sag or collapse even during moderate fires. Most types of masonry will withstand high temperatures for a long period before sustaining structural damage. Spalling of masonry is a condition brought about by long exposure to high temperatures or alternate heating then cooling by water whereby the surfaces exposed "pop-off" in chunks or scales. The thicker the masonry, the less likelihood there is of major structural damage from heat.

Heat causes glass to crack at relatively low temperatures, and at about 1600° F. it softens. Plaster tolerates fairly high temperatures before flaking and chalking on the surface, with a change in its chemical composition.

Paint softens and blisters in the presence of heat. Most woods shrink slightly as they lose moisture, and boards or timbers of wider dimension frequently check. Wood ignites at approximately 400° F. after 20 minutes exposure.

While it is impractical to record either the temperatures that are developed or their duration in fires of accidental origin, it is sometimes possible to estimate the degree of heat by observing the physical condition of glass, metals, and ceramics in the area after the fire. The following table lists some of the melting or fusing temperatures of materials in buildings.

APPROXIMATE FUSING TEMPERATURES

Material	F temperature
Tin	449°
Lead	621°
Zinc	786°
Aluminum	1217°
Copper	1981°
Iron (Cast)	2250°
Sand	2588°
Porcelain	2588°
Steel (structural)	2606°
Chromium	2750°

Heat rises during a fire, not only within a room, but also up stairways, elevator shafts, and other areas or vertical openings where its passage meets the least resistance. For that reason, concrete floors on the top of which fire is burning usually show very little and sometimes no damage in fires that may have caused serious loss to the walls and roof. Ceilings in rooms may be seriously injured by heat, and joists may be heavily charred, while the lower regions often escape with hardly more than discoloration of paint, or a blistering of the finish on wooden floors. "Look high for heat damage" is an expression that should be remembered.

SMOKE FROM FIRE

Smoke is the result of incomplete combustion. Heavy smoke develops when burning materials are shut off from adequate oxygen, preventing complete combustion.

There are two kinds of smoke that can occur during a fire. One is a *hot smoke*, the other is a cool or *cold smoke*. Hot smoke is that which is relatively close to the fire and carries the heat of the fire with it. Hot smoke causes the greater damage because it penetrates porous surfaces and, in contact with cool surfaces, it condenses, leaving heavy stains of tars and resins. Building decorations are readily damaged by such stains and require washing or redecorating. Wallpapers touched by hot smoke generally need replacing. Window glass becomes coated and must be washed. Ceramic tile normally has hairline cracks that are almost invisible and, under proper conditions, the hot smoke penetrates through the cracks in the glazed surface to the porous backing. This stain is impossible to remove and the tile must be replaced, or an allowance should be made for the damage. Enameled and porcelain plumbing fixtures will suffer less than ceramic tile. Test cleaning is the best way to determine whether there has been any permanent injury.

Stonework, brick, or cement surfaces are difficult and sometimes impossible to clean, once they have been exposed to hot smoke. Experts are frequently called in to test clean before an estimate of damage is made. Usually, if cleaning or sandblasting is impossible, cement washing, coating, or painting may be satisfactory to all interests.

Cold smoke is smoke that is carried some distance from the fire and has cooled and condensed in the air before it reaches the area or surfaces on which it deposits. Cold smoke is often in the form of soot. Vacuum cleaning or washing many times will restore a surface to its former condition. Cold smoke does not leave the characteristic pungent smoke odor that results from hot smoke.

Smoke odor in a building can be removed in nearly every case by specialists offering smoke odor removal service. They maintain service offices throughout the country. This treatment is particularly effective in removing smoke odor from inaccessible areas, under floors, or behind plastered walls and ceilings. Frequently exposed framing, sheathing, and like areas that are charred can be completely deodorized through the use of special liquids.

WATER

Water from automatic sprinkler systems and fire hoses, or from rain or melted snow entering through windows, doors, or openings in roofs and walls after a fire, hail, or windstorm, causes varying degrees of damage to finished millwork, flooring, plaster walls and ceilings, sheet rock, composition wallboards, and interior paint and wallpaper.

Electric cable and conduit in walls or floors may suffer damage from wetting. The extent of injury will depend on how well-sealed or impervious to moisture and water the cable shield is. It is often cheaper to replace them than to repair outlet receptacles, switches or electric fixtures that have been seriously water soaked. Electric motors that have been wet, even those submerged for a short time, can under some conditions be baked out.

Kiln-dried wood or well-seasoned woods absorb water up to about 25 percent by weight, during which time swelling occurs. After this "fibre-saturation" point is reached, the swelling ceases. Plaster softens in the presence of water, and wallboards frequently expand and warp, some of them permanently. The glue on wallpaper softens and eventually the wallpaper may peel off. Water in back of paint, particularly enamels and semi-glazed surface paints, forms small blisters full of water which ruin the decorations. Dirty water, which might be pumped from rivers or creeks, and dirt-laden water percolating through ceilings to floors below discolors decorations, carpeting and finished floors.

The chemical or physical properties of water used during a fire are seldom factors that increase damageability, except possibly salt water which has a corrosive effect on metals if left in contact for any length of time.

The weight of water (62.5 pounds per cubic foot) which has been absorbed by plaster and some types of contents such as paper or fabrics, may cause the collapse of wood beams, girders or sometimes entire floor structures.

Water expands various types of materials including newsprint, sisal, and cottonseed and, if it is closely packed in a building, it may swell and exert enough side pressure to push out masonry walls.

The freezing of water in the winter during or shortly after a fire causes further damage by expanding in plaster, water pipes, toilet bowls, traps, and hot water or steam heating systems.

ACTIVITY OF FIREFIGHTERS

A certain amount of damage is caused in gaining entrance to a building or to the rooms in it. Doors, door jambs and casings are sometimes broken or chopped, locks are forced, and glass in sash doors is broken. A common practice in good firefighting is to ventilate the premises. For that reason, firefighters break windows or skylights and occasionally chop through a roof to let out smoke and heat. Miscellaneous chopping and cutting may be done in tracing the course of a fire under floors, or in partitions and walls.

In fire-resistive multiple story buildings, toilet fixtures are sometimes broken or removed so that the accumulation of water on the floor can be squeegeed to the outlet. In buildings that have wood floors, holes may be bored or chopped to drain off water to lower floors.

CHEMICAL EXTINGUISHERS

Most towns and cities equip their fire trucks with chemical extinguishers for the average small fire. These extinguishers usually contain calcium chloride, foamite, carbon tetrachloride or carbon dioxide. While all but the latter of these chemicals may stain or discolor decorated surfaces, they are otherwise not harmful to building materials.

FALLING DEBRIS

During the more serious fires, there is always danger of weakened floor structures collapsing because of overloading with wet plaster and contents and the debris from floors above. Chimneys and brick walls may fall inward, crushing portions of the building that are yet unburned or otherwise uninjured. Charred wood girders, beams or rafters weaken and allow parts of the structure and debris to collapse.

Buildings adjacent to a burning building sometimes are damaged by walls collapsing outward, falling brick, or other debris.

EXPLOSIONS

In buildings of dwelling occupancy, the most common explosions are in furnance boxes of heating plants, caused by delayed ignition of accumulated fuel gases. The majority of these are low-pressure explosions and the damage is usually confined to the oil burner, fire box brick lining, insulated covering, and the smoke stack connecting the boiler to the chimney. Smoke, soot or fumes may follow such an explosion and be carried to other portions of the premises through open doors and hallways; when drawn into heating ducts they may reach into every part of the building through heat registers. The presence of oil in the smoke or fumes can cause substantial loss to decorations. If fire follows an oilburner explosion, more serious damage can be anticipated by burning of controls and by sooting. There have been occasions of violent explosions in the fire boxes of furnances, but they are more prevalent in the large industrial furnaces and also in furnaces that burn gas rather than oil or coal.

Explosions in manufacturing and other industrial plants originate from many different causes, the most common being ignition of the vapors of flammable liquids. There is also the hazard of dust explosions in plants that process starch, sugar, grains, cork, spices and numerous other products. Chemical explosions occur in plants experimenting with new products, and occasionally happen in munition plants, fireworks manufacturers, petroleum processors, fertilizers processors or storage plants. Such explosions can demolish buildings, or completely or partially damage them, depending on the force and the distance of the explosion from the risk. Masonry or plaster walls often crack. Windows and plate glass may also crack or shatter from ground shock or concussion.

The damage caused by blasting for roads, foundations, tunnels, and so forth follows a pattern similar to that mentioned above.

Examination of damage done by explosive forces requires great care. The estimating should be confined to explosion damage only; and the wear and tear or other conditions that may be chargeable to maintenance must not be included. (An excellent reference on this subject is *Blasting Claims—A Guide for Adjusters*, prepared by the American Insurance Association.)

A typical low-pressure explosion, not uncommon during fires, occurs when a heavy concentration of gases of incomplete combustion is ignited. These gases become "pocketed" in blind or concealed spaces, or at one end of a floor with the fire burning in the other end. If the premises are not vented to allow the gases to escape, they can be ignited by flames when the fire reaches the area. Windows, particularly display windows in store buidings, are frequently blown outward by such explosions. Because these explosions are low-pressured, they seldom cause serious structural damage.

LIGHTNING

Damage to buildings and structures by lightning occurs in all parts of the continent, but more frequently in the southeastern section of the United States, where the incidence of thunder storms is higher. Lightning is a high-voltage electrical discharge between the clouds and the earth. The voltage may run into millions, and the amperes into tens of thousands.

Once the discharge takes place, the effect on the object in its path on earth may be the result of a direct hit by a *lightning bolt*, or the result of an *atmospheric surge* of electricity induced in the area, usually by a bolt of lightning which has struck nearby. The latter is sometimes called a *ground surge*.

Damage caused by a direct hit can be trifling (such as knocking a few bricks from the top of a chimney), or devastating. The knowledge that the objective of lightning is the ground, will aid, perhaps, in understanding the unpredictable and sometimes perplexing path that lightning follows from the point of contact to the place of exit in a building. Where resistance to its path is encountered, instantaneous heat is created and, if the bolt of lightning is a powerful one, the result is explosive. Inspection of a building struck by lightning may reveal a few shattered clapboards, or a corner of a roof or a few bricks knocked off, giving the impression of minor, localized damage. A more thorough survey, however, often discloses a greater injury to the interior than anticipated. Electric receptacles may be scorched or blown out of the wall, cracks may appear in some rooms but not in others, and a joist or rafter may be split. When lightning strikes a structure, it frequently fans out rather than pursuing a direct course, making its path more difficult to trace.

Lightning that strikes behind a concrete block retaining wall can lay it flat on the ground. Chimneys struck by lightning require careful inspection from top to bottom, not only to include all damage caused by lightning, but also to detect and distinguish old cracks and damage caused by age and settlement.

Losses to pole transformers, appliances, motors and other equipment can be the result of atmospheric or ground surges, and not necessarily direct hits. In most instances examination will show either fusing of a part or parts or else explosive effects due to violent interior pressure. Because electrical appliances and apparatus are subject to inherent defects and wear, which bring about shorts and breakdowns, the inspection of damage attributed to atmospheric surges of lightning calls for a high degree of care. Expert advice should be sought when there is serious doubt that lightning is involved.

WINDSTORM

It has become a fairly general practice among insurance carriers to consider windstorm as any wind capable of causing damage without regard to a stipulated velocity. Gusts of wind in local areas, between tall buildings, or down alleys on relatively calm days may cause damage by slamming doors open or closed, tearing awnings and even breaking windows. Although this type of damage is contemplated under the policy, the windstorm provisions of a policy contain a number of exclusions not common to the hazard of fire or lightning.

Special Considerations

The person charged with preparing specifications for repairing a structure damaged during any type of windstorm either should be familiar with or instructed in the policy exclusion of certain kinds of losses and specific types of property.

Where windstorm is a *named peril* in the insurance policy the following exclusion is typical though the language may vary from one policy to another:

(a) loss caused directly or indirectly by frost or cold weather or ice (other than hail), snowstorm or sleet, all whether driven by wind or not;

(b) loss to the interior of the building(s), or the property covered therein caused by rain, snow, sand, or dust, all whether driven by wind or not, unless the building(s) covered or containing the property covered shall first sustain an actual damage to roof or walls by the direct force of wind or hail and then this company shall be liable for loss to the interior of the building(s) or the property covered therein as may be caused by rain, snow, sand, or dust entering the building(s) through openings in the roof or walls made by direct action of wind or hail.

Also, unless there is a specific endorsement, all policies exclude damage caused by flooding, surface waters, overflow of streams, wavewash, and tidal waters.

In situations where part of the damage has been caused by a covered peril and part by one not covered (or excluded) the estimate should be prepared to indicate the estimator's opinion of the damage *covered, not covered*, and that which is *questionable*.

In addition to excluding specific hazards, most forms exclude coverage on specified property unless liability is specifically assumed by endorsement. Property more commonly excluded consists of windmills, windpumps, and crop silos. Many forms also exclude such items as awnings and television antennas.

The windstorm provisions are not identical in all states, nor in all forms. The exact wording should always be checked before preparing an estimate of damage.

Tornadoes

The word tornado comes from the Latin word *tornare* meaning to turn. It is cyclonic, and is probably the most violent type of windstorm in nature. Approximately 150 to 200 tornadoes occur annually in the United States, the majority being in the midwestern area. The tornado moves generally from southwest to northeast with forward speeds ranging from a few miles per hour to as much as 60 miles per hour.

The path of a tornado is narrow, usually not much more than a quarter to one-third of a mile wide. The average length of the path of tornadoes is eight to ten miles. The destructive force in a tornado is the high velocity of the cyclonic winds, which often reach speeds up to 450 miles per hour. The counter-clockwise swirling action produces in the center a condition approaching a vacuum. The effect of this vacuum-like center is explosive when passing over any structure. Buildings are reduced to rubble and in many instances they are sucked up with other objects and carried for many miles. Probably no catastrophe so quickly and completely changes the appearance of a community. Trees are stripped of all foliage and many times completely denuded of their bark. Persons who have witnessed tornadoes close at hand describe their sound as "like a million bees" while in the air, and "like a thousand express trains" when it touches the ground. The characteristics of a tornado are its narrow and relatively short path and its destructive violence, which is almost atomic.

Lightning generally accompanies a tornado. Hail and rain also either precede, accompany, or follow this type of storm. Hailstones as large as baseballs have been reported. On one occasion disc-like hail, 6 to 10 inches in diameter and 2 to 3 inches thick, fell during a tornado.

About 70 percent of the tornadoes occur in the four months, March through June. Most of the others take place in the late summer; a few in other months. The states which have suffered most in property damage by tornadoes are, in order of the extent of damage: Oklahoma, Illinois, Missouri, Texas, Georgia, Ohio, Indiana, Kansas, and Iowa.

Hurricanes

A hurricane, like a tornado, is also cyclonic, with winds traveling in a counter-clockwise motion. However, the intensity is second to a tornado. Officially, hurricane winds are those with velocities in excess of 75 miles per hour. The path of destruction of a hurricane may cover several hundred miles. Our hurricanes generally start in the North Atlantic around the Cape Verde Islands. They travel westerly toward the West Indies, then northwesterly and as a rule curve to the northeast. They may travel inland toward the Gulf States or up the Atlantic Coast, either striking land or veering seaward.

The wind velocities of hurricanes may reach 150 to 200 miles per hour, but generally they do not exceed 100 to 125 miles per hour at the highest point. The width of the path of a hurricane may be 50 or 100 miles. The length of its path may be as much as several hundred or a couple of thousand miles. The forward speed of hurricanes varies from a few miles per hour to 60 miles per hour; as it passes over land, the forward and cyclonic velocities are retarded.

Hail seldom accompanies a hurricane, but extremely heavy rains are not uncommon. Along the coastal shores, high tides generally occur, inundating the low lands and causing serious damage.

The "eye" of a hurricane may be a few miles wide, or it may be as much as 40 or 50 miles wide. The suction effect, so common to tornadoes usually is present in a hurricane to a lesser degree.

Hurricanes occur most frequently in the months of June, July, August and September. There have been exceptions when hurricanes have occurred in October and November. The height of the season is considered to be the month of August and the first half of September. The states visited by hurricanes more than any others are those that lie along the eastern seaboard, and the Gulf states.

Where wind velocities reach over 100 to 200 miles per hour for any length of time, the roof structures on many buildings start to lift, show windows are sucked out or blown in, and total destruction—even disappearance of small lighter buildings—may take place. Wind velocities between 70 and 100 miles per hour usually tear off part or all of asphalt type shingles, and can cause extensive damage to wood, asbestos, aluminum, slate and roll roofing. Fences are overturned, depending on their construction and the condition of wooden posts in the ground. Trees, or their heavier limbs, when blown on top of or against buildings cause varying degrees of damage. Carports or other lightly built structures may be blown over or collapse. Flying shingles and portions of roofs or other debris cause damage to adjacent and nearby buildings.

Because windstorms are often preceded, accompanied or followed by rain, damage caused by rain entering openings made in the roof or walls (including broken windows) by the action of the wind should be included in an estimate of damage. The extent of such damage depends on the amount of water that gets into the premises, and the portions of the building with which it comes into contact. In most cases the damage by water is confined to ceilings, wallboard or plaster, insulation in attics, and decorations in those rooms affected. Occasionally, water on finished floors causes them to "cup," but in most cases they can be resanded and refinished. Wooden doors and windows may absorb water through bare or poorly painted surfaces, causing them to swell so that they will not open or close properly. Veneers may come loose on the lower part of the doors.

HAILSTONES

Hailstones are distinguished from ordinary hail, which consists of small grains of frozen rain or ice-like snow. Storms that produce hailstones ranging in size from 1/2 inch to 2 or 3 inches in diameter, are believed by some authorities to be associated with cyclonic disturbances in the upper atmosphere. The stones are comprised of laminations of ice, one layer being built up on top of another. They are extremely hard, the larger ones being difficult to break with a hammer.

Hailstorms may occur during a tornado or occasionally during a hurricane. They also take place alone or with rainstorms. They occur most often in the early spring months and are more prevalent in the central states than in other parts of the country.

The damage that can be done to buildings during a hailstorm depends on the quantity and size of the stones, and on whether rain accompanies or follows the storm. Large hailstones are capable of shattering slate, asbestos and Spanish tile roofs. They can break plate glass windows and puncture asphalt shingles and built-up roof coverings. Even the smaller hailstones are capable of perforating the asphaltic felt-type roofing. Aluminum roofing and siding, though not punctured, may be so badly pocked or dented that replacement is necessary. Wood siding also may be dented from hailstones, necessitating filling and painting, or, in the more serious cases, complete replacement.

Roofs that have been damaged by hailstones may admit rain during the storm or later before temporary or permanent repairs can be made. As a result, interiors are subject to further damage by water.

AIRCRAFT AND VEHICLE

Damage by vehicles, as a rule, pertains to those that run on land or tracks. The vehicle may be a passenger automobile out of control colliding with the side of a building to cause minor damage to an exterior wall, a show window, or a porch. Or it may be a large trailer truck with brake failure coming down a hill and crashing into a building to cause extensive damage. Railroad freight cars occasionally strike industrial structures while being switched in or out of a siding.

Aircraft occasionally crash into buildings, causing damage ranging from minor injury to serious damage, particularly if fire ensues.

SPRINKLER LEAKAGE

Buildings that are equipped with automatic sprinklers, stand pipes or other fire protective equipment can be damaged by leakage from within the system. Sprinkler heads may operate as a result of some inherent defect, or from excessive heat. Pipes freeze; heads may be injured manually, or leakage can result from corrosion of pipes; tanks may collapse, or any number of accidents might occur to cause discharge. The main damage to buildings by water leakage from the sprinkler system is to wall and ceiling surfaces and decorations. Sometimes plaster and wall treatments in lofts, offices, and showrooms are affected. Finished wood floors may have to be sanded or replaced, depending on their condition. Asphalt, rubber and plastic floor tile is frequently loosened by water seeping between the tile and the under-floor. Electric wiring and fixtures, wet down by water, must be checked for repair or replacement, as water may have penetrated the armored cable or pipe conduit and caused corrosion or potential shorting in the switches and other outlets.

FLOODING, BACKING UP OF SEWERS, ETC.

During heavy rainfalls, storm and sanitary sewers may not be adequate to carry off the surface waters fast enough to prevent backing up through drains and plumbing fixtures. Check valves installed in the sewer line will generally stop the backing up, but not all dwellings or commercial buildings have them.

Poorly drained lots or streets accumulate surface waters during heavy rains and basements become flooded when the water enters basement windows, seeps through walls, or comes in under cellar doors.

Creeks, rivers and tidewaters frequently rise in some localities when there have been excessive rains, or when hurricanes cause sudden high tides.

Hydrostatic pressure will develop under certain conditions whereby ground water percolating through the soil is forced through foundation walls and cracks in concrete floors. Under each situation mentioned, water may collect in basements of buildings from a depth of a few inches to several feet. Hydrostatic pressure may be strong enough to heave up parts of the basement floor or collapse foundations. Drains, capped inside the building, may burst, causing concrete floor to crack and heave. Sometimes, though rarely, footings and foundation walls can be undermined as a result of the displacement of subsoil by moving ground water.

When basements are inundated, wood flooring, plaster, trim and millwork will show only slight damage if the water recedes quickly. The longer the water remains, the greater possibility there is of serious injury. Electric wiring, outlets and fixtures, whenever exposed to water, should be checked for damage.

Furnaces that have been submerged will show more or less injury to asbestos insulation on the outside or insulation in the jacket. The fire clay bonding the brick in the fire box may soften and be so damaged that the fire box lining will need replacing. Oil burners, gas burners, and coal stokers will require cleaning and checking.

Electric motors of washing machines, clothes dryers, deep freeze boxes, refrigerators, compressors and elevators many times can be repaired by prompt removal for cleaning and baking out. Seriously damaged units may have to be replaced.

When wood beams and girders in older buildings have been under water for 24 hours or longer, they may swell, buckle and twist.

When sewage is mixed with the water, the problem of sanitation arises, and completely cleaning up a basement so affected can be costly. Also, silt and mud frequently are mixed in the water. This requires considerable labor to remove, sometimes being of such depth that shoveling is required to get it out.

SONIC BOOM

The coming of the jet plane has brought with it the mechanical phenomenon of "sonic boom," which is caused by shock waves in the air generated by planes at supersonic speed. It is said that the cracking of a whip accelerates the tip of it to the speed of sound, producing the same effect. Sound travels at approximately 760 miles per hour at standard atmospheric conditions. When objects move through space at less than the speed of sound, air particles are readily pushed aside. At supersonic speeds the air particles build up in front of the moving object and create atmospheric pressure waves, that is, shock waves. Planes traveling at supersonic speeds produce several shock waves.

Scientists are not certain of the exact nature of the phenomenon, but it is generally believed that the waves attach to the plane when supersonic speed is reached, and as the speed of the plane increases they are swept back forming the cone shape of a wide funnel. When the pressure waves reach the ground, they usually have joined into one, two, or three waves. The energy of the waves dissipates as they move toward the ground and also when they pass over it because of friction and obstructions.

The boom (sound) heard by a person standing in the path of the cone is caused by the instantaneous pressure of the plane passing over. If two or three waves are created, there will be a rapid, almost overlapping, repetition of the boom.

To what extent a sonic boom can cause structural damage to a building has not yet been fully determined. There is no question that under proper conditions a sonic boom can shatter brittle window glass and also break doors or their frames if they are slightly ajar or loose-fitting at the time of the boom.

The pressures necessary to crack plaster and brick walls, damage roof structures, and heave floors, are much greater than can be created by a sonic boom; it is most unlikely that such structural damage can be traced to that source. Nuclear explosion tests indicate the necessity of a free stream of pressure of 150 to 300 pounds per square foot to cause such structural injury. The highest recorded pressure induced by sonic boom is reported to be 33 pounds per square foot—and this from a jet flying 280 feet away.

Press, radio, or television reports on the occurrence of a sonic boom frequently invite residents for miles around to inspect their properties, often for the first time. This results in numerous claims for cracks in plaster, walls, basement floors and so forth, which allegedly were caused by the boom.

When inspection is made to estimate damage by sonic boom, it should be verified that one actually occurred, if possible, by reporting to the Claims Officer of the nearest air base, and by questioning surrounding property owners. The Claims Officer will usually conduct his own investigation, if a claim is filed, and his report should be obtained. When structural damage to walls, roofs and floors is claimed, it may be advisable to consult with a competent engineer, vibration or explosion expert.

EARTHQUAKE

The most frequent cause of earthquakes and earth tremors is the faulting of rock formations beneath the earth's crust. Where high mountains are in proximity to deep oceans, quakes are more numerous. Such a condition is found along the Pacific Ocean in a line running from Mexico to the Aleutian Islands. This is part of a so-called earthquake belt. While earthquakes have occurred in other sections of this country, their frequency and intensity are much less than along the Pacific Coast.

The effect of quakes on buildings and structures varies with the seriousness of the disturbance, the distance away and the type of earth formation through which the vibration waves travel. Major injury such as broken windows, large masonry cracks, or the collapse of all or part of a structure can be readily determined on inspection. Minor damage, however, such as hairline cracks in plaster and masonry are easily overlooked or difficult to distinguish from normal settlement cracks or deterioration that existed before the quake. Many times it will serve all interests best to have the property surveyed by a vibration expert or a structural engineer who can generally determine the damage and also the probable method and cost of repairs.

Earthquakes are sometimes followed by fire with the result that the damage is mixed, being partly due to vibration and partly due to fire. If earthquake damage is excluded in the policy of insurance, it will be necessary to separate fire and vibration damage. The best method of doing this is to prepare a two-column estimate which will show the cost of repairing the damage from both causes.

LANDSLIDE—MUDSLIDE

Landslides can be started on a dry hillside or a mountainside by some disturbance of the earth such as vibration. Most often, however, they occur when excessive rains follow a long dry spell. The porous soil permits water to percolate rapidly through it until it reaches an impervious layer of clay or rock. The water, following the slope of this impervious layer forms a slick medium on which the earth above begins to slip or slide.

The slide may cause cracking in walls and foundations due to earth movement, or may engulf or completely crush a building in its path.

When the soil which is sliding has been so completely saturated with water that it reaches the consistency of mud, the landslide is usually referred to as a mudslide. Mudslides also occur when the top stratum of soil is of a clay or adobe type which becomes saturated with water after a dry spell. When the underlay is rock on a slope, the wet clay begins to slide by gravity.

The early movement of a mudslide is usually slow and almost unnoticed. The first indications may be the appearance of cracks in retaining walls, foundations, or concrete drives and patios. The lateral movement of the slide depends on many factors. It may stop with the cracking of a foundation wall or it may continue downward taking the entire building along with it. It may also occur over a period of days, weeks or months.

When a mudslide is discovered in the early stages, steps can be taken to prevent greater damage by making tests to determine the depth and stability of the soil and the direction of the slide. When this data has been developed, measures can be taken to shore the building and establish drainage ditches to draw off the water in the soil and divert it away from the property. Only those thoroughly experienced in handling mudslides should be sought for advice as the work is highly specialized.

FREEZE-UP

A building without heat in freezing weather may suffer damage from frozen water pipes, sprinklers, tanks, traps, toilet bowls, and steam or hotwater heating systems. Not only will the plumbing and heating fixtures and piping be damaged by bursting but, if not discovered when they begin to thaw out, the leakage of water from within the system can cause serious loss to floors, walls, decorations, electrical wiring and so forth (see "Water Damage," page 3).

Preparing the specifications for repairing a freeze-up loss should not be started until the water has been shut off and the premises thawed and dried out. Tests will have to be made by competent plumbing and heating contractors to establish the extent of damage to piping, fixtures, and so forth. A careful inspection of the entire building area involved is required to determine where damage begins and ends.

DAMAGE BY HUMANS

Damage to buildings by people includes vandalism, malicious mischief, riot and civil commotion, and also damage done by burglars. While the intent of the individuals may not be the same, the general nature of damage is.

Window or plate glass may be broken, decorations defaced, plaster walls gouged, doors and locks broken and so forth. Not infrequently, vacant or unoccupied buildings are entered by children or youths who cause serious destruction. In addition to wanton damage to the property, plumbing, heating, and electrical fixtures or their parts may be removed.

Prior to preparing an estimate of this type of loss, a careful investigation of the circumstances is essential. Police reports should be obtained, and neighbors, former occupants, or others familiar with the physical condition of the property are generally questioned to develop, insofar as possible, the damage claimed as against any prior condition which existed as a result of wear and tear.

COLLAPSE

Collapse losses involving roofs, particularly on commercial properties, are not uncommon. Most of these occur when the roof, usually flat, has been subject to an unusu-

TRAPPED WATER

ICE DAM

WATER ENTRY

INSUFFICIENT
INSULATION

GUTTER

NO SOFFIT
VENTILATION

A

FLASHING
(ROLL ROOFING)

RAFTER

AIRWAY

SUFFICIENT
INSULATION

GUTTER

SOFFIT VENTILATION

B

M 134 787

Snow and ice dams: A, Ice dams often build up on the overhang of roofs and in gutters, causing melting snow water to back up under shingles and under the facia board of closed cornices. Damage to ceilings inside and to paint outside results. B, Eave protection for snow and ice dams. Lay smooth-surface 45-pound roll roofing on roof sheathing over the eaves extending upward well above the inside line of the wall.

Figure 1-1

ally heavy snow load, or an unusual accumulation of rain water. In either case, inadequate design is always suspect and the occasion generally requires the services of a structural or consulting engineering firm to investigate. Additionally, snowstorms and heavy rains are accompanied by a certain amount of wind. Consequently, when the peril of collapse is not covered under the insurance policy, there is frequently a tendency for the claim to be made under the peril of windstorm.

Because the person assigned to estimate the cost of repairs is many times the first or one of the first to arrive on the scene, he is in an excellent position to preserve evidence by photographs, diagrams and note-taking.

Collapse does not always involve a roof of a building. Floors of commercial buildings collapse partially or completely due to overloading. Basement walls of dwellings collapse inward due to hydrostatic pressure. Ceilings or parts of ceilings collapse for various reasons.

ICE DAMS ON ROOFS

Ice dams begin their build-up in gutters at the eaves or cornice line. The gutters collect snow and melting water which, freezing, fill the gutter and then snow and ice start building up on the roof from the gutter toward the ridge. Alternate freezing and thawing are factors increasing the build-up. Heat from the interior of the building melts the underside of the formation. The water cannot escape because of the frozen gutters. It eventually backs up under the shingles and enters the interior of the building causing damage to walls, ceiling and floors. An excellent illustration of how this occurs and corrective measures is shown in Figure 1–1. It is taken from "Wood-Frame House Construction," Agricultural Handbook No. 73, July 1970, U.S. Department of Agriculture, Washington, D.C.

CHAPTER 2

Fundamental Considerations

PURPOSES OF BUILDING LOSS ESTIMATES

When a building or structure has been damaged by fire, windstorm, explosion or some other hazard, the owner not only wants to know how much it will cost to repair the property but, if insured, an estimate of the loss will form the basis of claim. The insurer will want to know the amount of the damage in order to determine liability.

There are not only variations in methods of preparing estimates, but also differences of opinion as to extent of damage and cost of making repairs. The only way to reconcile differences in estimates is to analyze and check each item. This procedure requires that the estimate itself be in complete detail, showing all measurements, and dimensions, and listing the material and labor for each item of repair or reconstruction. *The estimate becomes, not only the foundation, but a fundamental device for the adjustment of a building loss.*

ESTIMATING NEW CONSTRUCTION VS. REPAIRS

The principles used to estimate the cost of restoring a damaged structure are the same as those used to estimate the cost of constructing a new building. Once the scope of the work has been established, and the specifications agreed to, the estimate is developed from the cost of materials and labor. Overhead charges and profit are added to arrive at a total cost.

Preparing a loss estimate differs in several respects from an estimate on new construction. A builder, for example, prepares a bid to erect a new house from plans and specifications which are furnished competitive bidders. All of the bidders base their estimates on the same measurements, materials and work to be done. Bids vary, usually because of errors on the part of the estimator, differences in opinion as to productivity of workers, use of labor-saving equipment or the ability of one builder to obtain materials at a lower cost.

Most building losses are estimated without plans and specifications being furnished. Each estimator inspects the property, takes measurements and decides on what has to be done to restore the building. The estimator decides on the kind and quality of materials, what must be repaired or replaced, and how much tearing out and carting is necessary. This procedure in loss estimating introduces a great many sources of error that are not inherent in estimating new construction.

ESTIMATING IS NOT AN EXACT SCIENCE

By definition "to estimate" means to approximate. We should clearly understand that it is not an exact science. So many chances of error exist, particularly in estimating building losses, so many misunderstandings, variations of opinion especially of the labor hours to perform operations. It is well known that the spread between contractors' bids on new construction work, with no loss involved, will range from 10 to 30 percent. This condition occurs in spite of the fact that each estimator uses identical plans and specifications. It is not unusual for even large contractors with a staff of experienced estimators to lose money on one job and make it up on another.

It is also rare that all of the detailed costs in the estimates of two contractors will be the same. The total amount for the job may check out reasonably close, but the individual items will vary up and down.

ACCURACY IS ESSENTIAL

While estimating is not an exact science, it nevertheless calls for the highest possible degree of accuracy in its preparation. The proper measurements, the identification of materials as to size, kind, and quality, the careful listing of items, the thoughtful consideration of labor requirements, and the double-checking of all arithmetical computations are of utmost importance. Shortcuts should be avoided, and outright guessing of dimensions or quantities will invariably result in errors and embarrassment.

QUALIFICATIONS OF AN ESTIMATOR

A general knowledge of materials and methods of construction is necessary in order to estimate costs. The greater knowledge a person has of building construction, the greater are the possibilities of becoming a highly competent estimator. One need not have been a carpenter, mason, bricklayer or painter. Many estimators employed by contractors have had little or no practical experience as mechanics in a building trade. They do, however, have a broad knowledge of building materials and construction which they have learned from study and observation. They are good mathematicians and detailers. They are able to visualize every operation that will be necessary to complete the work. Probably one of the most important qualifications that an estimator can have is that of making inquiry when uncertain. There are many excellent reference books that deal with construction methods, and no one who undertakes either to make an estimate or check one should be without good reference material.

SUBCONTRACTORS

The number of building trades which general contractors are equipped to estimate varies. It is fairly standard practice for them to figure all of the carpentry work. On the small or medium-sized jobs they may also estimate the cost of roofing, painting, plastering, concrete, or masonry. They frequently have their own people engaged in these trades and have sufficient work to keep them employed. The majority of general contractors sub-contract built-up roofing, plumbing, heating, electrical work, elevators, air conditioning, structural steel, and other mechanical or highly specialized building trades.

When subcontractors are asked to submit an estimate on losses, the person who has engaged the subcontractor should join in a survey of the damage, agree on specifications for repair or replacement, and instruct the subcontractor on how to prepare the estimate so that it can be audited if necessary.

Subcontractors generally add their own overhead and profit to their figures. A general contractor adds a charge for any supervision that is necessary, plus a normal profit percentage.

EXPERTS

In many cases, experts or specialists in certain building materials and operations are consulted or asked to inspect the damage and give an opinion on either the degree of damage or the cost of repairs. Damage involving building service equipment such as elevators, escalators, sprinkler systems, power plant boilers or generators, refrigerating apparatus, ventilating systems, and even such domestic appliances as electric ovens or ranges, dishwashers, dryers and air conditioners may require that either the manufacturer's representative or an expert inspect the property, before the extent of injury can be determined.

Sometimes it is prudent for the estimator to have a manufacturer's representative inspect certain specific materials such as brick of various types, marble or limestone facing and trim, cork insulation, rubber or asphalt tile flooring, or certain kinds of roofing. It is not uncommon to employ chemical or materials testing laboratories to determine whether certain materials have been damaged and, if so, to what extent repair or replacement is necessary. Among the experts frequently summoned are college professors in structural, mechanical, electrical and heating engineering. Also, professors of physics, chemistry and ceramics have been consulted on special subjects.

FUNDAMENTALS OF ESTIMATING

The principles of all estimating are identical, whether it is for new constuction or for repairing damaged buildings and structures. There are five essential factors to be considered: (1) Specifications, (2) Material, (3) Labor, (4) Overhead, (5) Profit.

Specifications

Specifications are determined *at the scene of the loss* by careful survey and inspection of each part of the property affected. All items of repair and replacement are noted. Measurements of rooms, story heights, damaged areas, and other dimensions needed to compute quantities are taken. Sizes and quality of materials are noted. This data when set up in an orderly manner in the finished estimate becomes the *specifications* for repair and replacement. The procedure for inspecting the property and taking off specifications is discussed in Chapter 3.

The ideal procedure for establishing the specifications of repair and replacement on building losses is to have the estimator representing the insured and the one who will represent the insurer get together at the loss before they prepare their estimates. If they can agree at least on the general scope of the work to be done, much will have been accomplished to avoid controversies after the estimates have been completed. It will be better, of course, if they are able to reach an understanding on all or almost all of the details. In many instances, as a result of such meetings the estimators agree on a specific expert, or experts, to advise or decide on questionable items such as how to repair or replace weakened floor structures, repair concrete floors and floor surfaces, structural steel members, etc.

Estimating Materials

In estimating materials, two important factors should be kept in mind; the *size and quality* of the particular material, and the proper *quantity* needed.

Quantities are sometimes fixed by room, wall or ceiling dimensions, or an entire roof is to be recovered. There are many instances, however, when the quantity of material required is a matter of judging how much area is involved and how badly it is damaged. Where a plaster wall has suffered some water damage as a result of fire on the floor above there may be a question as to how many square yards, if any, will have to be removed and replastered. Once the extent of damage has been decided upon, the quantities of material are obtained from measurements taken in the field.

Except in cases of special agreement with the insurer and insured, the kind, size, and quality of material listed in the estimate should be the same as that which existed at the time of the loss. If the roof shingles were 235-lb. asphalt strip shingles, the estimate should contemplate replacing that type, and not a 165-lb. or 265-lb. shingle.

Describing the material as one inch by three inch pine flooring would not be sufficient. There is edge-grain (comb-grain) pine flooring, and also flat grain. There are different grades of each, although they are difficult to determine when it is charred, dirty, painted or stained. Red oak flooring is cheaper than white oak, and quartered oak is more expensive than the latter, so that any description of oak flooring should indicate whether it is red, white or quarter-sawed and, whenever possible, the grade should be noted.

The following are common examples of why proper identification of materials is essential:

(a) Douglas fir is more expensive than hemlock.
(b) Calcimine is cheaper than latex or oil base paint.
(c) Standard gauge linoleum is cheaper than heavy guage.
(d) A 235-lb. asphalt shingle costs more than a 165-lb. asphalt shingle.
(e) A 3-ply built-up roof is cheaper than a 5-ply.
(f) Cedar shingle shakes cost more than sawed shingles.
(g) Double-strength glass costs more than single.
(h) 2-coat plaster is cheaper than 3-coat.
(i) Enamel plumbing fixtures are cheaper than porcelain.

The size or dimensions that are necessary to identify a particular material should be measured and noted. If a door is being described, it should include the number of panels (or lights of glass in a sash door), the kind of wood, the width, the height, and the thickness, otherwise it is not possible to obtain the proper price of the door from the local mill. The same rule applies to windows, cabinets, trim, and so forth. The type of hardware should be noted, as there is a substantial difference in price in the various materials, styles and grades.

Any dimension or special characteristic of the material which will affect the cost should be included in the description.

When there is a doubt as to the kind of material, and circumstances will permit, a sample should be removed and taken to a supplier for pricing. A loose shingle, a piece of flooring, a strip of wallpaper, or a piece of wood trim generally can be identified by a dealer as to its quality and cost.

The pricing of all material should be done at local sources, using lumber yards, mills, hardware stores, paint and wallpaper stores, or masonry supply houses. Where discounts are given to the trade either as a "trade" discount or a "cash" discount, they should be deducted from the cost. All customary charges for freight, cartage, or loading should be added to the cost. On large losses where materials can be ordered and delivered in carload or barge lots, the price of material should be estimated accordingly. In other words the price obtained should be the actual amount that a contractor would have to pay to have the material delivered to the job site, including all taxes.

Estimating Labor

Estimating labor consists of approximating the length of time that it will take a man, or a crew of men, to perform a specific job or operation. The cost of the labor is obtained by multiplying the number of hours by the hourly wage scale.

There are many factors which influence every estimator's judgment when considering the hours of labor required to install or apply materials or to perform specific operations. Most of these may be found within the following classifications which will serve as a check list, if the person making the estimate will ask these questions:

1. Are there any special or unusual conditions which the workers will encounter?
2. Is the physical area being worked in confined, or is it open so that the workers can move about freely?
3. Can labor-saving machines or equipment be used?
4. What experience have the workers with the particular type of work they are to do?
5. Is the quality of work required high or average?
6. Will the work have to be done from ladders or scaffolds?
7. Are the lighting conditions adequate?
8. Will the supervision of the workers be close or loose?
9. What will be the probable weather conditions?
10. What is the productivity of the individual worker?
11. Are the wages that are being paid standard or substandard?
12. Is smoking permitted on the job?
13. Are there any special union regulations limiting production?
14. Is the job near to or far from a source of labor?
15. Is the supply of labor good or not good?
16. Is it necessary to protect the contents or occupants of the premises while working?

1. *Unusual Conditions.* Special working conditions of a particular job require careful analysis and visualization by the person making the estimate. The estimate may specify the painting and decorating of a room of a given size. This could be a room which was vacant and unoccupied. The work needed to prepare the wall surface might be trifling or nothing, so that all a painter had to do would be to walk in and start painting.

A second situation might involve the living room of a palatial home which is furnished and occupied. The painter would be called upon to move furniture, protect the wall-to-wall carpeting, remove and rehang pictures, sandpaper woodwork, remove and reset a radiator, or even wash the walls and woodwork prior to decorating.

A third supposition might be that this room is an office in which two or three persons work who cannot be moved out. The painter may have to work around them and also be very careful not to spatter them or the contents with paint.

Each situation presents circumstances that are peculiar to the particular job. Actually none are serious *per se*, but additional time must be given the painters where their rate of work performed is reduced by physical conditions. If it is known that a painter will average a given number of square feet an hour under conditions similar to those outlined in the first situation, then it is possible to approximate the additional time that must be allowed to meet the conditions described in the other two situations.

2. *Physical Area.* The physical area in which a person is working will affect efficiency. In confined and cramped quarters such as closets or blind-attics, a mechanic's movements are hampered and consequently slowed down. In wide open areas, without restriction, efficiency reaches its highest point, everything else being equal.

3. *Labor-Saving Equipment.* The type of work tools and equipment that are used has an appreciable effect on the rate of speed with which work is done. The use of portable electric saws and drills, electric or pneumatic hammers, aluminum ladders, scaffold brackets, paint spray guns, electric wall paper removers, and other modern devices greatly improves working conditions and reduces cost through higher rates of production.

4. *Experience of Workers.* The experience and familiarity of the worker with the job affect the rate of production. A roofer who puts on asphalt shingles every day of the week, or a floorlayer who lays hardwood floors as a specialty, will turn out nearly twice as much work per hour as a carpenter who performs these tasks only occasionally. Experienced paperhangers will put on more rolls of paper per hour than a painter who has hung paper only a few times. A carpenter who has never laid out or built a winding staircase will take hours longer than a stair builder who has built many.

Workers unfamiliar with fire loss repairs, will perform slower than workers who are experienced in such lines.

There is always the new problem arising in loss repairs which confronts the most experienced workers such as repairing non-stock millwork, an ornamental church door, ornamental plaster, patching parquet flooring, matching paint colors inside or out, and repairing brick or stonework. Situations like these slow the workers down as they figure out the problem and perform the operation.

5. *Quality of Work.* The quality of work required directly affects the time required to do the work. The higher the quality, the more time will be needed, because it takes more time to do a good job. High grade painting on trim, where successive coats

are hand rubbed takes more hours than 2-coat work in less expensive residences. Cabinet work, or paneling of the highest grade, requires many more hours per unit or square foot than inexpensive cabinetmaking or paneling.

6. *Work Done From Ladders.* Operating from a scaffold, or from a ladder, slows workers down compared to the rate for similar work done on the ground. They not only work with regard for their personal safety when on a scaffold or ladder, but they are also hampered by lack of freedom of movement.

7. *Lighting Conditions.* The lighting conditions in the area where work is being done will affect performance. People work faster when they can clearly see at all times exactly what they are doing. When it is apparent that lighting conditions will be poor, or where portable extension lights have to be shifted about as the work is done, some consideration must be given to a reasonable increase in the time normally allowed.

8. *Supervision.* The type of individual who organizes and lays out the work, and who constructively supervises operations will have a decided effect on the cooperation and efficiency of the workers. The more experienced the supervisor, and the more capable in handling people, the lower will be the unit labor costs, compared with those of a loosely supervised job or one which is given little or no supervision.

9. *Weather Conditions.* Weather conditions will influence the length of time it takes to perform work operations. Where repairs are to be made to the exterior of a building, the workers will do more in pleasant weather than in cold winter months when they keep outside fires going to warm themselves, or go inside occasionally to escape the cold. They work with lighter clothing and can move about more freely. In winter months it is often necessary to provide temporary protection for materials, and allow time for shoveling away snow, especially in northern climates.

Conversely, in extremely hot humid seasons or climates, particularly some of the far southern sections, workers slow down, especially when working directly in the sun and heat.

10. *Productivity of Workers.* The individual productivity of workers varies. Some mechanics do excellent work but are slow; others do mediocre work but turn out large quantities. Many persons develop a fixed rate at which they work, others are naturally lazy, talkative, or clock watchers. Construction workers are no different from persons engaged in factories, offices, or mercantile establishments. They possess all of the human perfections and imperfections. A contractor who knows the individual characteristics of employees and estimates accordingly has an advantage over a competitor who is unfamiliar with the productiveness of the workers who will do the job.

11. *Wages Paid.* The old adage that "a laborer is worthy of his hire," has a certain amount of truth to it. Where a worker is aware that wages received are less than those paid others doing similar work, or less than the prevailing wage, there is a tendency to turn out the irreducible minimum of work.

12. *Is Smoking Permitted?* Where smoking is prohibited in such occupancies as hazardous manufacturing plants or department stores, workers who are making repairs will take time out to smoke outside of the prohibited area. Such time breaks, particularly on small jobs, can increase labor costs.

13. *Union Regulations.* Local union regulations, which limit or restrict the amount of work a person may do during a work day, naturally govern the basis for estimating labor costs. Such restrictions may stipulate the number of bricks a bricklayer can

lay, or may prohibit a plasterer from taking on a second job at another location, after completing one small job within one work day. Other limitations may require the presence of a foreman, even if only one worker is needed, or may involve the use of labor-saving equipment such as paint sprayers or rollers. Where such customs are known to exist, the person making the estimate should become thoroughly acquainted with the regulations, in order to give proper consideration when preparing an estimate.

14. *Source of Labor.* Where a job is located many miles from the source of labor, the estimate may have to include a provision for transportation. Workers are not expected, in certain jurisdictions, to travel great distances to and from work on their own time or at their own expense.

15. *Supply of Labor.* Labor is a commodity and follows the laws of supply and demand. Where the supply is ample or greater than the demand, the wage rate may be lower and the production is frequently higher. Where the supply is less than the demand, the opposite is often the case.

16. *Necessary to Protect the Contents or Occupants.* When it is necessary to protect the contents or the occupants of a building, the labor costs increase depending upon the degree of care required. Painting or plastering, for example, in an office building, or a retail store during working hours reduces considerably the rate at which workers perform, for they must be cautious not to damage the contents, and usually work around the occupants who are performing their own duties.

Labor Rates

The more progressive and competent contractors keep cost records and also time studies on each job for reference when estimating other jobs in the future. If a roofer, for example, keeps records of the hours it takes to apply asphalt strip shingles to a particular roof, and finds that it takes 40 hours for the roofers to lay 20 squares, he knows they are averaging 2 hours per square. Therefore, on a similar type of roof the labor would be estimated by multiplying the number of squares of shingles by 2 hours.

In the same manner a plumber or plumbing contractor knows from records that workers, under normal conditions, can rough in a bathroom and set fixtures in so many hours. A brick mason's records will note that bricklayers can lay a particular kind of brick in a type of wall at an average rate of so many hours per 1,000 brick. Contractors and subcontractors for different trades know from their own records, or those of others, what the average rate of work will be under given conditions. When the physical working conditions are different from those on which the records are based, the person making the estimate must adjust the labor to allow for more or less time. There are three sources from which to judge the hours of labor to perform operations and install material. Personal experience through having done the work; observation of workers actually engaged in doing the work; and time studies that have been recorded by others.

Very few people have performed all the operations of constructing a building. Many estimators have had personal experience in various trades and they can rely to some extent on that background.

Contractors, superintendents, and time-study engineers develop a knowledge of labor requirements, having been associated with and observed workers on many phases of construction.

A third source for estimating labor is through the records of others. Numerous books have been written by contractors, builders, teachers, and architects on the subject of estimating construction costs. One or more of them will be found on the bookshelves of every large and successful builder. They provide excellent references. No one looks upon them as the final word or as an indisputable authority, but since they provide the experience of the many, and frequently state the physical conditions under which the work was performed, they do provide an excellent basis for estimating labor. (See Bibliography.)

In each of the subsequent chapters, in which estimating of particular trades is discussed, the subject of labor is treated as it relates to each specific trade. Tables are given for estimating average hourly production rates for individual operations. *They are, in general, adequate for normal working conditions and include the time required to start, finish, and clean up.* Factors which affect the production rate, as outlined in this chapter, should be given careful consideration before increasing or decreasing the hours shown in the tables.

Local hourly wage rates should always be used when estimating labor costs. Union wage scales may be obtained locally or from the U.S. Department of Labor, Bureau of Labor Statistics. The latter, however, publish their statistics several months late and any subsequent increases are not reflected.

Average rates of $12.00 per hour for carpenters, masons, roofers, etc., and $9.00 per hour for helpers, have been used in the examples in this book. *They are for illustrative purposes only, and should not be considered as applying at any time in any locality.*

Overhead

Contractors usually consider two types of overhead to be included in their estimate.

1. Specific overhead, chargeable to each job for which an estimate is being prepared.
2. General overhead, not chargeable to each specific job.

Overhead Charged to Specific Jobs:

Foreman and/or Superintendent wages
Watchman wages
Rubbish removal, clean up costs
Temporary buildings and structures
    Office (trailer, etc.)
    Tool and/or storage sheds
    Toilet facilities
    Fences, guardrails, etc.
    Signs
Building permits
Utilities
    Temporary heat, light and power

      Temporary water
      Temporary telephone
      Out of town jobs
         Travel cost
         Premium labor rates
         Board and room
   Plant
      Rental of machines and equipment
      Scaffolding built on premises or rented
      Maintenance of machinery and equipment
      Depreciation
   Legal expense
   Insurance
      Public liability insurance
      Fire and Extended Cover insurance
      Workmen's Compensation
   Taxes
      Social Security taxes
      Federal Unemployment taxes
      State Unemployment taxes
      Sales taxes on materials
   Performance Bonds

Overhead Not Chargeable to Specific Jobs:

   Contractor's office, shop, etc.
   Rental
   Depreciation
   Heat, light and power for office, shop, etc.
   Stationery, postage, supplies
   Business machines
   Property and liability insurance
   Telephone
   Other
   Employees
      Executives
      Clerks
      Estimators
      Draftsmen
   Travel
   Advertising
   Donations

Overhead is not a theoretical charge. It is subject to analysis. Contractors who routinely do small jobs may operate from their homes. The wife may do the clerical work, answer the phone and do the book work. Equipment and surplus materials may be kept on the premises. The general overhead would be relatively small in dollar-cost when compared with that of a contractor routinely engaged in heavy construction, maintaining a separate office, shop, estimators, clerks, storage yard, equipment, etc. Yet, when annual overhead expenses of many contractors are related, *percentage-wise*, to income it is quite surprising that in many instances overhead charges against each job will range between 8 and 10 percent.

Profit

The profit percentage a contractor charges against the overall *job cost* varies with the size of the job and the competition. As the dollar amount of the job increases, the percentage profit a constructor is content with decreases, particularly when the bidding is competitive.

Small jobs of a few hundred dollars or more may well justify 10 to 20 percent profit. In other words, a $300 to $400 repair job cost could justify a profit of $60 to $80 whereas on a job cost of $5,000 or more the contractor would be satisfied with 10 percent. Unfortunately, most of the estimating on repair or replacement of buildings covered by property insurance is not competitive. So, the generally accepted profit allowed is 10 percent of the actual cost of repairs after general overhead.

OSHA (Occupational Safety and Health Administration)

The U.S. Department of Labor has published safety and health regulations for the construction industry pursuant to the Williams-Steiger Occupational Safety and Health Act of 1970.

These rules or regulations, while mostly applicable to heavy construction, are enforceable as respects small contractors and builders of residential properties. Authorized representatives of the U.S. Labor Department have the right of entry to any construction site to investigate compliance with the regulations or may designate the services, personnel, or facilities of any State or Federal agency to inspect. Many states have adopted the Plan and have primary responsibility on a reimbursable basis by the Government.

A complaint of the smaller contractors and builders is that the cost of compliance increases their operating overhead anywhere from 2 to 5 percent with no visible benefits in many instances. Nevertheless, OSHA should be taken into consideration in estimating, especially where substantial demolition and removal of debris are involved. Federal Register, Part 2, Vol. 37, No. 243 can be obtained from the Department of Labor. This document outlines the regulations. A brief resumé of *some* of the rules is given below for illustrative purposes. Note particularly Regulations pertaining to Demolition (Subpart T).

Basically, no contractor or subcontractor shall require any laborer or mechanic to work in surroundings which are *unsanitary, hazardous, or dangerous to health or safety*. It is the responsibility of the employer to initiate and maintain such programs necessary to comply.

The prime contractor and subcontractor(s) may agree on who shall furnish (for example) toilet facilities, first-aid services, fire protection or drinking water, but such agreement will not relieve either one of legal responsibility.

Machines or power tools not in compliance with the Regulations must be tagged unsafe, locked, or removed from the premises.

Employers must instruct employees in avoidance of unsafe conditions to avoid exposure to illness or injury. If harmful animals or plants (poison ivy?) are present, the employee is to be instructed regarding the hazards and in any first-aid. The same rules will apply where the employees are required to handle any flammable liquids, gases, or toxic materials.

Employers must insure availability of medical personnel for advice and consultation on matters of occupational health. Provisions shall be made for prompt medical attention in case of serious injury.

If no infirmary, clinic, hospital, or physician is reasonably accessible, a person certified in first-aid shall be available at the work site.

First-aid supplies shall be easily accessible. The first-aid kit must contain approved materials in a waterproof container and contents must be checked by the employer before being sent to the job and *at least weekly.*

Equipment must be furnished to transport any injured person to a physician or hospital, or else communication must be available for ambulance service.

Telephone numbers of physicians, hospitals, or ambulances must be conspicuously posted.

An adequate supply of potable water must be supplied in all places of employment. Containers shall be capable of being tightly closed and supplied with a tap. (No dipping of water from containers.) Container is to be clearly marked. A common drinking cup is prohibited. Where single service cups are supplied, a sanitary container for unused and one for used cups is to be provided.

Toilets are to be provided in accord with the Regulations.

Protection against the effects of noise-exposure shall be provided when sound levels exceed those set forth in a Table of "Permissible Noise Exposures." The protective devices inserted in the ears must be fitted individually by competent persons. Cotton is not acceptable.

All construction areas must be lighted to minimum intensities listed in a Table designating "foot-candles" for specific areas.

Hardhats (usually supplied by employer) are to be worn by employees working in areas where there is possible danger of head injury from impact, falling objects, or electrical shock or burns. Goggles and face protection are to be provided employees who are engaged in operations where machines or equipment present possible eye or face injury.

Firefighting equipment shall be furnished, maintained, conspicuously located, and periodically inspected. A fire alarm system (telephone, siren, etc.) shall be established so employees and the local fire department can be alerted. Alarm code and reporting instructions shall be conspicuously posted at entrance.

Employers shall not permit the use of unsafe hand tools.

Scaffolding requirements are extensive. They must be equipped with guardrails and also toeboards on all open sides and ends of platforms over 6 feet above ground or floor. If persons are permitted to work or pass under the scaffold, a 1/2" mesh screen must be installed between guardrail and toeboard.

Subpart T—Demolition

(1926.850) Preparatory operations.

(a) Prior to permitting employees to start demolition operations, an engineering survey shall be made, by a competent person, of the structure to determine the condition of the framing, floors, and walls, and possibility of unplanned collapse of any portion of the structure. Any adjacent structure where employees may be exposed shall also be similarly checked. The employer shall have in writing evidence that such a survey has been performed.

(b) When employees are required to work within a structure to be demolished which has been damaged by fire, flood, explosion, or other cause, the walls or floor shall be shored or braced.

(c) All electric, gas, water, steam, sewer, and other service lines shall be shut off, capped, or otherwise controlled, outside the building line before demolition work is started. In each case, any utility company which is involved shall be notified in advance.

(f) Where a hazard exists from fragmentation of glass, such hazards shall be removed.

(g) Where a hazard exists to employees falling through wall openings, the opening shall be protected to a height of approximately 42 inches.

(h) When debris is dropped through holes in the floor without the use of chutes, the area onto which the material is dropped shall be completely enclosed with barricades not less than 42 inches high and not less than six feet back from the projected edge of the opening above. Signs, warning of the hazard of falling materials shall be posted at each level. Removal shall not be permitted in this lower area until debris handling ceases above.

(1926.851) Stairs, passageways, and ladders.

(a) All stairs, passageways, ladders and incidental equipment thereto, which are covered by this section, shall be periodically inspected and maintained in a clean safe condition.

(1926.852) Chutes.

(a) No material shall be dropped to any point lying outside the exterior walls of the structure unless the area is effectively protected.

(b) A substantial gate shall be installed in each chute at or near the discharge end. A competent employee shall be assigned to control the operation of the gate, and the backing and loading of trucks.

(1926.854) Removal of walls, masonry sections, and chimneys.

(a) Masonry walls, or other sections of masonry, shall not be permitted to fall upon the floors of the building in such masses as to exceed the safe carrying capacities of the floors.

(b) No wall section, which is more than one story in height, shall be permitted to stand alone without lateral bracing, unless such wall was originally designed and constructed to stand without such lateral support, and is in a condition safe enough to be self-supporting. All walls shall be left in a stable condition at the end of each shift.

The word *shall* means mandatory, the penalties can be severe for noncompliance so even small contractors and subcontractors find it expedient to attend and/or have a foreman or superintendent attend training facilities set up by the Department. This can be expensive in time and travel for those long distances from such centers.

Since the safety and health rules and regulations of OSHA were first promulgated for the construction industry, it has been recognized that they were directed to, and applied to, heavy construction rather than to home builders. Investigation and enforcement, therefore have diminished substantially and the home builders are almost uneffected.

LOCAL CUSTOMS AND PRACTICES

One who estimates building losses or who checks estimates on a nationwide basis will need to adjust one's thinking to local customs and practices within the trade. Such customs will show variations in these four areas:

1. Nomenclature
2. Construction methods and materials.
3. Estimating methods.
4. Agreements of unit costs.

1. A wood sill for a frame building may be termed simply a sill in one section, or a sole-plate, shoe, or mudsill in other sections. Roof boards, or roof sheathing, are known as decking or roof decking in some sections. A leader pipe running from the gutters to the ground is frequently called a down-spout, and sub-flooring may be known as under-floor, or a porch may be referred to as a piazza, stoop, or veranda. A stud in some rural areas is termed a *scantling*, and a ridgepole is variously called a *ridgeplate* or *ridgepiece*. Edge grain, comb grain, and vertical grain refer, in different areas, to the same type of flooring.

2. Construction methods vary in many sections of the country. Some of these variations may be due to differences in climate. A northerner traveling in the southern states for the first time may be surprised not to see chimneys on many of the dwellings. Frame dwellings in the South are frequently built with single flooring and no exterior wall sheathing underneath the weather boards, brick veneer or stucco finish. The method of applying stucco in the South, and Southwest by placing chicken wire directly over the studs, is almost unheard of in northern climates. Light roof framing on low-pitched roofs, common in southern climates, would be inadequate in northern areas where snow loads must be considered in designing roof construction. Shallow footings or none at all, such as are found in frost-free climates, would violate the building codes in northern cities where depths of 3 to 4 feet are required.

It is essential for estimators to be familiar with local construction methods so that, when a building is completely destroyed, they will be able to figure the reproduction cost as it existed.

3. The different customs used in estimating are many and varied throughout the country. There may be a tendency toward standardization in local areas, but even here there will be found contractors who employ different methods to estimate the same thing. One estimator may figure the milling and cutting waste of $1'' \times 6''$ tongue and groove boards at one-sixth, while another will use 20 percent. One paperhanger, in computing the number of rolls of paper needed for a room, may allow 20 percent waste; another paperhanger may determine the waste by allowing one extra roll for two window openings, two extra rolls for three windows, and so forth. In calculating the quantity of wall sheathing, bevel siding, and similar materials, some contractors deduct the area of all window and door openings; others leave them in or deduct one half the area. In figuring interior plaster, or exterior stucco, there are differences in opinion as to whether or not openings should be deducted.

Many painters, when estimating the cost of painting a room, will measure the area of walls and the ceiling, ignoring the window or door openings, and multiply the area by a unit cost. In other sections of the country the same method will be used, but in addition a flat charge will be made for each door, window, and for the base board, on the theory that a different kind of paint or different color may be used on these units. Some painters estimate a job by the hours required, and figure a flat rate per hour for material, labor, overhead and profit. All of these systems seem to work, but whether they all are accurate is subject to analysis. The person who is called upon to make estimates in different parts of the country learns the various customs and practices, in order to discuss details intelligently when the occasion arises.

4. In some of the larger cities, contractors or estimators for the company, and those customarily figuring losses for insureds, tacitly have a general understanding or agreement on the approximate unit costs for such items as ordinary framings, roofing, painting, plastering, electric outlets, flooring, and so forth. Such unit costs are for the purpose of reaching an agreement in order that losses may be adjusted. They are not necessarily representative of a unit cost that might be used in a *competitive* estimate not involving the adjustment of an insurance loss.

EFFECTS OF ECONOMICAL FACTORS

As in other businesses, economic factors play a dominant role in construction job cost. Building material costs during the early part of the historic "Depression" of the 1930's were at rock bottom. Merchantable framing lumber was retailing in the northeast at an average of $20 per 1,000 board feet. Carpenters in some of the metropolitan areas were paid little more than a dollar an hour and laborers were paid less than that. In 1980, framing lumber averaged $350 to $400 per 1,000 board feet. Carpenters averaged $15 to $20 per hour in many metropolitan areas.

When materials are scarce, and prices change upward each month, contractors are reluctant to take jobs on a firm bid basis. More and more jobs are let on a cost-plus-basis with a condition that the total cost will not exceed a specified (safe) figure. If a job *is* taken on a fixed contract it can usually be assumed the figure is generous in order for the contractor to be protected against rising costs of labor or material, or both, before the work is completed. An understandable situation.

DEMOLITION AND REMOVAL OF DEBRIS

Before new materials can be installed in a structure that has been damaged, all of the burned, wet, broken or otherwise injured materials must be removed and eventually disposed of, either on the premises by burning, or off the premises by carting away to a public or private dump.

Damaged brickwork, lath and plaster, steel, framing lumber, roof shingles, finished flooring, and so forth, when not obliterated, will have to be removed to make ready for new materials. Generally, this work is done by common or unskilled labor if a substantial amount of demolition is involved. When a few rafters, floor joists or studs are to be taken out and replaced, a carpenter normally will do it. If a moderate amount of finished flooring, trim, electric wiring, plumbing pipes and fixtures are to be removed, the building trades journeyman who will install the new work generally takes care of the demolition. Much will depend on local working conditions and labor customs. In different sections of the country this operation appears in building loss estimates under one of the following various designations: Tear out and cart; demolition and trucking; debris; demolition; removal of debris; debris removal; and drayage. These terms all refer to the same operation, which is tearing out the damaged material and disposing of it.

There are no reliable tables, charts, or formulas for estimating the cost of this item. It is usually all labor and expense, plus any specialized equipment or material that must be purchased or rented to do the work. Experience, analysis, the ability to visualize each step, plus good judgment are needed to estimate the cost.

In most cases it will be found at the end of an estimate in the form of a lump sum amount. Many estimators make an educated guess at the over-all cost of the item, rather than trying to analyze each operation. Such over-all guesses are usually on the high side and, in fact, it is not uncommon for contractors to use the "debris" item to bury contingencies.

In some sections of the country, it is an accepted practice to add a few cents to the square foot price of shingles to allow for the cost of removing and carting away the old ones. Other unit prices will sometimes be increased to include removing and carting of the debris based on the cost per square foot, square yard, or cubic foot.

Because it is an operation precedent to reconstruction, and is almost always a labor, trucking, and expense item, it should be carefully studied, just as any other construction cost is analyzed. The only way in which this can be done is to visualize each individual operation and estimate the number of hours it may take to do the work. If oak or finished pine floor, subfloor, plaster, or brickwork must be removed, the estimated number of hours for each should be noted. If wood joists, studding, or rafters are to be removed, the operation is visualized and the probable number of hours is noted. After recording the estimated labor for each type of work to be performed, a summary should be made up by building trades, or by floor, room, or some other suitable classification. In this way omissions are frequently discovered and it also provides a sound basis for double-checking and for any later discussion during the adjustment.

Illustrative Example
Demolition and Removal of Debris

Remove 6 square wood shingles	9 hours
Remove roof boards 20′ × 30′	6 hours
Remove 8 rafters	4 hours
Remove 200 sq yds plaster and wood lath	10 hours
Tear out 150 sq ft 1″ × 3″ flooring	8 hours
Total labor	37
	@ $9.00
	$333.00
Loading and carting to public dump	100.00
Total	$433.00

It should be remembered that many times *undamaged* materials may have to be torn out because they are contiguous to materials that are damaged. Badly charred attic rafters or floor joists are a typical example where the roofing, in the case of charred rafters, is undamaged, and the finished floor, in the case of charred joists, is undamaged. The cost of removing the undamaged finished floor and roofing must be taken into account, as well as the cost of replacing them.

Machinery and equipment of a tenant may have to be moved and later replaced, in order to make the repairs to a floor. Charred joists under a bathroom floor may require the removal and replacement of all of the tile floor and plumbing fixtures above, to make proper repairs. Where a tenant has abandoned the contents debris, the cost of moving any portion of it to repair the building may have to be included in the estimate.

When subcontractors are employed to estimate for a particular building trade, a clear understanding should be reached as to who will estimate the cost of removing the debris and carting it away. In this way omissions or duplications can be eliminated.

If the debris cannot be satisfactorily disposed of on the premises by burning, or by dumping, an item of trucking must be added. A contractor who has trucks and drivers, generally knows what the proper charge should be. If it is required to hire a truck and driver, the local rate per day multiplied by the number of estimated days the truck is in service will determine the cost. The capacity of the truck, the time to load, and the estimated number of loads that can be hauled per day are considerations to be taken into account.

When preparing an estimate of the cost of removal of debris the answers to the following questions will aid in determining the probable cost.

1. What kind of material is to be removed?
   (a) How is it installed, or attached to the structure?
2. Is there contents debris either blocking or mixed in with the building debris?
   (a) Will the cost of removing contents debris be borne by the tenant? (or)
   (b) Must the cost be included in the building debris item?
3. What degree of care must be used in removing the debris?
   (a) To save and preserve it for reuse.
   (b) To avoid damage to other property.

(c) To avoid injury to workers.

(d) To avoid injury to occupants.

(e) To avoid injury to passersby.

4. Are protective barricades necessary?

(a) Partitioning, sidewalk fences, canopies.

5. Is special equipment needed?

(a) Scaffolding, debris chutes, etc.

(b) Bulldozers.

(c) Cranes or derricks.

(d) Pneumatic compressor and tools.

(e) Electric drills or hammers.

(f) Acetylene torches.

6. What type of labor will be used?

(a) Common.

(b) Skilled.

7. Is the damaged area easily accessible?

(a) Are elevators, stairways, and so forth, intact?

(b) Is the working space confined, or in open?

8. Can trucks load and park readily in the area?

9. How close is the building to a public dump, or disposal site?

(a) Is there a charge for dumping?

10. Can the debris be burned on the premises?

11. Is shoring of floors or walks necessary for removal of debris?

12. Can any material be sold for salvage?

(a) As scrap metal.

(b) As second-hand building materials.

(c) If contents have been abandoned, do they have any value?

13. Are specialists or subcontractors needed to perform any particular operation?

14. Must the tenants' property be shifted or moved?

(a) Machinery and equipment.

(b) Furniture and fixtures.

(c) Stock.

15. Are there special ordinances relating to wrecking? Special permits?

ESTIMATING SALVAGE

There are two kinds of salvage to be considered in estimating building losses.

1. Units or materials that, though undamaged, must be temporarily removed and later will be reinstalled.

2. Units or materials that are damaged or partially damaged, and which are to be removed and sold for salvage.

The first category would include such items as trim or millwork removed and stored during replastering, also service units, i.e., oil burners, plumbing and electrical fixtures, refrigerators, laundry equipment, which have to be removed mainly for protection while repairs are being made. Considerable care and skill are required to remove many such items so as not to damage them.

In the second category, salvage to be sold for value, there are very few building materials to be considered. Structural steel, and copper cable, wire or roofing in quantities, provide fair salvage depending on the scrap market, and the cost of getting it to the market. Common brick at one time was considered good salvage when labor wage rates were low enough to warrant the cost of cleaning and handling. In some sections of the country, second-hand bricks are now premium-priced as veneer for new homes. This architectural oddity is limited and confined to certain geographical areas.

Partially damaged service units of heating or plumbing fixtures bring small salvage, if any. Rough and finished carpentry salvage for resale is extremely rare.

KINDS OF ESTIMATES

There are several kinds of estimates, each of which has its special place in *approximating the reproductive cost* of buildings, depending on the degree of accuracy desired. The following are those most commonly used:

1. Cubic foot estimate
2. Square foot estimate
3. Semi-detailed, or panel method
4. The sectional, or bay method
5. Original cost method
6. Detailed estimate

All but the *detailed estimate* are used, primarily, to obtain the total replacement cost of a building rather than to determine partial repairs.

Cubic Foot Estimate

When the new-replacement cost of a building is desired on an approximate level only, one of the best methods used is the *cubic foot system*. The cost of building a dwelling, church, factory and so forth is either known or estimated in detail. The cost is divided by the total number of cubic feet contained within the outer walls, roof, and lowest floor of the structure. The quotient is the cost per cubic foot for the particular building under consideration.

Illustrative Example

$$\frac{\text{New replacement cost}}{\text{Number of cubic feet}} = \frac{\$70,000}{20,000} = \$3.50 \text{ per cubic foot}$$

The cost per cubic foot may then be applied to similar buildings by measuring the cubic foot volume and multiplying by the unit cost.

Though the theory behind cubic foot estimating is sound, there are a number of inherent weaknesses in the method which, unless understood and compensated for, will diminish its reliability. Very few buildings are identical in specifications, except in some housing developments, and it is therefore important to recognize construction and architectural differences between the building from which the unit was developed and the one to which it is being applied. The cubic foot cost of identical buildings decreases as the size (cubic volume) increases and vice versa. The ratio of the length to the width of a specific type of structure affects the cubic foot cost. The number and the height of stories in buildings of apparently similar construction and layout changes the cost per cubic foot. Any special or unusual interior finishes, built-in features, heating or air conditioning systems add to the difficulty of determining the proper cubic foot cost to be applied. The cost of porches, breezeways, and patios, which are not included in the volume computation, will have to be estimated and added.

While cubic foot appraisals sometimes come within 10 percent of the actual replacement cost, it must be considered a coincidence rather than a rule. Their main value is in quickly and inexpensively approximating building replacement costs. A builder in competition would not contract to build strictly on a cubic foot estimate.

There are several books and manuals published, with tables, charts and pictures, which are excellent guides for selecting an appropriate cost per cubic foot or square foot.

Measuring Cubic Foot Volume

The proper method of calculating the cubic foot volume of a building depends solely on the method used by the person who developed the cost per cubic foot.

A recognized procedure is to compute the square foot area of the building using the outside dimensions and multipying that area by the height in feet from the underside of the lowest floor (the basement if there is one) to a point halfway between the ceiling and the ridge or weather surface of the roof. Figure 2–1 shows several shapes of buildings and how the height is usually measured. The principal variation in methods used is that some systems measure from the top of the basement or lowest floor rather than from its underside. (See page 36.)

Square Foot Estimating

The replacement cost of buildings can also be approximated by measuring the square foot area of each floor in a building and multiplying by a predetermined cost per square foot. The cost per square foot is developed, as in the cubic foot method, from known replacement costs which are divided by the square foot floor area.

The area included in calculating the square foot cost of a building includes the first floor area, using outside dimensions, and similar living areas above the first floor. Open attics and basements are measured and priced separately in accordance with *their* cost per square foot.

The square foot method is more reliable when applied to buildings in which the construction and layout of the various floors are similar, such as schools, hotels, office buildings and hospitals. Most of the weaknesses of the cubic foot method are inherent also in the square foot method.

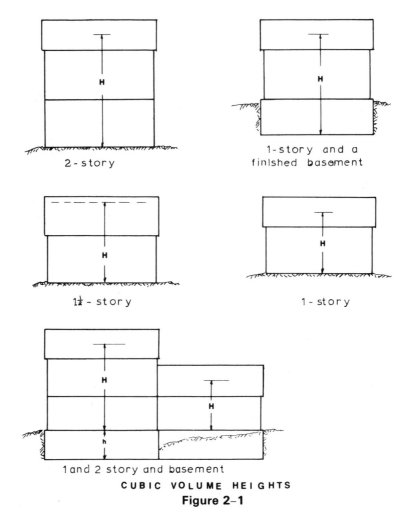

CUBIC VOLUME HEIGHTS

Figure 2-1

Recommended heights to use in computing the cubic-foot volume of a building.

Semi-Detailed or Panel Method

The areas of the exterior walls, partitions, floors and roof area are measured and computed. Knowing the construction materials through the walls, partitions, floors and roof, a unit cost per square foot is developed and *applied to the respective areas.*

The cost of partitions, floor and roof construction can be determined in the same manner. Interior and exterior doors and windows are counted and priced according to the individual cost for material and labor to install. Electric outlets and fixtures may be counted and priced per unit for labor and material. Built-in features, chimneys, fire places, stairways, heating and plumbing are added, generally in lump sum amounts, to complete the estimate, overhead and profit are added.

This method, in the hands of an experienced and thoughtful estimator, ranks next to the detailed method for accuracy. It is shorter than preparing a detailed estimate, but takes more time than the cubic foot or square foot system. It is especially suitable for office or factory buildings, apartment houses and warehouses.

The Sectional or Bay Method

A number of types of buildings are built in bays or sections of practically identical construction. By detail-estimating the new replacement cost of a single bay, the entire cost of the building can be computed by multiplying by the the number of bays. End walls must be added, and any dissimilarities in construction of one bay from others should be noted and taken into consideration. Such things as partitions, stairways, window or door openings, special equipment or services are added.

This method is useful in estimating foundry-type buildings, lumber sheds (either the open-front type or closed), barns, and also some warehouses and factories. Inspection of a building will quickly disclose whether the method should be employed. It is equally as accurate as the semi-detailed method, and for many buildings, is much faster.

Original Cost Method

When the original construction cost of a building is known or available, it can be brought up to date by applying a reliable cost index. The cost indexes are published by the American Appraisal Company, Engineering News-Record, publishers of appraisal books and construction trade journals. Building cost indexes are based on average conditions and indicate the percentage change in construction costs from one year to another, using a particular year as a base.

When using this method, it is important to know whether the original cost of the building included excavations, grading, foundations, footings, and so forth. If the cost of the lot is included, its approximate value at the time the building was built must be deducted before applying the cost index. If any additions have been made to the building, outside of maintenance items, since it was built, proper allowance should be made for them.

This method of estimating is approximate, the degree of accuracy depending on how recently the building was built, the dependability of the original cost figures, the reliability of the cost index, and the experience and judgment of the person applying it.

The Detailed Estimate

The detailed estimate is an itemization of all of the material and labor costs required to repair or replace a building or structure. It includes everything that a contractor will be required to provide and do to restore the property. It is the most accurate and dependable method of estimating because of its detailed build-up. It is also the most useful and satisfactory kind of an estimate for negotiating the adjustment of a building loss. Its main advantages over the other types are that:

(a) It identifies the room or area involved.

(b) The sizes, dimensions, and measurements are noted.

(c) The kind and quality of material are shown.

(d) The work to be performed is specified.

(e) The quantities of material are shown.

(f) The hours of labor allowances for each operation are shown.

(g) The material prices can be verified.

(h) The wage scales can be verified.

(i) All arithmetic calculations can be double-checked.

(j) It is orderly and therefore easy to follow.

(k) Comparison with other estimates is made easy.

(l) Differences are readily spotted.

(m) It provides a sound basis for discussion of details.

A detailed estimate is divided into two parts. The first part consists of methodically listing each operation and all of the materials that are necessary to complete the repairs or replacement of a structure. It follows a logical sequence of reconstruction. The second part of a detailed estimate consists of computing the cost of the materials and the cost of labor for each of the items listed. The total cost of the work is obtained by adding together the cost of the materials and the cost of the labor. When a building or structure has been totally destroyed, the listing of work and materials may be developed from plans or more often from sketches. The majority of losses, however, are partial and therefore the itemization is based on notes taken in the field at the scene of the loss.

MODULAR HOUSE CONSTRUCTION

The principal of modular house construction is a method of designing and erecting a building with its component parts manufactured in sizes which are multiples of a predetermined basic unit or module. The term "modular" applies to a broad spectrum of ideas in construction. There is no precise definition of the term. By definition "module" is a standard or unit of measurement—"the part of a construction used as a standard to which the rest is proportioned." According to the principle, when put together all parts of the house will join with a minimum of cutting and fitting on the job site.

Like the Unicom method of construction, developed by the National Forests Products Association, it utilizes "panelized" construction, with the modular aspect or feature being in the design of house plans. The Unicom method maintains the principle of a 4-inch module, but in practice house plans are drawn on the basis of 16, 24 and 48-inch modules. Wall sections consisting of complete units with doors and windows (also in modular units) are manufactured in a central production facility and installed with a minimum of cutting and fitting. First floor room heights are 8 ft 1 1/2 inches from floor to the ceiling framing, which accommodates standard 4 × 8 ft wall boards with space allowed for finish flooring and ceiling. Second floor wall heights may be planned in the same manner or, in the case of a 1 1/2 story house, a 7 ft 7 1/2 inch wall height.

Another type of construction provided by Continental Homes, called "core" construction, uses a pre-assembled Kitchen/Bath core, or module, in combination with a panelized package for the balance of the building.

"Modular housing is not a type of house; it is a building technique . . . In modular construction, buildings are designed in sections, or modules, in each of which sections the framing, wiring, plumbing, windows, doors, kitchens, baths, sheetrock, carpeting, siding, roofing, painting, and most of the trim work are completed in a central production facility. These sections are then transported to the job site, where they are installed on a previously prepared foundation, after which an assortment of mechanical connections and trim-out jobs are performed to complete the structure.

Usually, two to four sections will be used to form a single family house; office complexes, schools, motels and condominium projects may be comprised of a hundred or more sections per building, in two-three-or four story configurations.—With Pre-Site Building, more than 90 percent of the home is completed before it reaches the building site . . . The direct labor and materials cost that a builder would experience using conventional building techniques will be comparable to his costs for building the same structure using modules . . . But, the indirect construction costs which must be charged to conventional construction will decline substantially."*

Estimating Guidelines

There appear to be no special procedures necessary in estimating value and loss in modular or panelized constructed buildings . . . It would be prudent, however, to contact the manufacturer when confronted with other than relatively minor losses.

---

* Dennis C. Conlan, Special Projects Manager Continental Homes, A Marley Company Roanoke, Va.

CHAPTER 3

Fieldwork and Notes

The first and most important step in estimating building losses is the meticulous taking of measurements and notes in the field, because they form the basis for computing quantities, and pricing all materials and labor. The degree of care and thoughtfulness used in itemizing each operation is a major control of the accuracy of the final cost. The taking of notes is simply the process of identifying and listing all of the items of construction that are to be repaired or replaced, together with a word or statement to describe the work to be done.

EQUIPMENT FOR ESTIMATING

The estimator should carry a standard wood, metal, or plastic, 6-foot folding rule, and a suitable notebook. A strong flashlight is recommended for inspecting darkened rooms, basements, closets, or other poorly lighted areas. Some estimators use a 6- or 8-foot steel tape, and many carry a 25- or 50-foot steel tape for longer measurements. A small awl, carried in the brief case, is convenient for securing one end of a long tape when working alone. It can be stuck into a wood surface or brick-joint. Many estimators prefer using a clipboard and pad in place of a notebook; some also carry a camera in their car, as well as a collapsible aluminum ladder; others are experimenting with tape recorders to take notes.

PRELIMINARY SURVEY

When the estimator arrives at the place of loss, before taking any notes, the first consideration is to obtain an over-all mental picture of extent of damage, the scope of the work involved in making the estimate, and any specific or unusual problems. To do this one should make a general preliminary inspection of the premises.

If the damage is slight and confined to a small area, there is usually no need for an extensive survey, since a brief examination of the surroundings will disclose all of the information required to begin taking notes. If the damage is wide-spread and serious, a careful reconnaissance should be made in order to formulate a plan of procedure. During the preliminary survey a great many observations will be made. Some may suggest a special method, pattern, or sequence for taking-off the loss. Some will disclose problems that require special treatment. Most of these observations will include the following:

A. The Physical Characteristics of the Building
   1. Its use and occupancy.
      (a) Private or multiple dwelling

      (b) Offices
      (c) Hotel, motel, etc.
      (d) Manufacturing
      (e) Mercantile, institutional
      (f) Other occupancy

2. Number of stories, basement, sub-basement.
3. General plan of building and layout of interior.
4. Type of construction.
      (a) Frame
      (b) Brick, cement block
      (c) Brick-veneer
      (d) Mill
      (e) Fire-resistive
      (f) Metal clad on wood or steel frame
      (g) Non-classified
5. Special architectural or construction features.
6. Quality of materials and workmanship.
7. Age of building.
8. Physical condition of building prior to loss.
9. Obsolescent features.
10. Structural defects prior to the loss.
11. Unrepaired damage prior to the loss.
12. Damage resulting from an uninsured peril.
13. Building code violations.

B. Need for Emergency or Temporary Work
1. Removal of building or contents debris.
      (a) For safety
      (b) For accessibility
2. Removal of building equipment.
      (a) For its protection
      (b) For immediate repair
3. Drainage of plumbing, sprinkler system, and heating facilities to prevent freezing.
4. Shut off utilities—gas, water, electric.
5. Install temporary or permanent roof.
6. Temporary heat required to dry out building.
7. Close openings in walls, windows, roof.
8. Pump out basement, elevator pits, etc.
9. Mop up water on floors—spread sawdust.
10. Install temporary lights to prepare estimate.

C. Accessibility for Inspection
1. Need for ladders or special equipment.
2. Need to obtain tenants' keys for admission.

D. Need for Plans, Specifications, Photographs or Other Evidence
1. Availability of plans or specifications.

    2. Need for their preparation by the estimator.

    3. Need for photographs.

    4. Obtain description of construction and lay-out from tenants.

    5. Obtain description of construction and lay-out from owner.

    6. Obtain description of construction and lay-out from neighbors.

    7. Need for preserving physical evidence.

    8. Need for consulting manufacturers of special or unusual materials or equipment.

E. General Extent of Damage

    1. General area or part of building affected.

       (a) Parts in general needing repair

       (b) Parts in general needing replacement

    2. Have any repairs or replacements been made?

    3. Allow waiting period before deciding on damage to

       (a) Finish floors, asphalt tile, etc.

       (b) Plaster or wallboards

       (c) Paint and wallpaper, etc.

F. Question Ownership or Interest of Tenants and Landlord

    1. Improvements by tenant.

       (a) Store fronts, painting, etc.

    2. Equipment used by tenant.

       (a) Air conditioning, heating, light fixtures, etc.

       (b) Duct work, power wiring, etc.

G. Need to Engage Experts

H. Need to Employ Subcontractors

I. Specific Problems Due To Location

    1. Congested or business area.

       (a) Accessible for trucks loading and unloading.

       (b) Need for sidewalk canopy.

       (c) Need for debris chute.

    2. Proximity to labor and material sources.

    3. Existence of union regulations.

    4. Need to conform to special building ordinances.

    5. Possibility of condemnation of part or all of building.

J. General Area or Parts of Building *Not* Involved

ORDER IN TAKING NOTES

Following the preliminary survey of the damage, a methodical take off of the loss is begun. The starting place on fire repairs may be at the point of origin or area of the most severe damage. It may be either on the exterior, or interior of the building. When taking notes of damage by windstorm or hail, it is customary to start with the exterior and then go to the interior. Where the damage is caused by sprinkler leakage, the obvious place to begin taking notes is on the floor where the sprinkler leakage occurred and work downward. The conditions encountered in each instance will disclose the most likely place to commence taking the measurements and notes.

Experience has taught that progressing from one room to the next, and from one floor to the next, is a satisfactory plan to follow. One should never take off part of the loss in one area, move to another, and then return to the first area. Nor should one take notes on the interior for a while, go to the exterior, and then return to the interior and continue note taking. Not only will the finished estimate appear disorderly, but there is always a possibility of omissions and duplications. Insofar as it is practical, the work should be completed in one area, or one room on a floor, before moving to the next area or room on that floor.

Many estimators follow a rather fixed order of listing damage. This is particularly adaptable where a number of rooms are involved. A typical pattern for taking notes in a room might be as follows:

1. Rough carpentry items.
   (a) Studding
   (b) Floor joists
   (c) Rough flooring
2. Finish carpentry items.
   (a) Windows
   (b) Doors
   (c) Trim
   (d) Finish floor
3. Plastering (or wall treatment).
4. Painting and decoration.
5. Plumbing pipes or fixtures.
6. Heating pipes or fixtures.
7. Electric outlets and fixtures.
8. Demolition and removal of debris.
9. Special or miscellaneous items.

This order of listing becomes fixed in the mind of an experienced estimator, who tends to follow it automatically, always alert, however, to any special items.

There are no rules for the order of taking notes other than, whatever the procedure, they should conform to a logical, easy-to-follow sequence. The following outlines provide suggested systems for taking measurements and notes on various types of losses:

FIRE LOSSES

1. *Partial Losses (All parts of structure not involved):*

   (a) Start taking measurements and notes at point of origin, or the area of most serious destruction, and follow the course of the fire proceeding from room to room, floor to floor, to the portion of building not affected. Go to exterior and record necessary repairs on all sides, and to roof.

(b) Start taking measurements and notes at basement or lowest floor level. Proceed upward, going from room to room, floor to floor, to the highest floor involved. Go to exterior, recording necessary repairs on all sides and to roof.

(c) Start at roof, attic, or highest floor involved. Proceed downward, going from room to room, floor to floor, to lowest floor involved. Go to exterior, recording necessary repairs to all sides and to roof.

2. *Total Losses (or substantially all parts involved)*:

(a) Start taking measurements and notes at basement or lowest floor level. Proceed upward room by room, floor by floor, to highest floor. Go to exterior and record damage to all sides and roof.

(b) Start taking measurements and notes, at basement or lowest floor, by building trades, for entire structure, i.e. masonry, framing, sheathing, flooring, roofing, plastering, painting, plumbing, heating, electric, and so forth.

WINDSTORM AND HAIL LOSSES

1. *Partial Losses (all parts of structure not involved)*:

(a) Start taking measurements and notes at the exterior, recording necessary repairs on roof and all sides. Proceed to the interior, working from highest floor to lowest floor involved, going from room to room on each floor.

2. *Total Losses (or substantially all parts involved)*:
Proceed as in 2(a) and 2(b) under "Fire Losses."

EXPLOSION, BLASTING OR SONIC BOOM LOSSES

1. *Partial Losses (all parts of structure not involved)*:
(a) Start taking notes and measurements at lowest floor involved and proceed upward room by room, floor by floor, to highest floor involved. Go to exterior, recording necessary repairs to all sides and roof.

2. *Total Losses (or substantially all parts involved)*:
(a) Proceed as in 2(a) or 2(b) under "Fire Losses."

SMOKE AND FURNACE BOX EXPLOSION LOSSES

(a) Start taking measurements and notes at furnace room. Proceed from room to room, floor to floor, to highest floor involved. Go to exterior to note any damage.

VEHICLE DAMAGE LOSSES

(a) Start taking measurements and notes at point of collision. Complete exterior and proceed to interior, going from lowest floor involved to the highest floor involved.

LIGHTNING LOSSES

(a) Start taking measurements and notes at point where lightning struck. Complete examination of all sides and roof of structure. Go to interior and note damage beginning at lowest floor involved to highest floor involved, or follow path of lightning where it entered building.

SPRINKLER LEAKAGE AND WATER DAMAGE LOSSES

(a) Start taking measurements and notes at floor on which the leakage originated. Proceed downward room by room, floor by floor, to lowest floor involved.

FLOOD, WAVE-WASH, BACKING UP OF SEWER LOSSES

(a) Start taking measurements and notes at basement or lowest floor. Proceed, room by room, to floors above that are involved.

DESCRIPTIVE NOTES

The actual recording of notes in the field should be brief, simple, and readily understandable to anyone reviewing them, even though they may never have visited the scene of the loss. They should convey the following information:

1. The *location on the premises* where each item of repair or replacement is to be made.
2. All *measurements* required to compute the quantities of material, unless the quantity is determined in the field, in which case it should be noted.
3. The *kind, size, and quality of material* to be used.
4. A descriptive word or statement of the *work to be performed*.

1. *Location on the Premises.* The notes should show in what building the repairs are contemplated. They should identify the floor, and the room, or part of structure involved, as: *Main Residence, Second Floor, Master Bedroom*, or *Factory Warehouse, First Floor-Rear, Shipping Office*; or *Dormitory Building, Third Floor, South Wall.*

Notes made in this manner will make it clear to anyone the precise location to which the estimator is referring.

2. *Measurements.* Since very few loss estimates are prepared from plans from which measurements can be scaled or read, it is necessary to show in notes all measurements that are necessary to compute areas and quantities. The size of each room to be plastered, painted, or papered, or where floors are to be laid or scraped, should be indicated, giving the length, width, and height. The measurements of roof areas to be repaired or recovered, walls to be repaired or replaced, and other measurements from which quantities are computed, should be noted.

Illustrative Example

Main residence
  Second floor
    Master bedroom 14′ × 16′ × 8′
             (or)
Repair wall plaster 6′ × 8′
             (or)
Repair 210-lb. asphalt shingles 16′ × 20′
             (or)
Replace 8″ common brick wall 12′ × 32′
             (or)
Repair 1″ × 3″ oak floor 6′ × 10′
             (or)
Main residence
  Southwest side
    Replace bevel siding 10′ × 16′

In each instance, measurements that are noted as illustrated will provide the basis for computing the quantity of material needed, the labor to apply it, and the total cost of the operation.

3. *Kind, Size and Quality of Material.* The cost of a particular kind of material varies with its size and quality and it is impossible to price it without adequate information. Almost all materials used in the construction of a building are available in different grades, and the field notes should include sufficient data to properly identify them.

Illustrative Example

Install 4 lights double-strength glass 18″ × 24″.
             (or)
Replace 1 pair 3′ × 4′ × 13/8″ — 8-light pine sash.
             (or)
Plaster ceiling 2-coat sand-finish on gypsum lath.
             (or)
Apply Certigrade No. 3 Black Label — 16″ wood shingles.

4. *Work To Be Done.* Estimators use descriptive verb-words or their abbreviations to tell what work or operation is to be performed. They should be clear and informative. The following are examples of those most commonly used.

Build	Install	Plaster
Cart or truck	New (this or that)	Rebuild
Decorate	Paint	Refinish
Erect	Paper	Remove
Hang	Patch	Repair
Replace	Sand	Tear out
Reset	Scrape	Wash down
Resurface	Straighten	Whitewash
Rewire	Take down	

Much time can be saved by becoming familiar with these words and by accumulating others. They should be employed when recording notes. A few abbreviations are frequently useful, such as

Ref. . . . . Refinish
Rem. Remove
Rep. . . . . Repair
P&D. . . . . Paint and decorate
T&C. . . . . Tear out and cart
Pl. . . . . Plaster
WW. . . . . Whitewash

MEASUREMENTS AND DIMENSIONS

In measuring the length, width, and height of rooms, and in measuring surface area, except where scale plans or sketches are being made, it is considered sufficiently accurate to measure to the nearest whole foot or half foot. A room with actual dimensions of $12'2'' \times 18'5'' \times 8'2''$ may be recorded in the notes as $12' \times 18'6'' \times 8'$.

Dimensions taken for building materials, however, which are sold in specified lengths, widths, or thicknesses, should always be scaled accurately. The sizes of doors and windows, and their thicknesses, make a considerable difference in the price at the lumber yard. Baseboard and finish flooring are sold in different widths. Glass comes in single, double, semi-plate and plate thicknesses. Wood shingles are made in 16-, 18-, and 24-inch lengths. The dimensions of trim material vary for interior millwork, exterior cornices, water table, and corner boards, and they must be scaled accurately.

SKETCHES AND PLANS

Line drawings to record outside measurements, story heights, roof shape and pitch, interior layout of rooms or service facilities, can be extremely useful. On serious or total losses, they may provide a basis for eventually drawing formal plans. They are also a means for obtaining or double-checking quantities, locating windows, doors, and special building fixtures or equipment.

Sketches showing the details of certain construction, or architectural features, may also be useful later when the amount of material and labor is being estimated. A line sketch of a cornice construction or a framing detail of a building, takes only a few moments to make, but it may well serve to avoid a return trip to the premises should the written notes be inadequate, or if memory fails. As a matter of practice, many experienced estimators tape the outside of a building and draw a line sketch showing the dimensions before they enter the premises on all but the smaller losses.

When a loss is very serious, or a constructive total loss, plans of some kind will almost always be necessary to prepare an estimate. In some cases original plans may be obtained from the local Building Department, from the owner, from the architect, or contractor who built the building. In the majority of cases, however, the plans are either not available or incomplete, and new ones must be prepared. Very often the person who makes the estimate draws such plans.

Figure 3–1

A typical "field sketch" to record data for later use.

Much of the data for plans can be obtained at the scene of the loss. It usually is necessary to get a complete description of the building from persons who have been closely associated with it. As many persons as necessary should be interviewed to verify details. The following persons or firms are suggested as likely sources of information:

Owner or members of the family.
Friends of the owners who have visited premises.
Tenants or employees of tenants.
Superintendent of building.
Employees of owner.
City inspector.
Insurance inspector or engineer.
Guard.
Caretaker.
Builder of property.
Original architect.

Realty management firm.

Former appraiser of property.

Neighbors.

Real estate sales people familiar with the property.

Maintenance or repair firms who have worked in the building.

Any persons possessing photographs taken before the loss.

The questioning of persons concerning layout and construction is best done at the premises in order that their memories are refreshed by inspection and examination of the remaining portion of the building. It is well to have a list of questions ready to save time and to make certain that nothing is omitted. The list should follow the sequence of actual restoration from the ground, or foundation, to the roof. Circumstances in each case will govern the precise pattern of questioning to follow.

EMERGENCY OR TEMPORARY REPAIRS

After a loss has occurred, there usually are measures which have to be taken promptly to protect and preserve the property from further damage by the elements, trespassers, corrosion, dampness, water, and so forth. Most of the steps to be taken will fall into one or more of the following five classifications:

1. *Protection Against the Elements*
    (a) Close up openings in roofs and walls. Board up windows and doors. Permanently glaze windows and sash. Temporarily or permanently repair roof and skylights.
    (b) Drain plumbing and heating pipes, boilers, etc.
    (c) Provide temporary heat.
    (d) Install temporary partitions or barricades.
2. *Protection Against Water and Dampness*
    (a) Air and ventilate premises.
    (b) Mop floors, scatter sawdust, bore holes in floors.
    (c) Shut off water.
    (d) Pump out basement.
    (e) Clean, oil, dry equipment.
3. *Protection Against Trespassers*
    (a) Close openings in walls and roof.
    (b) Provide locks for doors and windows.
    (c) Provide guard.
4. *Protection Against Unsafe Conditions*
    (a) Take down hazardous masonry walls and chimneys.
    (b) Brace.
    (c) Remove excessive weight of debris or equipment from weak floors.
    (d) Shore up weak floors or roof structures.
    (e) Remove contents that swell in water (newsprint, cottonseed, etc.).
    (f) Shut off gas and electric.
5. *Provide Emergency Facilities*
    (a) Temporary electric lighting.

(b) Temporary electric power.

(c) Temporary repairs to elevators.

On arrival at the premises, the estimator may discover that part or all of the temporary protection and emergency repairs has been completed. In that case, notes should be taken in sufficient detail to check any bills or details which may be submitted later.

If none of the temporary or emergency work has been started, the estimator should include the necessary specifications in the notes.

DAMAGE THAT EXISTED PRIOR TO THE
OCCURRENCE OF THE LOSS

When a person is inspecting a building loss and making notes, the objective, in the majority of cases, is to determine the extent of damage which has been caused by a specific hazard or peril covered by the policy. It is necessary, therefore, to be alert to any conditions which indicate a need for repair, but which are in no way related to the peril involved. Buildings that are poorly maintained often contain defects which may consist only of deteriorated paint, broken glass or cracks in the plaster, or they may be far more serious, including such things as rotted timbers, sagged roofs, or bulged masonry walls. Frequently, these conditions are in such close proximity to the damage caused by the peril involved that they are inadvertently included in the estimate of loss. Many times, in fact, it is difficult to separate the two.

The more common defects and types of deterioration will be evident by the following.

1. Sagging roofs.
2. Sagging floors.
3. Cracks in exterior brick walls.
4. Cracks in exterior stucco.
5. Cracks in foundations or basement floors.
6. Cracks in interior plaster.
7. Stained or soiled wallpaper and paint.
8. Efflorescence on interior plaster.
9. Rust and corrosion on metal building materials, equipment or apparatus.
10. Rotted timbers or other woodwork.

These defective conditions may be the result of one or more of the following common causes:

1. Earth movement or subsidence, due to improper drainage, poor fill, shrinkage from drought, tunneling operations below, etc.
2. Poor foundation design—not below frost line, too narrow a footing, improper reinforcing or concrete materials and mixtures.
3. Inadequate size of timbers or their spacing to safely support load intended.
4. Shrinkage of wood timbers and floors.
5. Overloading by tenants.
6. Poor mixture or materials in stucco or masonry mortar.

7. Poor mixture in wall plaster.
8. Plaster or stucco improperly applied.
9. Inadequate bracing and bridging; or ties to veneer brick.
10. Fungus or insect injury to wood supports.
11. Seepage of rain through brickwork or other masonry.
12. Leakage of rain or melted snow through roofing, behind flashing, around windows and doors, or directly down chimney.
13. Hydrostatic pressure under basement floor, or backing up of storm and sanitary sewers.
14. Abuse by use, especially by tenants.
15. Vibration.

An experienced estimator is generally able to recognize most of these conditions and will include notes concerning their existence, their probable cause and the cost to repair them.

DAMAGE FROM A PREVIOUS LOSS NOT REPAIRED

An insured does not always make repairs to a building that has sustained a loss. There are several reasons. The damage may have been excluded or not covered under the insurance policy at the time of the loss. Or, it might have been less than the deductible. It may be in a location where appearances are not important. The damage may have been partially repaired, patched or covered over to conceal it.

VIOLATIONS OF LOCAL BUILDING CODES
OR ORDINANCES

Most municipalities have building codes and ordinances that are designed to regulate the alteration, repair, and construction of buildings. They also stipulate certain materials and construction for buildings that may be damaged by fire, lightning, windstorm, and so forth.

The field notes should indicate any violations, and the estimate should include only the *repairs* necessary to restore what existed at the time of loss, because the standard fire policy is limited to repairing and replacing with like kind and quality, without any increase of loss "occasioned by ordinance or law regulating construction or repair of buildings."

REPAIRS MADE PRIOR TO INSPECTION

It is not uncommon for the estimator to find that repairs were partially or completely made prior to inspection. When such repairs are minor, and readily explained as to the scope, there is no particular problem. The estimator obtains a copy of the estimate or invoice for making the repairs and prepares specifications and costs, frequently having the owner or owner's representative along to answer any questions.

If, however, the repairs are extensive, the estimator may need considerable time and ingenuity to establish with any degree of accuracy what occurred and develop a set of specifications on which to base the estimate.

Principal sources of information will be:

1. The insured or a representative.
2. The general contractor's estimate or records of payroll and materials used.
3. Subcontractor's records, estimates, etc.
4. Tenants in the area (if any).
5. Other insurers if they are involved.
6. Any plans, pictures, building permits, etc.

CONSTRUCTIVE TOTAL LOSSES

Estimators are confronted many times with the question of whether the building involved is a constructive total loss. Where valued policy laws are in effect, it is an extremely important question if the amount of insurance is in excess of the cost of rebuilding. Where valued policy laws are not in effect it is important if a reasonable part of the structure remains upon which to begin rebuilding.

The general rule applied by the courts seems to be:

Whether a building is an actual total loss (by fire) depends upon whether a reasonably prudent owner, uninsured, desiring to rebuild, would have used the remnant for restoring the building.*

ESTIMATING LOSSES CAUSED BY MIXED PERILS

Occasionally damage is caused by two perils—one covered, the other not covered. For example, during hurricanes along the seacoast, properties close to the shore are frequently damaged by wind (a covered peril), and by wave-wash (an excluded peril). Similarly, if a fire policy excludes "explosion" and one occurs followed by fire, the damage done by the explosion is not covered. Or, if a fire policy does not cover earthquake, and one occurs, only ensuing fire damage is covered.

Damages by mixed perils such as those described, require the estimator to resort to a three-column estimate which lists the repairs to damage covered, the damage not covered, and the damage which is questionable as to its being covered. (See Estimating in Catastrophes, Chapter 5)

However, if a covered peril such as fire occurs and during the fire there is an explosion, all of the damage is covered because fire is the proximate cause.

---

*Glens Falls Ins. Co. v. Peters, Texas Supreme Ct. Jan. 13, 1965. Royal Ins. Co. v. McIntyre, 90 Tex. 170 (1896). Fire Ass'n. v. Strayhorn, 211 S. W. 447 (1919).

CHAPTER 4

Pricing and Completing the Estimate

After the damaged property has been inspected and all the measurements and specifications for repair have been compiled, the final step to complete the estimate and obtain the total cost of repairs is the pricing. It consists of computing the cost of the materials required for each item listed, and estimating the number of hours and cost of the labor necessary to install the material or perform the operation indicated. Material costs are based on the local prices, delivered to the job. Labor costs are based on prevailing local wage scales for the building trades involved.

PRICING METHODS

As explained under "Fundamental Considerations" (Chapter 2), there are different customs and practices among contractors for figuring the cost of material and labor. Whatever procedure is followed, if it is to be considered reliable, must take into consideration these two factors.

1. Quantity of Material $\times$ Price = Cost of Material
2. Hours of Labor Required $\times$ Wage Scale = Cost of Labor

There are three general methods of pricing and setting up material and labor costs in an estimate.

1. The material is priced in detail for each item, and the labor cost is bulked in a lump-sum amount.
2. The material and labor are individually priced for *each* item, and shown separately.
3. A unit cost is applied to each item to obtain a single sum for the combined cost of material and labor.

METHOD 1: MATERIAL PRICED, LABOR LUMPED

Many contractors and builders employ this method. The materials are listed for each item of repair, quantities are extended and properly priced. The cost of labor is then determined by approximating the hours or days required to perform all of the work. Where minor repairs are involved, the method can be fairly accurate, but when it is used on losses where extensive repairs are required, it can take on the character of *guessing* rather than *estimating*. Carpentry work is often figured as shown in the following example.

Illustrative Example*

Framing	Dimensions	FBM		
42 Joists	2" × 12" × 16'	1,344		
48 Studs	2" × 4" × 9'	288		
19 Plates	2" × 4" × 12'	152		
12 Joists	2" × 8" × 16'	256		
2 (Girder)	2" × 12" × 14'	56		
2 (Girder)	2" × 12" × 16'	64		
38 Studs	2" × 4" × 8'	202		
18 Plates	2" × 4" × 12'	144		
15 Rafters	2" × 4" × 18'	180		
8 Rafters	2" × 6" × 18'	144		
Blocking	4" × 6" × 8'	16		
		2,846	@ .40 =	$1,138.40

Roof Sheathing				
1,000 FBM (1" × 6")			@ .25 =	250.00
Subflooring				
1,200 FBM (1" × 6")			@ .25 =	300.00
Finish Floor				
1,500 FBM (1" × 3" C.G. Pine)			@ 1.20 =	1,800.00
			Material =	$3,488.00
Carpenter Labor 25 days @ $96			=	2,400.00
			Material and Labor =	$5,888.40

It should be noted that some contractors who use this method may set up the labor cost in the finished estimate as a lump sum, but *actually* they have worked out the labor cost for each individual item which can be verified from their working papers.

METHOD 2: MATERIAL AND LABOR PRICED
AND SHOWN SEPARATELY

In this method the material cost is computed for each item and extended to a "Material Cost" column. Labor for each item is determined and the cost is extended to a "Labor Cost" column. A third column indicates the "Total" cost for both material and labor.

Illustrative Example

Item				Material Cost	Labor Cost	Total Cost
Framing						
12 Studs	2"×4"×10'	80 FBM				
16 Joists	2"×6"×16'	256 "				
20 Rafters	2"×6"×14'	280 "				
		616 "	@ $.40	$246.40		
20 Carpenter hours			@ 12.00		$240.00	$486.40
Sheathing						
864 FBM			.25	216.00		
18 Carpenters hours			@ 12.00		216.00	432.00
						$918.40

*Wages and prices are assumed for illustrative purposes

The form in which the estimate is set up varies with individual preferences of contractors. Printed forms are available from firms like Frank R. Walker, Chicago, Illinois, that specialize in contractors' stationery. Some of them provide columns for both quantities and unit prices for material and labor. Examples of this and other forms appear on pages 61, 62 and 63.

METHOD 3: UNIT COSTS FOR MATERIAL
AND LABOR COMBINED

Considerable estimating is done by the *unit cost* method. A unit cost is the combined cost of the material and labor needed to install a unit of material; a unit of material being a square foot, a cubic foot, a lineal foot, a board foot, etc.

If a roofing contractor's workers can apply a square (100 square feet) of 235-lb shingles per hour on a plain one-story dwelling roof, it is a simple matter to estimate the cost of applying 235-lb shingles to similar roofs. It can be done in two ways. If the roof requires 20 squares of shingles the roofer figures the time required by multiplying 20 by 2, equals 40 hours. If the price of the shingles is $20 per square and the wages are $12 per hour, the estimate would be:

20 squares of shingles	@	$ 20.00 =	$ 400.00
40 hours labor	@	12.00 =	480.00
	Cost of material and labor		$ 880.00

The unit cost method for estimating the shingles would be

1 square of shingles	@	$20.00 =	$ 20.00
2 hours labor		12.00 =	24.00
	Cost per square		$ 44.00
The cost of laying 20 squares would be			
20 × $44.00 =			$ 880.00

The advantage of this method of estimating is apparent. The unit cost may be reduced to the cost per square foot, $.44 (1 square = 100 square feet). In this illustration the roofer can quickly estimate similar jobs by obtaining the area and multiplying by $.44. Using the unit cost of $44 per square or $.44 per square foot is a matter of choice.

This illustrates the principle of developing unit costs. It can be used for most building materials, as well as for certain pure labor operations, such as removing old shingles, washing down walls or exterior brickwork.

DEVELOPING UNIT COSTS

A unit cost may be developed by starting out with a quantity of material for which the average labor rate of application or installation is known. Quantities and symbols that are usually used to *express* unit costs are listed in Table 4–1. The basic formula for developing unit costs and several examples are shown on page 57. The local price of the particular material is inserted in the formula, the hours of labor are multiplied by the local wage scale.

TABLE 4–1

Quantity Symbols and Abbreviations in Estimating

CONCRETE
Forms	sq ft
Foundations, walls, slabs, etc.	cu yds or cu ft
Gravel, stone, sand	cu yds or ton

MASONRY
Brick	per 1000
Cement	bag or sack
Concrete block	ea. or per 100
Coping	lin ft
Flue lining	each or lin ft
Gypsum block	per 100
Stone work	CWT (100 lbs)
Terrazzo floors	sq ft

CARPENTRY, Rough
Framing, dimension lumber	per bd ft or 1,000 FBM
Studding, sometimes per stud	
Bridging, wood and metal cross bridging	per set
Solid bridging	per pc
Furring	per 100 lin ft
Sheathing, subflooring, roof decking	
Composition	per sq ft
Plywood	per sq ft
Shiplap, T&G boards, square edge boards	per sq ft or 1,000 FBM
Trusses	each or 1,000 FBM

CARPENTRY, Finish
Doors	each
Flooring, wood finish	sq ft or 1,000 FBM
Sanding floors	sq ft
Screens and storm doors	each
Screen wire	per sq ft
Shutters	each
Stairs	per tread or riser
Trim, base boards, various moldings	per lin ft
Casings, window and door	per side
Shelving	sq ft or bd ft
Wood gutters, water table, cornice boards	lin ft or bd ft
Windows, sash	each
Wood panelling	sq ft or bd ft

INTERIOR WALL FINISH
Dry wall, gypsum board, etc.	sq ft
Lath and plaster	sq yd
Painting	sq ft or sq yd
Tile, floor and wall	sq ft
Wall paper	sq ft or roll
Paneling, wood and composition	sq ft

EXTERIOR WALL FINISH
    Shingles, wood, composition and metal   Sq ft or square (100 sq ft)
    Stucco, exterior plaster   sq yd
    Waterproofing   sq yd
    Siding, wood, composition and metal   sq ft

ROOFING   sq ft or square (100 sq ft)

ELECTRIC
    Outlets   each
    Wiring   lin ft or per outlet
    Fixtures   each

INSULATION
    Bats   sq ft or 100 sq ft
    Loose fill   bags or cu ft per thickness
    Vapor barriers   sq ft

KITCHEN CABINETS AND BUILT-INS   per unit or lin ft

FLOOR COVERINGS, Linoleum, vinyl, carpeting, etc.   sq yd or sq ft

MISCELLANEOUS
    Flashing   sq ft or lin ft
    Glass   light or sq ft
    Gutters and leaders, metal   lin ft
    Piping   lin ft
    Plumbing fixtures   each
    Steel beams and girders   per ton
    Weatherstripping   per opening

Basic Formula

Quantity of material	@ $	=	**$**
Hours of labor	@	=	_____
Unit cost of material and labor		=	**$**

Illustrative Example
Framing Lumber

1,000 FBM lumber	@	$400.00 =	$ 400.00
10 Lbs nails	@	.60 =	6.00
25 Hours labor	@	12.00 =	300.00
Unit cost per 1,000 FBM		=	$ 706.00
Unit cost per FBM		=	$ .706

One-coat Interior Paint
(Assume 1 gal paint covers 450 sq ft)

1 Gal paint	@	$ 14.00 =	$ 14.00
3 Hours labor	@	12.00 =	36.00
Cost per 450 sq ft		=	$ 50.00
Unit cost per sq ft	$\dfrac{\$50.00}{450}$ =		$ .11

An estimate that is priced by the unit cost method might appear in final form as follows (prices and wages assumed):

Repairs to Garage

Framing
10—2″×4″×12′ studs	80 FBM		
6—2″×6″×16′ rafters	96 FBM		
	176 FBM	@ .70 .	$ 123.20

Roof Boards
20′×24′	480 sq ft.		
Waste 1/6	80 sq. ft.		
	*560 FBM	@ .60 .	336.00

Bevel Siding
10′×16′	160 sq. ft.		
Waste 1/3	54 sq. ft.		
	214 FBM	@1.40 .	299.60

Exterior Paint
2,400 sq. ft. (2 coats)		@ .25 .	600.00
Total material and labor			$ 1358.80
Add overhead and profit 20%			271.76
			$ 1,630.56

A VARIATION FOR SETTING UP UNIT COSTS

Some estimators use a unit cost for the material and a separate unit cost for the labor, rather than a combined unit cost for material and labor. For example, an estimate for 2,000 square feet of 235 lb asphalt shingles would be set up in the estimate as follows:

2,000 sq ft shingles	@ $	.20 =	$ 400.00
Labor	@	.24 =	480.00
		Total =	$ 880.00
(or)			
20 squares shingles	@	$20.00 =	$ 400.00
Labor	@	24.00 =	480.00
		Total =	$ 880.00

CHARACTERISTICS OF UNIT COSTS

Although theoretically a unit cost should consist only of the cost of the materials and the labor cost to install them, there are variations in practice. Some contractors include overhead and profit. Others include removal of damaged material preparatory to installing the new. Still others may include waste. In checking a unit cost, if the percentage of overhead and profit is known, it can be deducted. But since demolition or tearing out and waste are difficult to extract from a unit, *it is recommended that they be treated separately in estimating.*

---

*When milling waste is added to the square foot area, the total becomes *board feet.*

It is recommended that waste *be added to the quantity of material* to be installed. Demolition, except for a few materials such as shingles, is best estimated under the general debris item. A unit cost should contemplate that the debris and damaged material have been removed preparatory to reconstruction, and that the conditions under which new materials are to be installed are comparable, to a large extent, to those normally encountered in *new construction* work.

While unit costs are developed from such quantities as 1,000 bricks or 1,000 FBM of lumber, they can be applied readily to smaller quantities of material where the working conditions are fairly normal. For example, a unit cost for finished 1″ × 3″ flooring developed from 1,000 FBM can be applied to the quantity of flooring required in a room 10 × 10 feet.

It would not be proper, however, to apply such a unit cost to a 2′ × 8′ closet where the area worked in is confined and difficult as compared to an open area contemplated in estimating the production rate for the 1,000 FBM. Here, as in all estimating, judgment must be exercised. The unit price can be increased to compensate for the additional labor required or the material and the time needed to lay the floor can be estimated separately.

Unit Costs Not Applicable to Patch Jobs

Unit costs, as a general rule, cannot be applied to small patch jobs requiring only a few yards of plaster, or several square feet of flooring or roofing. The reason is that the labor cost in patch work is very large compared to the material cost. It will distort the combined material and labor unit cost, making it difficult, if not impossible, to judge one that is adequate. In such cases the cost of the material should be obtained, and the hours of labor figured separately for the entire patch. In other words, the estimate should be made on a time and material basis. A roofer may spend an entire day patching in a square of shingles. If the shingles cost $20 a square, and labor is $12.00 and hour, the total cost per square would be:

1 square of shingles	$ 20.00
8 hours of labor @ $12.00	96.00
Total cost per square	$116.00

This is considerably different from a unit cost developed for average straight work involving an entire roof.

1 square of shingles	$20.00
2 hours of labor @ $12.00	24.00
Total unit cost per square	$44.00

Extracting Labor from Unit Costs

Estimators will frequently increase a unit cost to take care of more difficult working conditions than those contemplated when originally computing it. There is no difference in the amount of material in the unit, therefore *the entire increase must be charged to labor.* In an analysis of unit costs this characteristic must be constantly remembered, otherwise any indiscriminate increasing of unit costs *will automatically raise the hours of labor* to unrealistic quantities. If a predetermined unit cost for asphalt roof shingles is 44 cents a square foot, but is raised to 50 cents, the additional 6 cents a square foot is entirely labor.

Analyzing Unit Costs. It is recommended that the reader be well grounded in analyzing unit costs in order to reduce them to their basic components of labor and material, thereby extracting removal of debris, excessive labor loading, and other hidden items.

REFINEMENT OF UNIT COSTS

The question is often asked, "Should unit costs be in whole numbers or carried out to decimal places?" The answer primarily is one of judgment and, to a large degree, centers on how much confidence the estimator has in the hours of labor in the unit. There are times when the estimator analyzes the operation that is to be performed and concludes that the unit cost may be too high or too low and adjusts it accordingly. But as a general rule, unit costs are used in whole numbers except on large jobs and when bidding is highly competitive.

MINIMUM CHARGES FOR SMALL JOBS

Frequently small repair jobs are subject to minimum charges for labor and material, either combined or individually. For example, one cubic yard of ready-mix concrete to be delivered over several miles may cost the same as two yards. One lite of glass may cost as much as two or three. Several shingles to be replaced; a few hours of painting; checking electric appliances, a furnace, plumbing, etc., often are subject to minimum charges. The reason may involve the distance to be traveled, or, the fact that the best part of a day has been taken up and it is too late to go to another job. High overhead may also be a consideration in small jobs. (See also Union Regulations, page 22).

These are important considerations for the estimator particularly in periods or in areas where builders and subcontractors are extremely busy. Small jobs can be an annoyance because of having to take a worker away from a large job for a few hours or days.

In inflationary periods, the cost of making an estimate is frequently billed to the customer if the job is not given to the firm or individual who made the estimate.

THE FINISHED ESTIMATE

Since the real purpose and usefulness of an estimate of damage is to provide the groundwork for discussing the adjustment of a loss, the final form should set forth every item of repair or replacement. The work to be done should be so itemized and clearly described that even the uninitiated would be able to understand it. Quantities and the cost of labor and materials should be indicated.

Bulking, grouping, or lumping two, three, or several items into one sum, only tends to confuse and introduce unnecessary areas of misunderstanding. It raises differences of interpretation as to exactly what the estimator meant or intended. Such prac-

GENERAL ESTIMATE

ESTIMATE NO. _____

BUILDING _____

SHEET NO. _____

LOCATION _____

ESTIMATOR _____

ARCHITECTS_____

CHECKER _____

SUBJECT_____

DATE _____

DESCRIPTION OF WORK	NO. PIECES	DIMENSIONS			EXTENSIONS	EXTENSIONS	TOTAL ESTIMATED QUANTITY	UNIT PRICE MT'L	TOTAL ESTIMATED MATERIAL COST	UNIT PRICE	TOTAL ESTIMATE LABOR COST

Form 4–1

FORM 420 5 74

 GAB General Adjustment Bureau, Inc   **Detailed Estimate**

DATE

BRANCH   OFFICE TELEPHONE NO

OWNER		FILE NUMBER		BUILDING	CONTENTS
LOCATION		TELEPHONE NO			

$ _____ ESTIMATED REPLACEMENT COST $ _____ ESTIMATED REPLACEMENT COST

| POLICY NO | CLAIM NO | DATE OF LOSS | TYPE OF LOSS |

$ _____ LESS DEPRECIATION $ _____ LESS DEPRECIATION

| TYPE OF BUILDING | | BUILDING AGE | TOTAL SQ FT |

$ _____ ACTUAL CASH VALUE $ _____ ACTUAL CASH VALUE

| ROOF TYPE | | TYPE OF CONSTRUCTION | | NO ROOMS | NO BATHS |

DESCRIPTION	CHECK APPLICABLE ITEM	BUILDING → CONTENTS →	LABOR HOURS	LABOR RATE	TOTAL ① LABOR	❷ MATERIALS
			AGE	ITEM COST	DEPRECIATION	ACTUAL CASH ❸ VALUE
(CONTINUE ON FORM 448)			TOTAL			

TOTAL LABOR (BUILDING)	① $
TOTAL MATERIALS (BUILDING)	❷ $
TOTAL REPAIRS (BUILDING)	$
LESS DEPRECIATION	$
ACTUAL CASH VALUE (CONTENTS)	❸ $
LESS DEDUCTIBLE OR CO-INSURANCE PENALTY	$
ORDERED BY (APPRAISALS ONLY) TOTAL CLAIM	$

AGREED TO BY (APPRAISALS ONLY)   PREPARED BY

Form 4-2

Kemper
INSURANCE

BUILDING REPAIR ESTIMATE

	CLAIM NO.

INSURED

LOCATION OF PROPERTY

TYPE OF LOSS	LOSS DATE	TYPE OF BUILDING

REPLACEMENT VALUE	ACTUAL CASH VALUE	ESTIMATOR		DATE OF ESTIMATE

WHEN COMPLETING THE DATA BELOW, TRY TO DEVELOP AND USE "UNIT PRICES"; OTHERWISE, ESTIMATE "MATERIAL COST" AND "LABOR COST" SEPARATELY.

DESCRIPTION OF WORK REQUIRED (IDENTIFY MATERIALS – SHOW ROOM OR AREA MEASUREMENT)	QUANTITY	UNIT PRICE (WHEN USED)	MATERIAL COST	LABOR COST	TOTAL COST

EK432 2-73 1M PADS

PRINTED IN U.S.A.

Form 4–3

tices are without good reason. An estimator, to be as accurate as possible, must break down each operation of repair. The final estimate should reflect such a breakdown. Each item should lend itself to analysis because most of the differences in estimates can be traced to errors in measurements, arithmetic, material costs, labor allowances, or judgment.

Follow Field Notes

Generally, when field notes have been taken off as suggested in Chapter 3, there is a logical starting point, and continuity flows throughout these notes. The sizes of rooms, descriptive specifications, dimensions, and measurements need only be transcribed to the final estimate. Areas and quantities can be computed, priced, and extended.

Frequently, it may be desirable or convenient to keep certain types of work or trades together. Framing, for example, may involve the roof structure, some of the sidewalls, and perhaps a few of the floor joists in the rooms below. These items may be pulled out of the notes and listed under one heading, "Framing," in the final estimate. When subcontractors are used on such trades as plastering, painting, electrical or steelwork, the work contemplated by each should be specified in the bid or estimate in detail to permit checking. A subcontractor's estimate in a *lump sum* amount has little or no value.

Bulking Labor

This practice is not recommended in the final form. It is meaningless to anyone except possibly the person who prepared the figures. Not many estimators can guess at the overall number of hours that a job of any size will take. Many very capable estimators show labor in bulk, but they obtain the final labor figure by a careful analysis of each operation. Such analysis is usually in their notes or worksheets. Because building loss estimates require explicit details for negotiating, the labor as well as the material should be itemized for each operation.

Showing Unit Costs

The final form need not necessarily show individual unit costs because they may be computed by dividing the cost of the item by the quantity. For simplicity, however, it is desirable that unit costs be shown for each item.

Overhead and Profit

It is also recommended for uniformity that contractor's overhead and profit be added at the end of the entire estimate. Unit costs, or labor and material costs, for each item are better shown on a "net" basis; that is, one that excludes overhead and profit. The cost of each item should be made up solely of the material and the labor necessary to perform the operation or install the material. One exception to this recommendation may be made in catastrophe losses where unit costs, which sometimes include overhead and profit, are agreed to by the local contractors' association and representatives of the insurance companies (see Chapter 5).

Other exceptions may be made where the pricing of certain items is based on a unit cost quotation of a subcontractor. This may be a roofer, painter, mason, electrician or tiler who quotes a price per unit which the estimator uses, and which includes the subcontractor's overhead and profit.

Recapitulation

Estimators who figure a great many building losses recognize the value of recapping their estimates by trades. A comparison of the recapitulation of two estimates will quickly reveal whether the differences are in one or two items, or are scattered throughout. Recapping an estimate is done after the entire estimate is completed and priced. The amounts of money shown for every item in a particular trade as carpentry, painting, plastering, or electric are added together. The various trades are listed on a separate sheet entitled "Recapitulation" and the total is shown opposite each trade. This should be done with great care, with a check-off of each item as it is listed from the estimate. The total of the recapitulation must always equal the total for the detailed estimate.

Examples of loss estimates are shown on pages 66 to 78. They have been selected from a number submitted by contractors with considerable experience in estimating building losses and in making repairs.

COMPUTERIZED ESTIMATING

A relatively recent innovation in estimating building construction costs, as to both loss and damage and replacement cost, is what is known as *computerized estimating*. In its application to building losses under insurance contracts it is gaining recognition as a time and expense saver.

When using this system an individual takes off the specifications—the measurements, scope of work to be done, kind and quality of materials,—and notes these details on specially coded Work Lists or Work Sheets that have been designed and are furnished by the particular computerized estimating service being used. These coded work sheets are planned separately for interior rooms and for the exterior of the building and may consist of a few thousand to as many as ten thousand so-called *action items* which are classified by each building trade and by the specific operation to be performed, i.e. *Remove, Install, Repair, Refinish, Paint,* and *Clean.* Provision is made for the person taking off the specifications to override or bypass the computer by means of special coded notations on the worksheets.

The function of the computer is to extend measurements into areas and quantities and *price-out* the items on the coded worksheets. Material prices, wages and such things as unit costs are programmed into the computer for each locality and are kept current. In other words, the computer is told what has to be done to build or repair a structure through the coded work sheets, and then does all of the arithmetical calculations. In this respect the term *Computerized Estimating* is a misnomer, though the role of the computer in the finished estimate is as important as the role of the person who has charge of preparing the specifications.

DRIVER-EDDY *construction company*

☒ 2727 SOUTH SHANNON ST. • SANTA ANA, CA 92704 • TEL. 714/545-7189
☐ 5801 WEST JEFFERSON BLVD. • LOS ANGELES, CA 90016 • TEL. 213/870-7821
☐ 4930 NAPLES STREET • SAN DIEGO, CA 92110 • TEL. 714/275-3230

World Insurance Company
1000 Wilshire Blvd.   DATE   January 12, 1983
Los Angeles, CA

OWNER: Mr. and Mrs. John Smith

LOCATION: 000 Victoria Ave.   ESTIMATOR Robert Driver
            Los Angeles, CA

                                       JOB NO.   00001

PROPOSED WORK TO BE ACCOMPLISHED	QUANTITY	UNIT PRICE	AMOUNT
Fire Damage - Residential			
1) Building permit and plan check			$ 210.00
2) Demolition			2,188.00
3) Cartage	3 bins	$125.00	375.00
4) Structural lumber and roof sheathing			3,676.40
5) Rough hardware			270.00
6) Carpenter labor framing			4,868.00
7) Roofing - 20 squares heavy shake			4,000.00
8) Sheet metal, gutters, downspouts, flashings			590.00
9) Electrical wiring and 100 amp service			2,375.00
10) Electrical fixtures			260.00
11) Plumbing - vents, gas test			485.00
12) Heating and air-conditioning - replace ductwork and thermostat			610.00
13) Deodorize attic areas not replaced			305.00
14) Insulation			860.00
Living Room (12x18x9)			
1) Lath and plaster ceiling	24 yds	15.00	360.00
2) Replace 2'6"x6'8" door complete			155.00
3) Prepare and paint walls and ceiling	84 yds	2.50	210.00
4) Paint trim	8 openings	8.00	64.00
Dining Room (9x12x9)			
1) Lath and plaster ceiling	12 yds	15.00	180.00
2) Replace 3'0"x4'6" DH sash complete			185.00

FIRE   AND   RECLAMATION . . . . REMODELING . . . . NEW   CONSTRUCTION

Estimate

ESTIMATE CONTINUATION

PROPOSED WORK TO BE ACCOMPLISHED	QUANTITY	UNIT PRICE	AMOUNT
Dining Room continued			
3) Sand and refinish hardwood floor	12 yds	$ 7.00	$ 84.00
4) Recase 2 openings	2 openings		42.00
5) Prepare and paint walls and ceiling	54 yds	2.50	135.00
6) Paint 6 openings	6 openings	8.00	48.00
Hall (3x9x9)			
1) Lath and plaster ceiling	3 yds	15.00	45.00
2) Sand and refinish hardwood floor	3 yds	7.00	21.00
3) Remove wallpaper	8 rolls	8.00	64.00
4) Furnish and hang wallpaper	8 rolls	20.00	160.00
5) Prepare and paint ceiling	3 yds	2.50	7.50
6) Paint trim	4 openings	8.00	32.00
Bath (6x8x9)			
1) Clean ceramic tile counter top and shower and regrout			150.00
2) Clean fixtures			35.00
3) Replace bath accessories			37.50
4) Prepare and paint walls and ceiling	33 yds	2.50	82.50
5) Paint trim	3 openings	8.00	24.00
Bedroom (12x12x9)			
1) Prepare and paint walls and ceiling	64 yds	2.50	160.00
2) Paint trim	5 openings	8.00	40.00
Bedroom (12x15x9)			
1) Replace 2'6"x6'8" door complete			155.00
2) Remove wallpaper	12 rolls	8.00	96.00
3) Furnish and hang wallpaper	12 rolls	22.00	264.00
4) Prepare and paint 2 walls and ceiling	47 yds	2.50	117.50
5) Paint trim	6 openings	8.00	48.00
Kitchen (9x12x9)			
1) Heavy clean ceramic tile and fixtures			110.00
2) Replace 5'0"x4'0" sash complete			198.00
3) Prepare and paint walls and ceiling	54 yds	2.50	135.00
4) Refinish natural cabinets			215.00
5) Paint trim	4 openings	8.00	32.00

Estimate (continued)

67

NAME   __Mr. & Mrs. John Smith__

JOB NO. __00001__

DATE   __January 12, 1983__

PAGE NO.   __3__

PROPOSED WORK TO BE ACCOMPLISHED	QUANTITY	UNIT PRICE	AMOUNT
__Bath (4x6x9)__			
1) Clean fixtures			$ 20.00
2) Prepare and paint walls and ceiling	22 yds	$ 2.50	55.00
3) Paint trim	2 openings	8.00	16.00
__Exterior__			
1) Prepare and paint front and south elevation stucco and trim			498.00
2) Replace 4 screens, aluminum frame			48.00
Final clean-up windows and final debris removal			150.00
Job Estimate			$25,551.40
10% Overhead			2,555.14
10% Profit			2,555.14
TOTAL ESTIMATE			$30,661.68

Estimate (continued)

GUNN & GUNN *Construction Company*

161 N. W. 52 STREET. MIAMI. FLORIDA 33127

January 23, 19 3

Universal Insurance Company
111 S. W. 8th Street
Miami, Florida

Re: John Doe
      000 S. W. 8th Street
      Miami, Florida

Gentlemen:

We propose to furnish labor and materials to repair fire damages to the captioned C.B.S. single family residence according to the following specifications:

Kitchen 8'6"x9'x8'6"

Rocklath & plaster ceiling & drop soffit over cabinets	18 yds	@	15.60 yd	$280.80
Replace sheet vinyl floor	76 sf	@	1.75 sf	133.00
Replace 23 size awning window complete	1	@		72.30
Replace 3620 formica wall cabinet over exhaust hood		@		136.04
Replace 108" formica form top w/18" backsplash		@		168.48
Replace tarnished wall & base cabinet hinges & pulls		@		52.50
Replace 36" Nutone hood & fan reusing existing ductwork		@		133.48
Labor removing & setting cabinets & form top		@		98.82
Replace wallpaper on drop soffit, minimum job		@		85.00
Plumbing & electrical connections for sink & cook top		@		165.00
Enamel doorsides & trim 2 coats	2	@	9.50 ea	19.00
Paint new plaster ceiling 2 "	76 sf	@	.22 sf	16.72
Clean & paint interior & polish exterior of wall & base cabinets	66 sf	@	2.25 sf	148.50

Dining Room 9'x10'x8'6"

Sand & refinish oak floor	90 sf	@	1.15 sf	103.50
Replace Levelor blinds	26 sf	@	4.10 sf	106.60
Clean & paint 3 plaster walls & ceiling 2 coats	336 sf	@	.24 sf	80.64
Enamel doorside & trim 2 "	1	@		9.50
" baseboard & shoe molding	29 lf	@	.24 lf	6.96
Clean awning windows & screens	2	@	6.00 ea	12.00
Chemically clean hanging light fixture	1	@		4.50

Estimate

Living Room 15'6"x17'x8'6"

Sand & refinish oak floor	263 sf @	1.15 sf	302.45
Replace Levelor blinds	52 sf @	4.10 sf	213.20
Clean & paint plaster walls & ceiling			
2 coats	815 sf @	.24 sf	195.60
Enamel doorsides & trim 2 "	3 @	9.50 ea	28.50
" baseboard & shoe molding	65 lf @	.37 lf	24.05
Clean windows & screens	4 @	6.00 ea	24.00

Hallway 3'x19'x8'6"

Clean & paint plaster walls & ceiling			
2 coats	411 sf @	.24 sf	98.64
Enamel louver doorsides & trim	2 @	12.00 ea	24.00
" doorsides & trim 2 coats	3 @	9.50 ea	28.50
" baseboard & shoe molding	42 lf @	.37 lf	15.54
" scuttle hole cover, frame & trim	1 @		8.00
Clean, wax & buff oak floor	54 sf @	.38 sf	20.52

Hall Closet 2'6"x5'x8'6"

Clean & paint plaster walls & ceiling			
2 coats	140 sf @	.24 sf	33.60
Enamel doorside & trim 2 "	1 @		9.50
" baseboard & shoe molding	15 lf @	.37 lf	5.55
" shelves & hang rod 2 "	3 pcs @	4.50 ea	13.50
Clean, wax & buff oak floor	13 sf @	.38 sf	4.94

Bathroom 5'x8'x8'6"

Clean & paint plaster walls & ceiling			
2 coats	161 sf @	.24 sf	38.64
Enamel doorside & trim 2 "	1 @		9.50
Clean tile floor, wainscot, fixtures, medicine cabinet, window & screen	@		55.00
Clean light fixture	1 @		2.50

Sub Contractors & General Conditions

Electrical: kitchen, rewire circuit to hood fan, check all outlets, switches, receptacles to electric panel, connect new hood fan, furnish & install 8" drum type ceiling fixture   @   210.00

Labor to tear out damaged plaster   @   56.12

Job cleanup, hauling debris to dump & repairs permit   @   90.00

	3345.19
Contractor's O. H. & Supervision 10%	334.52
	3679.71
Contractor's Profit 10%	367.97
Total	$4047.68

Thank you for the opportunity of submitting this proposal.

Sincerely,

GUNN & GUNN CONSTRUCTION COMPANY

JOHN GUNN

JG/lg

Estimate (continued)

BUILDING APPRAISALS
—
SENIOR MEMBER
AMERICAN SOCIETY OF APPRAISERS

GEORGE A. PASQUALUCCI, A.S.A.
40 SUMNER STREET
P.O. BOX 514
QUINCY, MASS. 02169
472-1452

Lumbermens Mutual Casualty Company
196 Main Street
Brockton, Massachusetts   02402

Attention of Mr. Charles Gillon

	Re:	Insured:	Vicky L. Smith
		Claimant:	Martin P. Jones
			112 Elm Street
			Boston, Mass.
		Damage to dwelling by vehicle	
		Date of Loss:	January 4, 1983

Dear Mr. Gillon:

   The following is an appraisal of damages at the above captioned reference and the cost to complete the repairs as outlined:

	LABOR	MATERIAL	TOTAL

<u>Damaged Area</u>

<u>Front Porch</u>

Dimensions -
   25'0" x 7'9" x 8'8"
   height from deck to
   ceiling line

<u>Temporary Repairs</u>

Note - Work completed.

Shore up roof - use of second hand
lumber
2 x 10"   1/14'
2 x 6 " 1/14'   1/6'   1/5'

| Lump Sum Allowance - Labor and Material | | 70.73 | 70.73 |

Estimate

	LABOR	MATERIAL	TOTAL
Demolition			
Cleaning & Carting to include -			
Floor			
Framing			
Underpinning			
Brick			
Debris			
Total	428.40	48.00	476.40
Shoring and Bracing			
Jack up entire porch and realign same	148.96	60.00	208.96
Repairs			
Masonry			
Rebuild 4 No. brick piers Dimensions 10" x 10" x 2'4" height			
Brick - hard burned common type　　　320 No.	148.92	130.00	278.92
Framing - Hem - Fir			
2 x 4"　2/8'　2/10'　　24 BF	7.68	8.64	16.32
Corner Post　　　　　3 No.			
7" x 7" x 7' - Reset	62.79	-	62.79
Balustrades			
2 sections - Reset	41.86	-	41.86
Lattice - pine　1 3/8" Area to cover 18'3" + 7'9" x 2'3" height			
Total　　　　　　630'	94.50	100.80	195.30
Trim - Pine			
Base			
1 x 5"　2/10'　2/8'　　36 LF	12.60	14.40	27.00
Fascia			
1 x 8"　2/10'　2/8'　　36 LF	12.60	23.04	35.64

Estimate (continued)

		LABOR	MATERIAL	TOTAL
Molding				
1 3/4" cove molding				
2/10' 2/8'	36 LF	10.80	15.12	25.92
Corner boards				
1 x 7" 1/6'	6 LF	2.10	3.36	5.46
1 x 6" 1/6'	6 LF	2.10	2.88	4.98

Flooring

Matched fir 4" wide
to cover entire deck
SF + 25% = 250 SF   112.50   367.50   480.00

Painting

All new work to match existing
Application - two coats   240.00   240.00

Nails and Rough Hardware

Lump Sum Allowance   20.00   20.00

Dumping Fees (Minimum Load)   7.00   7.00

Permit   21.00   21.00

Trucking

Includes pickup and delivery of
materials and equipment   48.00   48.00
                                  1085.81 1180.47   2266.28
         5% Mass. Sales Tax on Material   38.69
                                                         2304.97
         10% Overhead   230.50
                                                         2535.47
         10% Profit   253.55
         TOTAL $2789.02

If you should have any questions, please call me.

Very truly yours,

George A. Pasqualucci, A.S.A.

GP:ja
Enc.

Estimate (continued)

DEWEY C. WHITE   LEN CAMPBELL

WHITE REPAIR & CONTRACTING CO.

——————————— GENERAL CONTRACTORS ● RESIDENTIAL—COMMERCIAL ———————————

FIRE DAMAGE REPAIRS A SPECIALTY

452-8778   *Serving Atlanta Since 1960*   3922 North Peachtree Road
                                                                                  Atlanta, Georgia 30341

February 2, 1983

Kemper Insurance Group
1401 Peachtree Street N.E.
Atlanta, Georgia 30309

Attention: Mr. John F. Hardman

RE: John Doe
    1122 Moreland Place S.E.
    Atlanta, Georgia

Dear Sir:

We propose to furnish labor and materials to accomplish the following work at
the above address:

Tear out & haul away all charred & damaged material & debris	160.00

Roof

Replace 6 pcs. 2 x 6 x 18 roof rafters 108 b.f. @675M	72.90
Replace 1 pc. 2 x 8 x 10 ridge board 13 b.f. @675M	8.78
Replace 260 b.f. 1 x 6 roof decking @550M	143.00
Replace 8 pcs. 2 x 4 x 16 roof braces 85 b.f. @675M	57.38
Replace 4 squares 15 lb. roof felt @6.95	27.80
Install new roof covering on low side of split level only, where roof structure is being replaced matching existing color 4 squares @36.00	144.00
Spray paint & deodorize attic area complete with aluminum paint	145.00
Replace 1400 sq.ft. of 4" rockwool insulation in attic area complete to eliminate smoke odor	420.00
Rebuild 2'0" x 21'0" of overhang, soffit, facia, and frieze board across rear of house @3.25 per l.f.	68.25
Replace 21 l.f. of O.G. type gutter on rear of house @1.60 per l.f.	33.60
Sandblast all smoked brick veneer on rear of house	85.00

Interior

Kitchen 11'0" x 22'0" x 8'0"

Replace 4 pcs. 2 x 6 x 12' ceiling joists 48 bd.ft. @675M	32.40
Replace one 3'0" x 3'2" aluminum window unit complete	48.25
Replace one 2'8" x 6'8" rear exterior door unit complete with weatherstripping & hardware.	82.50

Estimate

WHITE REPAIR & CONTRACTING CO.

Replace 38 pcs. 2 x 4 x 8' wall studs & plates 203 bd.ft. @675M	137.03
Replace one 2'6" x 6'8" cased opening complete	36.00
Replace one 2'6" x 6'8" interior door unit complete	49.00
Replace 80 sq.ft. 5/8" plywood underlayment where damaged only @.55	44.00
Replace one 4'6" x 3'2" aluminum window unit complete	68.60
Apply gypsum wall board to ceiling only 242 sq.ft. @.40 per ft.	96.80
Replace ¼" prefinished birch paneling on sidewalls complete 352 sq.ft.	193.60
Replace one 4'0" x 6'8" cased opening complete	42.00
Replace 2 member base molding 66 l.f. @.85 per lin. ft.	56.10
Replace 4 sides door & window trim @8.00 each	32.00
Replace 4" crown molding complete 66 l.f. @ .75	49.50
Replace inlaid vinyl floor covering complete 242 sq.ft. @1.15	278.30
Replace 8 lin. ft. 3'6" high birch wall cabinets complete @36.00 l.f.	288.00
Replace 8 lin. ft. 2'0" high birch wall cabinets complete @28.00 l.f.	84.00
Replace one 2'6" x 8'0" high birch oven cabinet	125.00
Replace 11 lin. ft. 3'0" high birch base cabinets complete 30.00 l.f.	330.00
Replace one 2'0" x 11'0" formica counter top complete with 4" backsplash @ 12.95 lin. ft.	142.45
Replace one 3'0" x 5'0" bar double thick complete with formica counter top	185.00
Replace one 36" range hood complete with vent	94.50
Replace one Westinghouse extra wide oven with glass door (Harvest Gold)	359.95
Replace one Westinghouse surface unit (Harvest Gold)	134.50
Stain 7 openings complete @3.00	21.00
Paint ceiling only complete 242 sq.ft. @ .14 per sq.ft.	33.88
Stain & varnish wall & base cabinets	78.00

Living Room 12'0" x 22'0" x 8'0"

Replace 7 sides of door and window trim @8.00	56.00
Replace one 11" x 48" window glass	13.20
Replace two 11" x 30" window glass @ 8.25 each	16.50
Sand, strip & varnish 4 openings @6.00	24.00
Clean & paint walls & ceiling complete 808 sq.ft. @ .16 per ft.	129.28
Machine sand & refinish hardwood floor 264 sq.ft. @.60 per ft.	158.40
Replace one heat thermostat control	38.50
Repair front door	16.00

Hallway 6'0" x 6'0" & 3'0" x 6'0" x 8'0"

Replace three 2'6" x 6'8" interior door units leading to bedrooms @49.	147.00
Replace 5 sides of door trim @8.00	40.00
Replace attic entrance door	6.00
Repair all damaged gypsum wallboard on sidewalls & ceiling as needed	35.00
Clean & paint walls & ceiling complete 390 sq.ft. @ .16 per ft.	62.40
Sand & stain 5 openings complete @6.00	30.00
Machine sand & refinish 6 step treads & risers	18.00
Machine sand & refinish hall floor 36 sq.ft. @ .60 per sq.ft.	21.60

Estimate (continued)

WHITE REPAIR & CONTRACTING CO.

Page Three
Mr. John F. Hardman

Electric
Replace all wiring where damaged by fire and water. Replace all light
fixtures where damaged only.   400.00

Labor & Material	$5699.95
10% Insurance & Taxes	$ 569.99
10% Supervision	$ 569.99
TOTAL	$6839.93

Prices quoted include Workmens Compensation, General Liability Insurance and
all taxes required by State and Federal Authorities pertaining to Old Age
Benefits.

Trusting that we may serve you, we are

                              Very truly yours,

                              WHITE REPAIR & CONTRACTING CO.

                              Dewey C. White

DCW/mci

                    Estimate (continued)

N. Frank Vought Inc.

BUILDERS AND APPRAISERS · SINCE 1882

ROBERT B. ROGERS, III

26 EAST 1st STREET
MT. VERNON, N. Y. 10550
(914) 668-2420
OR (212) 823-3600

February 3, 1983

Tri-City Adjustment Company
126 - 9th Avenue
New York, New York 10032

RE: FIRE LOSS TO BUILDING - 1980 Union Street
                              Bronx, New York
        Assured: Blank Realty Corp.
        Date of Fire: January 19, 1982

Gentlemen:

Our estimate of the cost to repair and make good the damage
caused by fire at the above mentioned location, using materials
of a like kind and quality as before the fire, is as follows:

FIRST FLOOR (10' Ceiling)
CANDY STORE (19'x30')
White Plains Road Side

Replace one 4'6" x 4' double hung rear window with wide jambs and trim		$ 475.00
Remove and reset iron window guards		60.00
Plaster ceiling and repair side walls	1,253' @ $1.55	1,942.00
Wire lath and plaster two columns		225.00
Test and repair gas piping and risers		175.00
Repair and scrape flooring	600' @ 65¢	390.00
Remove and reset radiator		22.00
Ten ceiling outlets)		
Four double wall receptacles)		
Three switches)		
Entry outlet and fixture)		
Fuse box)		590.00
Check and repair water lines and branches		115.00
One plate glass - 34x72)		
One plate glass - 60x72)		
Two plate glass in doors - 24x62)		
Two dt. transom glass - 60x22)		
Two dt. transom glass - 34x22)		
Two dt. transom glass - 32x18)		
Two dt. glass - 26x48 in moveable sash)		
One dt. glass - 28x36)		506.00
Replace one window sash and trim		85.00
Repair store front mouldings		160.00
Three exterior outlets over show windows		90.00
Repair metal cornice (fire damage only)		85.00
Replace framing and metal panel under south show window		110.00
Paint store front and new rear window (No interior painting)		175.00

(cont'd)

Estimate

RE: FIRE LOSS TO BUILDING - 1980 Union Street
 Bronx, New York
 Assured: Blank Realty Corp.
 Date of Fire: January 19, 1983

Rear Passage
One door complete $ 125.00
Scrape and repair flooring   40.00
Plastering and patching   150' @ $1.55   233.00
Ceiling outlet and fixture)
Wall outlet and receptacle)
One switch   )   130.00
Repair and ease two doors   36.00
Paint   105.00

Storeroom and Lavatory
Repair and glaze one window   36.00

VACANT STORE
Paint   175.00

UPHOLSTERY STORE
Touch up painting at door and
passage   60.00

BASEMENT
Test and check building wiring
under Candy Store   85.00

EXTENSION ROOF
Repair roof boarding   80.00
Repair rubberoid roofing and
flashing   115.00

GENERAL
Demolition, debris removal and
cleaning up   425.00
                                                        $6,850.00
         Builders Overhead and Profit - 15%   1,027.50
                                                        $7,877.50

-We Do Not Include-
Emergency or temporary repairs
Store fixtures
Kitchen equipment and fixtures
New York State sales tax

NOTE: If painting of store is part of
building loss add $380.00 net to our
estimate.

APPROXIMATE CUBIC CONTENT   -   957,500'
Unit Cost   - $2.20
APPROXIMATE REPLACEMENT COST - $2,106,500.00

                Yours very truly,

                N. FRANK VOUGHT, INC.

RBR:hvg

Estimate (continued)

Though there are only a handful of computerized estimating services presently operating, there will, no doubt, be more of them since pilot tests have demonstrated that they can reduce the time and the cost of making estimates. CompuClaim Corp, domiciled in Richardson, Texas, is designed, as its name implies, for insurance claim personnel, adjusters and estimators of building losses. COMPU U.S.A., located in Dallas, Texas, also provides services for insurance claim personnel, adjusters and remodeling contractors. The Olender System, R. J. Olender, Inc., Bensalem, Penn., while designed basically for the insurance industry, has been used by remodeling contractors and subcontractors. The NIKA Corporation, operating out of Chicago, Ill., is used by the insurance industry, financial institutions, and government agencies such as HUD and Veterans' Administration. NIKA also provides for initial and interim field inspections by a staff experienced in construction and evaluation of building damage.

The four computerized estimating services mentioned above are operating on a national level or are capable of doing so. All have operated for insurers in major catastrophes. Each furnishes instruction manuals and will provide the training for client personnel in field take-off and office input operators. Claim personnel can telephone from the location of the loss and obtain on-the-spot complete estimates on small losses. Self-contained mobile units are available when and where needed, especially in catastrophes.

Services offered are flexible enough to meet most situations. Clients may tie directly into the services' computers through mini-computers or through their own hardware. They may subscribe to a part or to the complete service.

The efficacy of computerized estimating depends, as with any other system, on (1) the care and accuracy in taking off the specifications and, (2) on applying the appropriate material prices, and labor charges. Labor must take into account all the factors that effectuate the time required to install and apply materials and perform specific operations as described in Chapter 2, pages 20 to 23. In traditional estimating, the individual who takes off the specifications for building or restoring a structure is usually the one who judges the hours of labor required. In computerized estimating, the pricing is programmed into the computer by a person or by persons who, generally, have not seen the building to be repaired or constructed. Therefore, well-balanced *averages* of labor hours, work and materials quality are essential. The person who takes off the specifications should know or at least have some idea of the basis for the pricing mechanism of the computer so that one will know when and to what extent to by-pass it.

While the theory underlying *computerized estimating* is sound and may mark a major advance in the estimating of building losses and construction, unless the data processing dovetails with the field specifications there is the possible danger of a hodge-podge of errors.

CHAPTER 5

Estimating in Catastrophes

When a peril such as fire, explosion, hailstorm, windstorm, earthquake, flood, or freeze-up, results in a number of claims under insurance policies greatly in excess of normal, and requiring the recruiting of additional adjusters, roofers, etc., it is generally classified as a catastrophe in the industry. Damage may be confined to one city or town, as in a tornado, or it may extend over several states as do hurricanes.

Conflagrations in cities today are infrequent due to better construction and improved firefighting services. There still are brush fires and forest fires which, burning out of control, destroy buildings and structures that lie in their paths. There are occasional violent explosions in chemical and other industrial plants which result in damage to adjacent structures and frequently to buildings at considerable distance.

The most common catastrophes confronting insurance companies are caused by windstorms occurring over a wide area, hailstorms, flooding, hurricanes and tornadoes. (See Causes of Loss—Chapter 1). The number of claims arising from a single catastrophe may be a few thousand or as many as a million as, for example, when hurricanes of severity travel over several states before dissipating.

CATASTROPHE "PLANS"

As a result of their experience in numerous catastrophes the insurance companies have formulated *catastrophe plans* to better organize and supervise the handling of claims. The "Objective" stated in the "Plan for Handling Catastrophe Losses" adopted by the American Insurance Association, quite clearly and very briefly outlines the basic problem:

> Following the occurrence of a major catastrophe, conditions in the area involved present a serious handicap to the orderly and efficient adjustment of losses. The sudden necessity for handling expeditiously a multitude of claims places a severe strain upon regular adjusting facilities and there is a tendency to use personnel without due consideration of qualifications, and to close losses without proper inspection of the properties involved. Combined with the fact that the labor and material markets experience excessive demands at the same time, the orderly adjustment of losses, vital to the interests of the public as well as to our business, requires that we establish a routine for the efficient handling of such claims.

Catastrophe plans have been adopted also by the Property Loss Research Bureau, The National Association of Independent Insurance Adjusters, several state and regional agents associations, and also by companies that maintain their own staff of adjusters. All of the plans are flexible and designed for cooperative action with each other.

The purpose of each plan is to move adjusters, builders, experts, supervisory personnel, supplies and equipment into a stricken area as soon as it is practical, following the disaster. By so doing, damaged property can be quickly protected from the elements, uniform procedures can be put into operation, claim handling and processing are fast-tracked, and claimants can be provided with ready funds through the prompt payment of claims.

CONDITIONS TO EXPECT

An adjuster or a builder who has been assigned to a catastrophe area for the first time will usually be frustrated in the early stages by the confusion that exists. The degree of the confusion will depend a great deal upon the violence of the disaster and the area that it covers.

Residents are often in shock during the first few days. Their main concern is naturally for the well-being of other members of the family, or for friends who may have been injured or made homeless. After a tornado, hurricane, or flood, people are occupied in searching for articles of personal property that have been blown or washed away from the premises. There is cleaning up to be done. Many will have moved temporarily or permanently to other quarters, and they may be difficult to locate. When found, they may be too upset and distracted to discuss their damage calmly.

Following the more serious catastrophes, the National Guard, Army, or State Police are in charge of the town or city, patroling, issuing emergency passes, or prohibiting admission entirely to certain areas until order has been restored and the hazard of looting has been eliminated.

Communications, power and transportation may be partially or completely out of service. Many times the drinking water becomes contaminated. Hotel, motel and restaurant facilities usually are inadequate to accommodate the influx of the vast number of people.

The adjuster or builder generally is expected to handle the same number of claims in a day that were normally handled in a week, and it is not uncommon to work from early morning to nightfall. Claims people are constantly under such pressure from insureds, agents and their own organizations that it is almost impossible to devote the necessary time to each loss to prepare a careful, orderly, estimate of damage. There is a tendency under such pressure to grow careless, and take short cuts, and to make guesses and compromises in specifications in order to move on to the next loss.

Knowing, or anticipating many or all of these conditions, the adjuster or builder is well advised to sacrifice a small amount of speed in order to perform as efficiently as possible. One should not be influenced to compete with some other adjuster or builder who is out to set a record of fast claim handling. It is not uncommon to find it necessary to reopen losses that have been handled in haste because of overpayment or underpayment—particularly the latter.

INSPECTING PROPERTIES

The procedures that been outlined under "Fieldwork and Notes" in Chapter 3, should be followed in catastrophe losses just as in any other loss. The methods of estimating, the damageability of materials, and methods of repair have been discussed in the other chapters. However, because of the special problems in catastrophes, the following suggestions are offered:

1. A set of sample policy forms used in the area should be acquired and carried in the brief case. The form used on the particular risk should be known and its conditions well understood in order that the estimate of damage will include all of the property covered but will not encompass excluded property or perils.
2. The inspection of the premises is always best made in the presence of the insured or his authorized representative. The insured may then point out the damaged portions of the property, and the occasion presents an opportunity to discuss specifications and cost of repairs.
3. If temporary repairs to roofs, walls, or windows have not been made, provisions for immediate attention should be made. There are numerous instances of subsequent damage by rain to the interiors of buildings long after the loss has been agreed upon. Frequently it is necessary to reopen and readjust.
4. Although no damage is claimed for the interior of the building it should nevertheless be carefully inspected for such things as wet or stained walls and floors, and cracks in plaster. The insured or tenant often overlooks damage of this type, and may discover them at a later date and make supplemental claim.
5. A roof that has been exposed to strong winds or hail is best inspected from the roof rather than from the ground by eye or with field glasses. Asphalt shingles may appear undamaged from the ground but on close examination may be broken or cracked underneath the tabs where they have been flapped up and then laid back again. Perforations by hail are not always visible from the ground, and if necessary a ladder should be obtained for close inspection on the roof.
6. It is good practice to walk around the exterior of the building to look for any damage that might have been overlooked by the insured.
7. Outbuildings, yard fixtures and other outside property that is covered under the policy should be inspected.
8. Previous damage such as that done by hail or wind, which has never been repaired, should not be included in the estimate. (See Chapter 3.)

MIXED DAMAGE—THE FOUR COLUMN ESTIMATE

In some catastrophes part of the damage may be caused by a peril covered under the policy while other damage has been caused by an excluded peril. For example, when hurricanes strike coastal areas such as in the Gulf of Mexico or along the Atlantic seaboard, properties adjacent to the coast may suffer damage from both wind, a covered peril, and from wavewash, tidewater or other flooding which is excluded. The estimator is then required to make a careful and reasonable separation of the covered and uncovered damage.

A sample of an estimate sheet used in such cases is shown on page 84. This is a four-column estimate showing total cost of repairs; estimate to repair the water damage; estimate to repair debatable items; and the estimate to repair wind damage.

NATIONAL FLOOD INSURANCE PROGRAM

The National Flood Insurance Program (NFIP) is administered by the Federal Insurance Administration (FIA). Their headquarters is 6430 Rockledge Drive, Bethesda, Maryland 20034.

There are two NFIP flood insurance policies covering property.

1. Dwelling Form
2. General Property Form

The Dwelling Form covers residential buildings of one to four families and the contents of such residences. The General Property Form covers the buildings and contents of residential structures consisting of five or more units. It also covers buildings and contents of properties occupied for commercial, manufacturing and farming purposes. The General Conditions and Provisions of both policies closely follow those of the 1943 New York Standard Fire Policy. Under both policies the Definition of Flood states:

Wherever in this policy the term "flood" occurs, it shall be held to mean:

A. A general and temporary condition of partial or complete inundation of normally dry land areas from:
   1. The overflow of inland or tidal waters.
   2. The unusual and rapid accumulation or runoff of surface waters from any source.
   3. Mudslide (i. e. mudflow), a river or flow of liquid mud proximately caused by flooding as defined in subparagraph A-2 above or by the accumulation of water under the ground.

B. The collapse or subsidence of land along the shore of a lake or other body of water as a result of erosion or undermining caused by waves or currents of water exceeding the anticipated cyclical levels.

Servicing agents for underwriting and for claim handling are approved and designated by NFIP. A comprehensive claim manual for adjusters is furnished by NFIP. It contains 49 questions and answers. These are not intended to be all-inclusive but do provide answers to the most frequently asked questions about coverage and about perils and property excluded. While designed for claim personnel, it should be read, understood and carried by anyone who prepares building estimates on NFIP claims. It may be obtained by writing to the National Flood Insurance Program in Bethesda, Maryland.

Form 5–2 (page 85) is the NFIP "Work Sheet Building" form. It requires the separation of *wind damage* and *water damage* as well as the full cost of repair on each

General
Adjustment
Bureau, Inc.

SPECIAL WAVEWASH STATEMENT

TYPE OF LOSS	GAB FILE NO.
DATE OF LOSS	ADJUSTER

INSURED _____

LOCATION _____

ITEM NO. _____ TYPE OF CONSTRUCTION _____

ADDITIONS _____ DIMENSIONS _____ SQ. FT. AREA _____

NO. ROOMS _____ NO. BATHS _____ AGE _____ GENERAL CONDITION _____

EST. REPLACEMENT COST $ _____ DEPRECIATION $ _____ A.C.V. $ _____

DETAILS OF LOSS	TOTAL REPAIR OR REPL. COST	WATER	DEBATABLE	WIND
FOUNDATION				
DIMENSION LUMBER				
EXTERIOR SIDING				
SHEATHING				
ROOFING				
INT. WALLS & CEILINGS				
FLOORING				
MILLWORK				
HARDWARE				
INTERIOR PAINT				
EXTERIOR PAINT				
ELECTRICAL				
PLUMBING - AIR COND.				
DEMOLITION - DEBRIS				
OTHER				
TOTALS	$	$	$	$

COMPROMISE OF DEBATABLE TOTAL ADDED $ _____

TOTAL COST OF WIND REPAIRS OR REPLACEMENTS $ _____

LESS DEPRECIATION $ _____, DEDUCTIBLE $ _____, COINSURANCE $ _____ $ _____

INSURANCE $ _____ AND INSURED CLAIMS $ _____

APPLICATION OF DEDUCTIBLE OR COINSURANCE CLAUSE:

_____

_____

_____

_____

_____

610 - (1-67)

Form 5-1

WORKSHEET - BUILDING

(See Reverse Side for Privacy Act Statement)

Date of Report

Insured and Location

Policy No. FL

Adjusting Firm and Address

(ADJUSTER)

Adjuster's File No.

Date of Loss

Type of Building

Age

Measure Dimensions and Draw Diagram of Ground Floor Area.
Attach Snapshot.

Building Dimensions | Total Sq. Ft. | No. Rooms

Type of Foundation | Finished Basement Yes ☐ No ☐ | No. Stories

Interior Wall Construction | Exterior Wall Construction | No. Baths

Estimated Repl. Cost _____

Less Depreciation _____

Actual Cash Value _____

Quantity	Detailed Description	Full Cost Repair	Depreciation	Flood Loss	Wind Loss If Any

593-109 (8-79)

HUD 1695 9-78)

Form 5–2

item involved. Provision is made for attaching a photograph and making a sketch of the ground floor area. An estimate of the replacement cost and actual cash value is also a requirement.

ADVISORY PRICE LISTS

One of the first steps taken by the administrators of a catastrophe plan is to arrange for a conference with local builders' and contractors' associations to discuss and agree when possible on unit costs for various building items. These unit costs are published and then used as a guide by the adjusters and builders in estimating the losses in the area. If no agreement can be reached, an outstanding, competent contractor is asked to prepare such a list which is used for the same purpose.

These advisory lists are necessary in a catastrophe because most of the claims are handled by adjusters who take-off the specifications and apply the published unit costs. There are never enough contractors available to appraise the damage to each property. The unit costs agreed to are generally "full" and no flexibility is provided for in the list for grade and quality of material.

CHAPTER 6

Estimating Losses of Suspicious Origin

BACKGROUND

Arson-for-profit is a crime perpetrated by private property owners and by criminals, organized or individuals. The number of known or suspected arson fires in the United States has increased 724 percent since 1964, according to the National Fire Protection Association. The National Commission on Fire Prevention and Control, in its report entitled AMERICA BURNING, states, "In many large cities fire chiefs believe that almost half of all fires in their experience have been set." It is estimated that this crime is costing more than a billion dollars a year. There is an erroneous belief by the public and many law enforcement agencies that the cost of this crime is absorbed by the insurance companies. The facts are that every dollar paid out by the insurance companies on fires set by arsonists comes out of the insuring public's pocket in the form of higher premiums.

The increased cost of insurance is relatively a small part of the total dollar cost paid by the public for arson fires. When commercial properties are involved, either directly or by exposure, there are unemployment insurance payments, and frequently a loss of tax revenue from real estate, income, and sales. To these costs must be added the serious injuries and the loss of life of firefighters as well as innocent persons who are caught or trapped in burning buildings that have been deliberately bombed or torched.

INSURANCE COMPANY ACTIVITIES
TO CURTAIL ARSON

The suppression of incendiary fires rests in no small measure with their detection and proper investigation. The insurance companies, through their underwriting, loss control, and claim departments, are focusing on the elimination of *potential* arson-prone risks through selective underwriting and inspection; and reinforcing their investigative function and capacity where arson is suspected. The insurance industry has responded to the alarming increase in arson by developing a broad and long-range program for controlling the crime. That response includes the following organizations.

The Insurance Committee for Arson Control (ICAC)

This Committee is an all-industry effort to control the growing problem of arson. Its purpose is to encourage the formation of local-committee-based Arson Control Task Forces; stimulate greater cooperation in the public and private sectors and assist in the development of measures to improve prevention, detection, investigation and prosecution of arson.

Insurance Crime Prevention Institute (ICPI)

This Institute is a nationwide organization whose objective is to stop insurance fraud. It represents over three hundred (300) Mutual, Stock and Independent insurers writing the majority of property and casualty insurance in the United States. The primary function of ICPI is to investigate fraudulent property and casualty insurance claims, except those occurring under Accident and Health Insurance and Workers Compensation lines.

Insurance Claims Services (ICS)

This is a non-profit affiliate of Property Loss Research Bureau (PLRB). It provides insurers with investigative services in suspected arson losses under fire and inland marine policies.

Property Insurance Loss Register (PILR)

This is a voluntary, self-sustaining, non-profit subscription service, administered by the American Insurance Association (AIA). Approximately four hundred eighty-six (486) companies or groups of companies support this organization. One of the primary objectives of the Property Insurance Loss Register is to accumulate records of property loss claims as they occur and make the information available to its subscribers. An important function is to serve as a "universal" Property Insurance Reporting Form, gradually replacing the present Fire Marshal Reporting and other Special Fire Incident Reports.

Property Loss Research Bureau (PLRB)

This is a non-profit association of one hundred (100) insurance companies. It acts as an educational investigative and consultant agency on property loss matters.

Property Claim Service (PCS)

This is a subscription service of the American Insurance Association. It is a non-profit association of one hundred seventy (170) insurance companies. It is an educational agency for subscriber companies.

National Committee on Property Insurance (NCPI)

This is a not-for-profit corporation. Its membership consists of the majority of the nation's property insurance writers. It provides advice and assistance to its members who write property and casualty insurance in distressed and residual markets. Its Anti-Arson Action Plan is a major contribution to the insurance Industry.

ARSON REPORTING IMMUNITY LEGISLATION

A very important factor that has effected the detection, investigation and prosecution of arson has been the exchange of arson-related information between the insurance companies and the public officials following a loss of suspected incendiary origin. Until recently a deterrent for such information sharing on the part of insurance companies has been the exposure to the threat of civil suit.

Following the enactment of a law in 1978 by the Ohio Legislature granting immunity from suit to insurers when they share arson-related information with law enforcement officials, the Alliance of American Insurers (AAI) drafted model legislation entitled ARSON REPORTING IMMUNITY BILL. Under this legislation the law requires insurers to inform state fire marshals and other such persons of fires of suspicious origin, and it permits insurers and arson investigators to exchange information developed during their separate investigations. Forty states have passed some form of arson reporting immunity legislation. Nine of these reflect all of the elements of the Alliance of American Insurers Model Law.

While Arson Reporting Immunity Laws will hopefully become more uniform, it is recommended that anyone representing an insurance company in any capacity on a loss of suspicious origin be fully informed of the arson immunity laws in the state in which he or she is operating.

STATE AND MUNICIPAL ACTIVITIES
TO CURTAIL ARSON

State and local governments have primary *jurisdictional* responsibility for investigating and prosecuting action. State and/or municipal authorities are generally active when arson appears probable. They may include the Chief of the Fire Department, Chief of the Police Department, State Police, municipal and/or state Fire Marshals, and Sheriffs office. If organized crime is thought to be involved, the Federal Bureau of Investigation (FBI) may enter the case to investigate and prosecute if there are violations of federal law. Additionally, the Bureau of Alcohol, Tobacco and Firearms (ATF), and the investigative division of the Treasury Department can enter an arson investigation any time a trailer or an explosive is used as long as interstate commerce is involved. Since gasoline is considered an explosive, this opens up many cases in which they may participate.

Over the past few years, with awareness of the increase in arson-for-profit, the federal, state and municipal agencies and authorities, working together with the insurance industry, have made much progress toward the coordination of effort to curtail and control this crime.

The Arson Resources Directory, published by the Office of Planning and Education, U.S. Fire Administration, January 1, 1980, lists hundreds of organizations and individuals throughout the United States working to fight the crime of arson. For example, there are 31 statewide anti-arson organizations; 14 Metropolitan Arson Task Forces; 6 cities with early warning systems; numerous insurance industry, fire department, college, federal and state educational programs; tipster-hot-line programs with rewards in 16 states.

THE BUILDING ESTIMATOR'S ROLE

The estimator of a building loss may be a building contractor or may be an insurance company claims representative capable of preparing estimates. If a builder, contractor or subcontractor, assignment to the loss may be made before the adjuster or company claimsperson is able to visit the loss. This is not unusual when workloads are heavy, the loss is not thought to be serious, or is in an outlying area inconvenient for the claimsperson to get to promptly. In these situations the building estimator is usually the first person representing the insurance company to visit the scene of the loss. If the origin is incendiary, the estimator will also be the first to have the opportunity to observe and discover physical evidence that the fire may have been deliberately set; evidence that could be removed or destroyed before the claimsperson, a trained arson investigator, fire marshal or other law enforcement officer or agent could be brought in.

The estimator, therefore, plays a very valuable role in providing early warning of possible arson in a situation which might otherwise be passed over as accidental. Even though estimators may not be the first on the scene, they often discover evidence, in the course of inspecting the damage, that was inadvertantly missed by others.

Estimators, recognizing that they have a specialized and an important place in an insurance company's claim function, must exercise great caution and not become overzealous when confronted with what appears to be evidence of arson. Persons who represent insurers on losses are not law enforcement officers or agents. A duly constituted legal authority may make a charge of arson, which is later proven unfounded, directly or by insinuation against a person without incurring any liability for such false charge. *This immunity does not extend to a person representing an insurance company.* The estimator must also avoid being identified with any criminal investigation that is being conducted by the authorities and should not make an appearance at the office of an investigative official during such investigation. Civil suits against insurers for alleged real or imagined accusations by their representative have increased over the years.

There are certain fundamentals regarding the crime of arson that will help guide a person to observe and identify evidence of incendiarism.

FUNDAMENTAL CONSIDERATIONS

As a prelude to discussing specific methods and procedures for observing and identifying evidence when the cause of fire and/or bombing appears to be other than accidental, some of the basic concepts relating to the crime of arson should be explained:

1. Definition of arson
2. Essential elements for proof of arson
3. Arson losses vs. accidental losses
4. Direct evidence
5. Indirect or circumstantial evidence
6. Motives for arson
7. Did the insured have an opportunity to set the fire?
8. Preservation of evidence

Definition of Arson

Arson may be defined as the willful and malicious burning of property of value, whether the property of the arsonist or the property of another. In most jurisdictions arson includes aiding, counseling, or procuring the burning of property. In any definition of arson, bombing of property should be included because one or both are used in the willful and malicious destruction of property.

Essential Elements For the Proof of Arson

There are three basic elements insurance company lawyers need to defend an action on an insurance policy where the insured is suspected of arson:

1. That the fire was of incendiary origin (incendiary—from Latin *incendere* to set on fire.);
2. That the insured had a motive to cause the fire (motive—purpose—objective);
3. That the insured had an opportunity, acting directly or through an agent, to cause the fire.

Arson, by legal definition, does not require a motive on the part of the arsonist because a mentally retarded person or an emotionally disturbed person could willfully and maliciously burn property without a rational motive. When arson is committed by or for any of the interests under an insurance policy, it almost always entails a motive— usually one of economic advantage. When such a motive can be proved, the position of the defense is greatly strengthened.

In criminal prosecution, arson must be proved *beyond a reasonable doubt.* In a civil suit the insurer need prove its defense only by a *fair preponderance of the evidence.* Direct evidence is generally the proof needed to establish guilt beyond a reasonable doubt. Circumstantial evidence is generally relied upon to develop a fair preponderance of evidence in civil cases.

Arson Losses vs. Accidental Losses

When suspicion surfaces to indicate the possibility that a fire is other than of accidental origin, the first step is to try to eliminate as many accidental causes as possible. Keeping in mind that an arsonist is capable of starting a fire in a way that will give the appearance of being accidental, the following more common causes should be considered and eliminated if found not to be applicable:

1. *Careless smoking.* Careless smoking as a cause of fire can frequently be traced to the point of origin: a person falling asleep in bed or in a chair while smoking; the emptying of ash trays that contain lighted smoking materials into waste recepticals; smoking materials falling unnoticed onto the seat of an automobile while it is parked in the garage.

2. *Electrical.* Defective wiring to or in lights, motors and appliances. Appliances inadvertantly left on; light bulbs left on adjacent to flammable material; overloaded circuits, and pennies placed back of fuses that have blown.

3. *Spontaneous combustion.* Destructive fermentation or chemical change that takes place in certain materials that results in a build-up of heat sufficient to cause fire. Wet hay stored in barns, ignition of linseed-oil paint rags kept in a confined area are examples.

4. *Lightning in the area*—either direct hit or a ground surge.

5. *Cooking and heating equipment and devices.* Grease fires while cooking in fat. Ovens left on and pans of food left on top of a stove with the burner on during occupants' absence. Backfire of oil burners, defective floor furnaces, leaking gas or oil lines. Faulty chimneys, sparks from an open fireplace, space heaters overturned. Clothing left too close to an open gas flame, a stove or a fireplace to dry.

6. *Trespassers, burglars or persons indulging in vandalism or malicious mischief.*

7. *Small children playing with matches or cigarette lighters.*

8. Brush or grass fires, incinerator fires and bonfires out of control of insured or neighbors.

Direct Evidence

Direct evidence of arson is the factual evidence that will prove that the fire is of incendiary origin; that it was set deliberately by someone. In most cases, however, such evidence does not necessarily prove that the *insured* was that someone unless a reliable witness testifies that he saw the insured in the act of setting the fire. Some of the more common examples of direct evidence are the following:

1. Two or more independent fires found burning in one room, different parts of a building, or in separate buildings.

2. Trailers used to spread or conduct the fire from one room to another, from floor to floor or throughout a large area. Trailers commonly used are rope, toilet tissue, paper towels, newspapers, rags, bedding and other materials saturated with flammables.

3. Ignition and timing devices that enable a person to get away from the premises before the fire starts and thus establish an alibi for his whereabouts. Timing devices similar to alarm clocks are common. Devices are often connected to telephone mechanisms to operate when an outside call comes in. Others may be connected to furnace thermostats. Slow-burning candles are common. Oven gas jets may be turned on to be ignited by a top-of-the-stove pilot light when the area is filled with gas fumes, or ignited when the door bell is rung—the buzzer being adjusted to emit a spark.

4. Chemicals are sometimes used to start fires, phosphorous on water being an example. When water is allowed to leak slowly from the container the phosphorous will ignite as it is exposed to the air, setting fire to flammable material placed nearby.

5. Accelerants, such as paint thinners, kerosine and gasoline are often used to saturate the contents of buildings or to pour over wood floors. Deep charring of wood floors or wood paneling in localized areas may indicate the use of accelerants. Empty and partially filled containers in or outside the building bear investigation, particulary if they have an odor of an accelerant.
6. Odors of solvents, gasoline, kerosine, or bottled gas are sometimes an indication of possible incendiarism, especially when such fluids or gas are not normally used on the premises.
7. Intensely burning fires in the early stages, particularly when accompanied by a large volume of smoke of a color which is not related to the building materials or the contents of the building give rise to the suspicion that the fire is not accidental. Any evidence of arson possibility through smoke coloration should be checked with the fire department that attended the fire and other on the scene witnesses.
8. Holes made in walls to expose wood studding or to increase draft are sometimes made by a person before setting a building on fire.
9. In commercial buildings, the automatic sprinklers may be shut off or the tank emptied if the system is a gravity type; automatic fire doors may be propped open to permit fire to spread from one area to another. Debris from a fire may disclose materials used by an arsonist, particularly mechanical devices, accelerant containers and trailers that failed to ignite.

Indirect or Circumstantial Evidence

Whenever direct evidence of incendiarism is uncovered, the first essential of arson is satisfied. Connecting the insured, an agent of the insured, or any other person under suspicion with starting the fire will depend to a large extent on the *quality* and *quantity* of circumstantial evidence that is uncovered in the investigation.

Circumstantial evidence is that "evidence not bearing directly on the fact in dispute, but on various attendant circumstances from which the judge or jury might infer the occurrence of the fact in dispute". The key word here is *infer*. The indirect evidence must be of sufficient quantity and quality that it may be reasonably inferred the insured is involved.

Typical examples of indirect evidence are the following:

1. The *time of the fire* may, when considered with other facts, support the suspicion that the fire was other than accidental. Late night fires and early morning fires that are not explainable are examples.
2. Unusual actions of the insured on the day of the fire may lead to explanation. Examples would be taking the family to dinner or a late movie; seeing that the family pet dog or cat is in a kennel or with a friend when the fire occurs; fire occurs shortly after insured and family go on trip or vacation; or business premises burns shortly after insured closes for the night.
3. Location where fire starts on premises may add to other evidence to support suspicion. Examples would be closet fires or store rooms where unusual high values are claimed.

4. An insured with a record of fires of unexplained origin is always suspect.
5. Evidence that personal property such as stocks, furniture, furs, or jewelry were removed from the premises prior to the fire adds to suspicion and circumstantial evidence.
6. Contradictory statements and misstatements by an insured regarding such things as cause of fire, place of origin, insured's whereabouts at time of fire and other matters lead to suspicion of a circumstantial nature.
7. Evidence of tampering with thermostatic heat controls oil or gas lines to furnace, or hot water heater, or overfusing of electrical circuits, can give rise to suspicions.

Motives of Arsonists

If a motive can be established for the insured to set the fire or have it set by an agent, circumstantial evidence is greatly reinforced.

Motives for incendiary fires may be broadly classified as follows:

A. Economic gain for the insured
B. Economic gain by others than insured
C. Revenge, intimidation or jealousy
D. To commit or conceal a crime
E. Sociopolitical reasons
F. Pathological persons

A. *Economic gain by insured*:

1. To collect insurance money.
2. Insured mortgagor desires to liquidate mortgage.
3. Over-insurance of urban properties of low market value but high replacement recently purchased.
4. To liquidate excessive, obsolete and hard-to-move stocks of merchandise.
5. Tenant desires to terminate long-term unfavorable lease.
6. Insured desires to relocate business that has outgrown existing facilities.
7. Insured desires to avoid pending business failure.
8. New highway will bypass insured's business location.
9. Partner desires to dissolve partnership.
10. Declining neighborhood threatens to or is lowering realty values.
11. Insurers threaten cancellation.
12. Insured's building is on leased ground and must be moved.
13. Land more valuable without building on it.
14. Insured desires to void unprofitable contract or franchise
15. Impending condemnation of insured's building for one reason or another.

B. *Economic gain by others*:

1. Estate being liquidated over long period and one or more heirs impatient for settlement.
2. Payment offered arsonist to set fire.

3. Mortgagee desires to liquidate mortgage.
4. One competitor seeks to eliminate another.
5. Building owner desires to terminate low rental lease and renegotiate a new lease or get another tenant at a more favorable lease.
6. Building contractor sets fire to obtain a profitable reconstruction contract.
7. Arsonist seeks to obtain employment as a security guard, firefighter, etc.

C. *Fires set for revenge, intimidation or jealousy:*

1. Husband or wife against spouse.
2. Male or female against lover.
3. Husband or wife against third party threatening their marriage, happiness or security.
4. Hateful revenge.
5. Strikers in labor-management dispute.
6. Racketeers intimidating business people.

D. *Fires set to commit or conceal a crime:*

1. Fires set to do away with a person.
2. Fires set to cover up evidence of crimes as burglary, theft or pilferage in warehouse and stores, murder, rape, etc.

E. *Fires set for sociopolitical purposes:*

1. Vandals, rioters, subversives, terrorists, and similar groups who bomb and set fires to focus attention on their cause or to force recognition of their demands.

F. *Fires set by pathological person:*

1. Pyromaniac (compulsive fire setter).
2. Persons who set fires for thrills.
3. Sex degenerates who set fires for sexual relief.
4. Fires set by mentally retarded, senile and insane persons.

Did Insured Have Opportunity to Cause the Fire or Have an Agent Cause It?

This is the most difficult of the essential elements of arson to prove. Such evidence as can be obtained is almost always circumstantial. Tracing the insured's movements just prior to the fire is usually accomplished by questioning the insured, relatives, neighbors and business associates. Then comparing stories told.

If it is confirmed that the insured was out of town, on vacation or nowhere near the property at the time of loss the alternates exist that a timing device was employed, the insured had some agent set the fire, or the insured, in fact, had nothing to do with the fire and it was accidental or set by an outsider.

If it is confirmed the insured was near or at the scene when the fire occurred and other evidence supports the suspicion of arson, circumstantial evidence of the insured's involvement is reinforced.

Preservation Of Evidence

There are two things to remember about preserving evidence.

1. Only a qualified expert should move or handle evidence of incendiarism. Until such an expert can be brought to the scene the evidence should remain undisturbed and under surveillance at all times.
2. If evidence must be removed, photographs should be taken by a professional under direction of an attorney, fire marshal, or other law enforcement officer familiar with arson investigation in order to get adequate pictures of the evidence from all angles. Removal from the premises should be done only by law enforcement officials or under their supervision. Evidence of tampering with gas or oil lines; electrical circuits, wiring, fuses, devices or appliances should be examined by qualified experts in those fields and in the presence of or by authority of law enforcement officials.

CHAPTER 7

Arithmetic for Estimating

A sound working knowledge of elementary arithmetic and simple geometry is necessary to calculate quantities and costs in estimating building losses. Unless a person is regularly or frequently called upon to add, subtract, multiply, and divide whole numbers, fractions, and decimals, the principles are forgotten, mistakes are easily made, and estimates become unreliable. A knowledge of the formulas used in calculating the areas of simple figures and the volumes of ordinary solids is of equal importance.

FRACTIONS

A fraction is part of a whole number, for example: 1/2, 1/3, 1/4, 1/8. When written on two lines, the number above the line is called the *numerator*; the number below the line is called the *denominator*. The line itself signifies that the top number is to be divided by the bottom number.

$$\frac{1}{2} = \frac{\text{numerator}}{\text{denominator}}$$

The fraction 3/4 means that we are dealing with three parts of a whole unit which has been divided in four equal parts. It indicates that 3 is to be divided by 4. When the division indicated by a fraction is performed, the result is the *decimal equivalent* of the fraction.

$$\frac{3}{4} = .75, \text{ or } \frac{1}{2} = .5$$

TABLE 7–1
Common Fractions with Decimal Equivalents

1/64	0.015625	1/7	.1428
1/32	.03125	1/6	.1666
1/16	.0625	1/5	.2
1/12	.0833	1/4	.25
1/10	.1	1/2	.5
1/9	.1111	1/3	.333
1/8	.125	1	1.0

Any fraction less than a whole unit as 1/2, 1/6, 4/5, and so forth is called a *proper fraction.*

A fraction equal to or greater than a whole unit, as 3/3, 5/2 or 8/6, is called an *improper fraction.*

Any number that includes a whole number and a fraction as 1 1/2 is referred to as a *mixed number.*

A mixed number may be converted to a fraction by multiplying the denominator by the whole number and adding the numerator. This result is placed over the denominator of the fraction.

Illustrative Example

$$3\tfrac{1}{2} = 2 \times 3 + 1 = 7/2$$
$$6\tfrac{2}{3} = 3 \times 6 + 2 = 20/3$$

A fraction may be reduced to its lowest terms by dividing both the numerator and the denominator by a common number, or numbers, until they no longer can be divided by a common number.

Illustrative Example

$$\frac{12}{30} \div \frac{3}{3} = \frac{4}{10}$$

$$\frac{4}{10} \div \frac{2}{2} = \frac{2}{5}$$

In this example both the top and bottom are first divided by 3. The top and bottom of the resultant fraction 4/10 is then divided by 2 reducing the fraction to its lowest terms. The same result is obtained if the numerator and denominator of 12/30 are divided by 6.

Adding and Subtracting Fractions

Fractions may be added or subtracted only if they have a common (the same) denominator.

Illustrative Example

Add:			Subtract:	

$$\begin{array}{rcl} \dfrac{3}{4} &=& \dfrac{3}{4} \\[2mm] +\dfrac{1}{2} &=& \dfrac{2}{4} \\[2mm] \hline & & \dfrac{5}{4} \end{array} \qquad\qquad \begin{array}{rcl} \dfrac{3}{4} &=& \dfrac{3}{4} \\[2mm] -\dfrac{1}{2} &=& \dfrac{2}{4} \\[2mm] \hline & & \dfrac{1}{4} \end{array}$$

When adding or subtracting fractions, the numerators only are added or subtracted, and the result is placed over the common denominator.

Multiplication of Fractions

To multiply two fractions, first multiply the numerators, then multiply the denominators. Reduce the resultant fraction to its lowest terms by dividing numerator and denominator by a common number.

Illustrative Example

$$3/4 \times 1/3 = 3/4 \times 1/3 = 3/12 = 1/4$$
(or by cancellation)
$$3/4 \times 1/3 = 1/4$$

To multiply mixed numbers, first reduce them to improper fractions.

$$1\ 1/2 \times 3\ 1/4 =$$
$$3/2 \times 13/4 = 39/8 = 4\ 7/8$$

Division of Fractions

To divide one fraction by another, invert the divisor and proceed the same as in multiplication.

Illustrative Example

Divide:
$$4/7 \text{ by } 1/2 =$$
$$4/7 \times 2/1 = 8/7 = 1\ 1/7$$
Expressed another way
$$4/7 \div 1/2 = 4/7 \times 2/1 = 8/7 = 1\ 1/7$$

Cancellation is useful in solving problems involving the multiplication or division of two or more fractions (it cannot be used for addition or subtraction).

Illustrative Example

Multiply
$$\frac{2}{4} \times \frac{3}{8} \times \frac{5}{10} \times \frac{4}{9}$$

(or)

$$\frac{3 \times 8 \times 4 \times 2}{1 \qquad 12}$$

$$\frac{\overset{1}{2}}{4} \times \frac{\overset{1}{3}}{\underset{2}{8}} \times \frac{\overset{1}{5}}{10} \times \frac{\overset{1}{4}}{\underset{3}{9}} = \frac{1}{24}$$

$$\frac{3 \times 8 \times 4 \times 2}{\underset{3}{\overset{1/2}{12}}} = 16$$
$$1$$

SOME APPLICATIONS OF FRACTIONS

IN ESTIMATING

Board Feet—To compute the quantity of board feet in lumber, divide the number of pieces times the dimensions by 12.

Illustrative Example

To obtain the number of board feet in 10 pieces of 2″ × 4″ that are 12 feet long.

$$\frac{10 \times 2 \times 4 \times \cancel{12}}{\cancel{12}} = 80 \text{ FBM}$$

Number of Rafters, Joists or Studs—To find the number of joists, rafters or studs required in a given space, divide the length of the space in feet by the distance in feet, center to center between joists, rafters or studs. Add one unit for the end.

Illustrative Example

Find the number of rafters placed 16 inches on center in a flat roof 36 feet long:

(16 inches = 4/3 feet)

$$36/\frac{4}{3} = \overset{9}{\cancel{36}} \times 3/\cancel{4} = 27 \text{ plus } 1 = 28 \text{ rafters}$$

Obtain the number of studs placed 20 inches on center in a partition 30 feet long:

(20 inches = 5/3 feet)

$$30/\frac{5}{3} = \overset{6}{\cancel{30}} \times 3/\cancel{5} = 18 \text{ plus } 1 = 19 \text{ studs}$$

SURFACE MEASUREMENT

A *polygon* is a plane figure bounded by straight-line sides. A three-sided polygon is called a *triangle*; a four-sided polygon is a *quadrilateral*. Various types of polygons are shown below.

In discussing the properties of any polygon in a vertical plane, the *base* is the bottom side, and the *height* or altitude is the perpendicular distance from the highest point to the base. The symbol for the *base* is "b" and the symbol for the height is "h."

The *area* of a polygon is the measurement of its surface without regard to its thickness. In estimating buildings, the calculation of areas of polygons has to do principally with the surfaces of floors, ceilings, interior and exterior walls, and of roof surfaces.

The *perimeter* of a polygon is the sum of the length of its sides. It is the distance around the figure.

RECTANGLES

The area of a rectangle equals the base multiplied by the height. If the rectangle is horizontal, as the floor or ceiling of a room, the area equals the length (l) multiplied by the width (w).

RHOMBOIDS, PARALLELOGRAMS

The area of a rhomboid equals the base multiplied by the height. All opposite sides of a rhomboid are parallel.

Area = b × h
Area = 10 × 8 = 80

TRAPEZOIDS

The area of a trapezoid equals one-half of the sum of the top and bottom multiplied by the height. Only the top and bottom sides are parallel.

Area = ½ (b + b′) × h

$$\text{Area} = \frac{8 + 10}{2} \times 8 = 72 \qquad \text{Area} = \frac{6 + 12}{2} \times 8 = 72$$

IRREGULAR POLYGONS

The areas of trapeziums or irregular polygons can only be determined by breaking the figures up into triangles, or regular polygons, and obtaining the areas separately.

Illustrative Example

Area ABDE = 10 × 15 = 150
Area BCD = ½(6 × 15) = 45
             Total   195

TRIANGLES

The area of any triangle equals one-half of the base times the height. (The height is the perpendicular distance from the base to the highest point.)

Area $\times \dfrac{12 \times 10}{2} = 60$ Area $= \dfrac{9 \times 10}{2} = 45$

Hero's Formula for Area of a Triangle

The area of any triangle may also be computed by the following formula where "s" equals one-half of the sum of the sides of the triangle.

Illustrative Example

$s = \frac{1}{2}(10 + 6 + 8) = 12$
Area $= \sqrt{s(s-a)\,(s-b)\,(s-c)}$
Area $= \sqrt{12(12-10)\,(12-6)\,(12-8)}$
$\quad = \sqrt{12(2)\,(6)\,(4)}$
$\quad = \sqrt{576}$
$\quad = 24$

This formula is advantageous where no angular or other measurements of the triangle are available except the length of the three sides.

Rule of Pythagoras for Right Triangles

The square of the length of the hypotenuse is equal to the sum of the squares of the lengths of the other two sides.

Illustrative Example

$5^2 = 3^3 + 4^2$
$25 = 9 + 16$
$25 = 25$

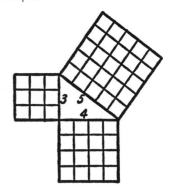

By means of this formula if any two sides of a right triangle are known, the other side can be computed.

$$c = \sqrt{a^2 + b^2}$$
$$a = \sqrt{c^2 - b^2}$$
$$b = \sqrt{c^2 - a^2}$$

Illustrative Example

$a = 3\ b = 4\ c = 5$
Known a and b—find c.
$c = \sqrt{3^2 + 4^2}$
$ = \sqrt{9 + 16}$
$ = \sqrt{25}$
$ = 5$

Illustrative Example

$a = 3\ b = 4\ c = 5$
Known: c and b—find a.
$a = \sqrt{5^2 - 4^2}$
$ \sqrt{25 - 16}$
$ = \sqrt{9}$
$ = 3$

Illustrative Example

$a = 3\ b = 4\ c = 5$
Known: c and a—find b.
$b = \sqrt{5^2 - 3^2}$
$ = \sqrt{25 - 9}$
$ = \sqrt{16}$
$ = \sqrt{4}$

Square Root from Table

Table 7–2 gives the square and the root beginning with number ten. The nearest root shown may be used or for greater accuracy interpolate.

TABLE 7–2

Square Roots of Numbers

Number	Square Root	Number	Square Root	Number	Square Root	Number	Square Root
100	10.00	150	12.25	200	14.14	250	15.81
101	10.05	151	12.29	201	14.18	251	15.84
102	10.10	152	12.33	202	14.21	252	15.87
103	10.15	153	12.37	203	14.25	253	15.91
104	10.20	154	12.41	204	14.28	254	15.94
105	10.25	155	12.45	205	14.32	255	15.97
106	10.30	156	12.49	206	14.35	256	16.00
107	10.34	157	12.53	207	14.39	257	16.03
108	10.39	158	12.57	208	14.42	258	16.06
109	10.44	159	12.61	209	14.46	259	16.09
110	10.49	160	12.65	210	14.49	260	16.12
111	10.54	161	12.69	211	14.53	261	16.16
112	10.58	162	12.73	212	14.56	262	16.19
113	10.63	163	12.77	213	14.59	263	16.22

TABLE 7–2

Square Roots of Numbers

Number	Square Root	Number	Square Root	Number	Square Root	Number	Square Root
114	10.68	164	12.81	214	14.63	264	16.25
115	10.72	165	12.85	215	14.66	265	16.28
116	10.77	166	12.88	216	14.70	266	16.31
117	10.82	167	12.92	217	14.73	267	16.34
118	10.86	168	12.96	218	14.76	268	16.37
119	10.91	169	13.00	219	14.80	269	16.40
120	10.95	170	13.04	220	14.83	270	16.43
121	11.00	171	13.08	221	14.87	271	16.46
122	11.05	172	13.11	222	14.90	272	16.49
123	11.09	173	13.15	223	14.93	273	16.52
124	11.14	174	13.19	224	14.97	274	16.55
125	11.18	175	13.23	225	15.00	275	16.58
126	11.23	176	13.27	226	15.03	276	16.61
127	11.27	177	13.30	227	15.07	277	16.64
128	11.31	178	13.34	228	15.10	278	16.67
129	11.36	179	13.38	229	15.13	279	16.70
130	11.40	180	13.42	230	15.17	280	16.73
131	11.45	181	13.45	231	15.20	281	16.76
132	11.49	182	13.49	232	15.23	282	16.79
133	11.53	183	13.53	233	15.26	283	16.82
134	11.58	184	13.56	234	15.30	284	16.85
135	11.62	185	13.60	235	15.33	285	16.88
136	11.66	186	13.64	236	15.36	286	16.91
137	11.70	187	13.67	237	15.39	287	16.94
138	11.75	188	13.71	238	15.43	288	16.97
139	11.79	189	13.75	239	15.46	289	17.00
140	11.83	190	13.78	240	15.49	290	17.03
141	11.87	191	13.82	241	15.52	291	17.06
142	11.92	192	13.86	242	15.56	292	17.09
143	11.96	193	13.89	243	15.59	293	17.12
144	12.00	194	13.93	244	15.62	294	17.15
145	12.04	195	13.96	245	15.65	295	17.18
146	12.08	196	14.00	246	15.68	296	17.20
147	12.12	197	14.04	247	15.72	297	17.23
148	12.17	198	14.07	248	15.75	298	17.26
149	12.21	199	14.11	249	15.78	299	17.29
300	17.32	350	18.71	400	20.00	450	21.21
301	17.35	351	18.73	401	20.02	451	21.24
302	17.38	352	18.76	402	20.05	452	21.26
303	17.41	353	18.79	403	20.07	453	21.28
304	17.44	354	18.81	404	20.10	454	21.31
305	17.46	355	18.84	405	20.12	455	21.33
306	17.49	356	18.87	406	20.15	456	21.35

TABLE 7–2 (continued)

Square Roots of Numbers

Number	Square Root	Number	Square Root	Number	Square Root	Number	Square Root
307	17.52	357	18.89	407	20.17	457	21.38
308	17.55	358	18.92	408	20.20	458	21.40
309	17.58	359	18.95	409	20.22	459	21.42
310	17.61	360	18.97	410	20.25	460	21.45
311	17.64	361	19.00	411	20.27	461	21.47
312	17.66	362	19.03	412	20.30	462	21.49
313	17.69	363	19.05	413	20.32	463	21.52
314	17.72	364	19.08	414	20.35	464	21.54
315	17.75	365	19.10	415	20.37	465	21.56
316	17.78	366	19.13	416	20.40	466	21.59
317	17.80	367	19.16	417	20.42	467	21.61
318	17.83	368	19.18	418	20.45	468	21.63
319	17.86	369	19.21	419	20.47	469	21.66
320	17.89	370	19.24	420	20.49	470	21.68
321	17.92	371	19.26	421	20.52	471	21.70
322	17.94	372	19.29	422	20.54	472	21.73
323	17.97	373	19.31	423	20.57	473	31.75
324	18.00	374	19.34	424	20.59	474	21.77
325	18.03	375	19.36	425	20.62	475	21.79
326	18.06	376	19.39	426	20.64	476	21.82
327	18.08	377	19.42	427	20.66	477	21.84
328	18.11	378	19.44	428	20.69	478	21.86
329	18.14	379	19.47	429	20.71	479	21.89
330	18.17	380	19.49	430	20.74	480	21.91
331	18.19	381	19.52	431	20.76	481	21.93
332	18.22	382	19.54	432	20.78	482	21.95
333	18.25	383	19.57	433	20.81	483	21.98
334	18.28	384	19.60	434	20.83	484	22.00
335	18.30	385	19.62	435	20.86	485	22.02
336	18.33	386	19.65	436	20.88	486	22.05
337	18.36	387	19.67	437	20.90	487	22.07
338	18.38	388	19.70	438	20.93	488	22.09
339	18.41	389	19.72	439	20.95	489	22.11
340	18.44	390	19.75	440	20.98	490	22.14
341	18.47	391	19.77	441	21.00	491	22.16
342	18.49	392	19.80	442	21.02	492	22.18
343	18.52	393	19.82	443	21.05	493	22.20
344	18.55	394	19.85	444	21.07	494	22.23
345	18.57	395	19.87	445	21.10	495	22.25
346	18.60	396	19.90	446	21.12	496	22.27
347	18.63	397	19.92	447	21.14	497	22.29
348	18.65	398	19.95	448	21.17	498	22.32
349	18.68	399	19.97	449	21.19	499	22.34
						500	22.36

APPLICATION OF SURFACE MEASUREMENT

TO ESTIMATING

Measurement of floor and ceiling areas

The floor or ceiling area of a room is obtained by multiplying the length by the width.

Illustrative Example

Area = 14′ × 16′ = 224 sq ft

Measurement of sidewall areas of rooms

The sidewall area of a room including the window or door openings is equal to the perimeter times the height from the floor to the ceiling.

Illustrative Example

Perimeter = 2(14′ + 16′) = 60 lin ft
Area = 60′ × 8′ = 480 sq ft

Another method is to obtain the area of the walls separately and add them together to get the total wall area.

2 walls = 2(16′ × 8′) = 256 sq ft
2 walls = 2(14′ × 8′) = 224 sq ft
480 sq ft

Measurement of exterior wall areas

The exterior areas of buildings may be obtained either by computing each wall area separately and adding them together, or by taking the perimeter of a building and multiplying by the height of the wall. Gable ends, dormers, extensions, and so forth, should be figured separately and added.

Illustrative Example

Perimeter = 2(20′ + 30′) = 100 lin ft
×10
1,000 sq ft

Plus Gables = $\frac{2(8′ × 20′)}{2}$ = 160

Total 1,160 sq ft

Measurement of Rafter Lengths

When the *rise* of a roof is known, and also the *run* (one half the span), the length of the rafter may be calculated from the "Rule of Pythagoras." (See also Table 14–2 pages 103–105.)

Rise = 6 ft
Span = 16 ft
Run = $\frac{16}{2}$ = 8 ft
Rafter AC or BC = $\sqrt{6^2 + 8^2}$
　　　　　　　= $\sqrt{36 + 64}$
　　　　　　　= $\sqrt{100}$
　　　　　　　= 10 ft

Note: To this length must be added any existing overhang.

PROPERTIES OF CIRCLES

The *circumference* of a circle is the distance around its exterior.

The *diameter* is a line from one side of the circumference to the opposite side and passing through the center of the circle.

The *radius* of a circle is a line from the center of the circle to any point on the circumference, and, therefore equals one-half the diameter.

In any circle the circumference when divided by the diameter will always equal the same number called pi with symbol π. Its value is approximately 3.1416, or 3 1/7. In other words, the circumference of a circle is approximately 3 1/7 times as long as its diameter.

$\dfrac{\text{Circumference}}{\text{Diameter}} = \pi = 3.1416$
　　　　　　　　　　$= 3\ 1/7$

The circumference of a circle equals π times the diameter. This is expressed as C = π D, or C = π 2R (also 3 1/7 × diameter).

The diameter of a circle equals the circumference divided by π. This is expressed as D = $\dfrac{C}{\pi}$

The area of a circle in terms of its radius is π times the radius squared. This is expressed as A = πR^2 or 3 1/7 × R^2.

Illustrative Example

The diameter of a sprinkler tank is 20 feet. What is the area of its base?

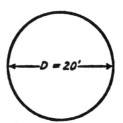

$$A = \pi R^2$$
$$= 3.1416 \times 10^2$$
$$= 3.1416 \times 100$$
$$= 314.16 \text{ sq ft}$$

If the height "h" of the tank is 30 feet, what is the exterior wall area? It would be the circumference of the base multiplied by the height.

Wall surface area of a tank	$= \pi D h$
Circumference	$= \pi D$
Circumference	$= 3.1416 \times 20'$
Circumference	$= 62.832 \text{ lin ft}$
Surface area	$= 62.832 \times 30'$
Surface area	$= 1,884.96 \text{ sq ft}$

VOLUME MEASUREMENT

In the measurement of volume three dimensions are considered, namely length, width and height. Volume is expressed in terms of cubic units of inches, feet, and so forth.

Rectangular Solids

The rectangular solid is commonly encountered in estimating.

The formula for the volume of a rectangular solid is:

$$V = LWH$$

Illustrative Example

V = 20′ × 16′ × 10′
  = 3,200 cu ft

Cylinders

The cylinder is another common type of solid encountered in estimating, and generally its volume in terms of gallons capacity is required.

The cubic volume of a cylinder may be found by multiplying the area of its base by the height or length of the tank. The formula would be expressed: $V = \pi R^2 H$

Illustrative Example

π = 3.1416
R = 10 ft
H = 40 ft
V = 3.1416 × 10² × 40′
  = 12,566.4 cu ft

To convert the contents in cubic feet to gallons multiply the number of cubic feet by 7.4805 the number of gallons in one cubic foot.

12,566.4 × 7.4805 = 94,003 gallons

TABLE 7–3

Table of Weights and Measures

Linear Measure

12 inches	= 1 foot
3 feet	= 1 yard
5½ yards	= 1 rod
320 rods	= 1 mile
5,280 feet	= 1 mile

Surface Measure

144 square inches	= 1 square foot
9 square feet	= 1 square yard
30 ¼ square yards	= 1 square rod
160 square rods	= 1 acre
43,560 square feet	= 1 acre
640 acres	= 1 square mile or one section

Volume Measure

1,728 cubic inches	= 1 cubic foot
27 cubic feet	= 1 cubic yard
128 cubic feet	= 1 cord
(A cord of wood	= 4′ × 4′ × 8′)

Liquid Measure

4 gills	= 1 pint
2 pints	= 1 quart
4 quarts	= 1 gallon
231 cubic inches	= 1 gallon
31½ gallons	= 1 barrel
63 gallons	= 1 hogshead
7.4805 gallons	= 1 cubic foot
32 ounces	= 1 quart

Dry Measure

2 pints	= 1 quart
8 quarts	= 1 peck
4 pecks	= 1 bushel

Weight Measure

16 ounces	= 1 pound
100 pounds	= 1 hundred weight
2,000 pounds	= 1 ton
2,240 pounds	= 1 long ton

Metric System

10 millimeters (mm)	= 1 centimeter	(cm) =	0.3937	inches
10 centimeters	= 1 decimeter	(dm) =	3.9370	inches
10 decimeters)	= 1 meter	(m) =	39.37	inches
100 centimeters)		=	3.28	feet
10 meters	= 1 dekameter	(dkm) =	393.7	inches
10 dekameters	= 1 hectometer	(hm) =	328.08	feet
10 hectometers	= 1 kilometer	(km) =	0.62137	miles
10 kilometers	= 1 myriameter	(mym) =	6.2137	miles

TABLE 7–3 (Continued)

Metric Conversion Factors

To Convert From	To	Multiply By
acres	square feet	43,560.0
acres	square meters	4,047.0
acres	square yards	4,840.0
board feet	cubic inches	144.0
centimeters	feet	0.03281
centimeters	inches	0.3937
cubic feet	cubic centimeters	28,317.0
cubic feet	cubic meters	0.028317
cubic feet	cubic yards	0.03704
cubic feet	gallons (U.S.A.)	7.481
cubic feet	gallons (Imperial)	6.22905
cubic feet of water	pounds	62.37
cubic inches	cubic centimeters	16.38716
cubic yards	cubic meters	0.764559
fathoms	feet	6.0
feet	centimeters	30.48
feet	meters	0.304801
gallons (U.S.A.)	cubic centimeters	3,785.0
gallons (U.S.A.)	cubic feet	0.13368
gallons (U.S.A.)	Imperial gallons	0.832702
gallons (Imperial)	gallons (U.S.A.)	1.20091
gallons (U.S.A.)	cubic inches	231.0
gallons (U.S.A.)	liters	3.78543
gallons (U.S.A.)	ounces	128.0
horsepower (U.S.A.)	horsepower (metric)	1.01387
inches	centimeters	2.54001
inches	meters	0.0254001
inches	millimeters	25.4001
liters	cubic feet	0.03532
liters	gallons (U.S.A.)	0.26418
miles (statute)	kilometers	1.60935
miles (statute)	miles (nautical)	0.8684
miles (nautical)	feet	6,080.204
miles (nautical)	miles (statute)	1.1516
pounds (avoirdupois)	grams (metric)	453.592
pounds (avoirdupois)	kilograms	0.453592
square centimeters	square feet	0.0010764
square feet	square centimeters	929.0
square feet	square meters	0.0929034
square inches	square centimeters	6.452
square yards	square meters	0.83613
tons (metric)	tons (short)	1.1023
tons (long)	pounds	2,240.0
tons (short)	pounds	2,000.0
tons (metric)	pounds	2,204.6
yards	meters	0.91442

Temperature Conversion

	To	Multiply By
degrees F less 32°F	degrees C	.5556
degrees C	degrees F − 32°F	1.8

Examples:

72° F − 32° = 40° × .5556 = 22° C

100° C × 1.8 = 180° F − 32° or 212° F

CHAPTER 8

Concrete

Concrete is a mortar containing aggregate coarser than sand, usually gravel, stone or cinders. Plain concrete, as distinguished from reinforced, is used when the stresses are in compression as in footings and certain types of foundations. When tensile stresses are involved as in beams, girders and floor slabs, steel reinforcing is placed in the concrete near the tension area, in the form of rods, or mesh, or both, depending on the design. Concrete is made by mixing together cement, sand and a coarse aggregate, and allowing the mass to harden in forms. Other additives may be introduced to produce hardness, to prevent freezing, or to make the concrete waterproof.

The problem of estimating any complicated or extensive concrete work in connection with building losses is not frequently encountered. The most common cases that arise will be those which involve replacing or repairing damaged sidewalks, driveways, basement or garage floors, and occasionally repairs to foundations, columns, slabs, and so forth.

In preparing an estimate for concrete, four factors are to be considered in addition to any earth handling: (1) Forms, (2) Concrete, (3) Reinforcement, (4) Finishing.

FORMS

The unit of measurement of forms should be the actual square feet of concrete *in contact with the forms*. A foundation wall 20 feet long, 8 feet high and 12 inches thick would require 8′ × 20′ = 160 square feet of forms for each side or a total of 320 square feet excluding the ends. Wood forms are conventionally constructed of 2 by 4-inch studs and sheathing or plywood with necessary bracing. The studding is spaced from 12 inches to 24 inches on center depending on the height of the form. Where the wall height is 3 or 4 feet, spacing of 24 inches is adequate. For higher walls the studs are placed closer together to withstand the outward pressure exerted by the plastic mass of wet concrete.

Estimating Material and Labor for Forms

For average estimating purposes the quantity of lumber (studding, sheathing and bracing) required *per square foot of concrete surface* is between 2 1/2 to 3 board feet. Using this basis, the board feet of lumber needed for the forms for a wall 8 feet high and 20 feet long, assuming 2 1/2 FBM per square foot of surface, would be:

$$2(8' \times 20') = 320 \quad \text{sq ft}$$
$$\underline{\times \ 2\frac{1}{2}} \quad \text{FBM}$$
$$800 \quad \text{FBM}$$

In pricing the lumber, unless quantities in excess of two or three thousand board feet are involved, an average price between the cost of sheathing boards and the studding may be used. If sheathing, for example is $350.00 per 1,000 FBM, and 2" $\times$ 4" studding is $400.00 per 1,000 FBM, an average price of $375.00 may be used.

Wood forms are usually built and erected on the job by carpenters. In some localities, carpenter helpers, or common laborers assist in the work. Laborers generally strip the forms. The number of hours required to build, erect, and strip concrete forms varies with the type of structure for which the form is being made.

Figures 8–1 and 8–2 show several common types of forms, and the method of estimating the board feet of lumber for each type, when an accurate quantity of form material is desired. (See pages 114–115.)

Table 8–1 shows the approximate board feet of lumber and the average hours of labor required to build, erect, and strip 100 square feet of forms of various types.

Forms for Slabs on Fill

Forms for concrete slabs that are placed directly on the surface of the ground such as sidewalks, garage or basement floors, or the floors of one-story dwellings without basements, will require a minimum of form work.

TABLE 8–1

Approximate Board Feet of Lumber and Hours of Labor Required to Build, Erect, and Strip 100 Square Feet of Wooden Forms

Type of Form	FBM Lumber Per 100 Sq Ft Surface	Hours Labor Per 100 Sq Ft Surface		
		Build and Erect	Strip	Total Hours
Footings	350	6	2	8
Foundations and walls	250	8	2	10
Floor slabs above grade	250	10	4	14
Columns	300	10	4	14
Beams and girders	300	12	4	16
Stairs and steps	300	12	2	14

After the area has been prepared, stakes are driven and side pieces to contain the concrete are put in place using braced boards, 2 $\times$ 4's, 2 $\times$ 6's or other stock of suitable dimensions. These curbing pieces are usually placed so that the upper edges are even with the top of the finished concrete slab. In this way they serve as leveling guides. When estimating the amount of material, and the labor to set forms of this kind, each job should be figured according to its individual circumstances and requirements.

TYPICAL FOUNDATION OR WALL FORM
(BRACING OMITTED)

MATERIAL FOR SECTION SHOWN

SHOE	2-2"×4"×8'	= 10⅔ F.B.M.
STUDS	14-2"×4"×6'	= 56 "
* WALES	12-2"×4"×8'	= 64 "
1" BOARDS	2(6×8)=96 S.F.	
	WASTE ¼ 24	
		=120 "
		250⅓ "

FORM SURFACE AREA
2 × 6'×8' = 96 SQ. FT.

F.B.M. PER SQ. FT.
250⅓ ÷ 96 = 2.8 F.B.M. PER SQ. FT.

* WHEN HEAVY WALES ARE REQUIRED, IT WILL BE NOTED THAT THE F.B.M. PER SQ. FT. IS SUBSTANTIALLY INCREASED OVER THE AVERAGE SHOWN IN TABLE 8-1

12"

8'-0"

6'

WIRE TIE

SPREADER

FOOTING

2"×4" STUDS 16" O.C.

DOUBLED 2"×4" WALES

3'

STRIKE BOARD

DIVIDER

2"×4" SIDES FOR 4" SLABS.

STAKES DRIVEN TO HOLD SIDE FORMS.

10'

TYPICAL SIDEWALK OR SLAB-ON-FILL FORM
MATERIALS FOR FORMS OF THIS TYPE SHOULD BE COMPUTED ACCORDING TO THE REQUIREMENTS FOR EACH INDIVIDUAL JOB.

Figure 8-1
Concrete forms.

Typical Footing Form.
Material for Four-Foot Section
SIDES 2-2"x8"x4' = 10⅔ F.B.M.
STAKES 4-2"x4"x2' = 5⅓ "
BRACE 1-2"x4"x3' = 2 "
Total = 18 "

FORM SURFACE AREA
ONE SIDE 8 x 4 = ⅔ SQ. FT.
TWO SIDES ⅔ x 2 = 14/⁴ "
= 5⅓ " "

F.B.M. PER SQ. FT.
18 ÷ 5⅓ = 3.38 F.B.M. PER SQ. FT.

.2 x 8 SIDES

2 x 8 SIDES

8"

2"x4" STAKES ⨍ BRACES
SPACED 4' O.C.

Typical Column Form
MATERIAL FOR EIGHT-FOOT SECTION
TIE BLOCKS 4(4-2"x4"x2') = 21⅓ F.B.M.
1" BOARDS 4'x8' = 32 SQ. FT.
WASTE ¼ = 1 " "
TOTAL 40 "
61⅓ "

FORM SURFACE AREA
10"x10"x8' = 26⅔ SQ. FT.

F.B.M. PER SQ. FT.
61⅓ ÷ 26⅔ = 2.3 F.B.M. PER SQ. FT.

2"x4" TIE BLOCKS

12" 12"

Figure 8–2
Concrete forms (footing and column).

A rule of thumb for determining the number of board feet of form lumber for slabs on fill is to use the same number of lineal feet of side curbing as the distance around the slab and add 25 per cent for stakes and braces.

Illustrative Example

Determine the board feet of lumber for a walk 4 inches thick, 3 feet wide by 30 feet long.

Perimeter = 2(30′ + 3′) = 66 lin ft

66 lin ft 2″ × 4″ =	44 FBM
Plus 25% Braces and Stakes	11 FBM
Total	55 FBM

A carpenter can place, level, drive the stakes and brace ground slab forms at the average rate of 20 lineal feet per hour. The number of hours of labor to install the forms in the above example would be:

$$\frac{66}{20} = 3.3 \text{ hours}$$

(and) $\dfrac{55}{3.3} = 16.7$ FBM per hour

Estimating Costs

Since the unit of measurement for concrete forms is a square foot of surface area in contact with the concrete, the unit cost of forms may be obtained using Table 8–1 for the number of board feet of lumber and hours of labor per 100 square feet of form.

Nails and wire are not included in the following examples as the quantity required per square foot is negligible when related to the total unit cost. If large quantities of form work are involved, an allowance of 30 pounds of nails per 1,000 board feet of lumber, and 40 feet No. 8 form wire per 100 square feet of form may be added.

Illustrative Examples of Unit Costs For Forms
(Assumed Prices and Wages)

Foundation and Wall Forms

Table
(8–1) 250 FBM lumber	@ $400.00 =	$100.00
(8–1) 8 hours carpenter, build, erect	@ 12.00 =	96.00
(8–1) 2 hours laborer to strip	@ 9.00 =	18.00
	Unit cost per 100 sq ft =	$214.00

Unit cost per sq ft $\dfrac{\$214.00}{100}$ = $ 2.14

Floor Slabs Above Ground

(8–1) 250 FBM lumber	@ $400.00 =	$100.00
(8–1) 10 hours carpenter, build, erect	@ 12.00 =	120.00
(8–1) 4 hours laborer to strip	@ 9.00 =	36.00
	Unit cost per 100 sq ft =	$256.00

$$\text{Unit cost per sq ft } \frac{\$256.00}{100} = \$ \quad 2.56$$

Columns

(8–1) 300 FBM lumber	@ $400.00 =	$120.00
(8–1) 10 hours carpenter, build, erect	@ 12.00 =	120.00
(8–1) 4 hours laborer to strip	@ 9.00 =	36.00
	Unit cost per 100 sq ft =	$276.00

$$\text{Unit cost per sq ft } \frac{\$276.00}{100} = \$ \quad 2.76$$

Beams and Girders

Table

(8–1) 300 FBM lumber (average price)	@ $400.00 =	$120.00
(8–1) 12 hours carpenter, build, erect	@ 12.00 =	144.00
(8–1) 4 hours laborer to strip	@ 9.00 =	36.00
	Unit cost per 100 sq ft =	$300.00

$$\text{Unit cost per sq ft } \frac{\$300.00}{100} = \$ \quad 3.00$$

Application of Unit Costs for Form Work

When the surface area of the concrete structure has been measured and computed in square feet, a unit cost is developed. The square foot area is then multiplied by the unit cost to obtain the total cost of building, erecting, and stripping the forms.

Using the unit costs shown in the foregoing example for *Foundations and Walls*, the cost of forms for a 12-inch wall 8 feet high and 20 feet long would be:

1 Side 8' × 20' =	160 sq ft	
2 Sides	× 2	
	320 sq ft	
Unit cost per sq ft	$ 2.14	
Total cost	$684.80	

As another example, using the unit cost shown, the total cost of the forms for 10 columns 12" × 12" × 10' high would be (perimeter of each column is 4 lin ft):

1 Column 4' × 10' =	40 sq ft	
10 Columns	× 10	
	400 sq ft	
Unit cost per sq ft	$ 2.76	
Total cost	$1,104.00	

Salvage in Forms

Ordinarily when forms are built with new lumber specifically for footings, foundations, and so forth, on moderate-sized concrete jobs, the value of the lumber salvage is seldom considered in estimating unless there is a definite reuse for the lumber in the balance of reconstruction. In that event an allowance of approximately 2 to 3 hours labor per 100 square feet should be allowed for cleaning, removing nails, and for breakage, in addition to stripping time.

Some contractors keep on hand old form lumber that has been salvaged from previous concrete jobs and can be used again. They may also have available prefabricated plywood panel forms, or patented metal forms that can be used many times, cleaned and stored for reuse. In estimating form work, consideration should be given to the availability of reusable forms as a means of reducing the cost. A reasonable charge should be included in the estimate for the rental of such forms, and also for cleaning and trucking them to and from the job.

CONCRETE MIXES

Concrete is a mixture of cement, sand and a coarse aggregate, usually gravel or crushed stone graded from 3/4 inch to 1 1/2 inches. The *mix* in concrete refers to the proportions of these three components and is stated in cubic feet of cement, cubic feet of sand, and cubic feet of stone. A mix of 1:3:5 means that the proportions are:

> 1 cu ft cement
> 3 cu ft sand
> 5 cu ft stone

In actual construction the proper proportions, the quality of the materials, and the amount of water used are important and controlled factors. The Portland Cement Association publishes considerable reference material on the subject of mixes for various kinds of work. It has been said that it costs very little more to make good concrete than it does to make concrete of a poor quality. For estimating purposes, materials may be figured on the basis of a mix of 1 part cement, 2 1/4 parts sand and 3 parts stone. The size of the stone should not exceed 1 inch. This proportion is generally considered satisfactory for good concrete to be used in footings, foundations, floors, walks, driveways, beams, girder, columns and swimming pools.

In terms of a cubic yard of concrete the quantity of materials required for a 1:2 1/4:3 mix would be:(Table 8–2, Mix No. 3)

> 6¼ sacks of cement
> 14 cu ft of sand (.52 cu yds)
> 19 cu ft of stone (.70 cu yds)

A sack of cement contains one cubic foot and weighs 94 pounds. A barrel of cement contains four sacks.

A yard of sand or stone contains 27 cubic feet, and weighs 2,600 to 2,800 pounds.

Where it is desired to use other proportions of cement, sand and stone, Table 8–2 shows the recommended quantity of each of these materials for one cubic yard of concrete.

Estimating Quantities

The unit of measurement for concrete is expressed in cubic yards, or sometimes in cubic feet (there are 27 cubic feet in a cubic yard). Coarse aggregate like gravel or stone contains voids between individual particles which are taken up with the sand. Similarly there are voids between the particles of sand which are taken up with the finer particles of cement. For this reason, in any prescribed *mix*, the combined cubic feet of cement, sand, and aggregate will exceed 27 cubic feet. When the materials are mixed, however, the voids are filled and the resultant bulk is 27 cubic feet, or one cubic yard. For example, a mix of 1:2 1/4:3 is composed of 6 1/4 cubic feet of cement, 14 cubic feet of sand and 19 cubic feet of coarse aggregate, a total of 39 1/4 cubic feet. After mixing, the mass will occupy 27 cubic feet.

TABLE 8–2
Quantities of Cement, Sand and Stone Required
for 1 Cubic Yard Concrete

| No.* | Mix | Sacks of Cement | Materials Per Cu Yd Concrete | | Max. Aggregate |
			Cu Yds Sand	Cu Yds Stone	
1	1:1:1¾	10	.37	.63	⅜ "
2	1:2:2¼	7¾	.56	.65	¾ "
3	1:2¼:3	6¼	.52	.70	1"
4	1:3:4	5	.56	.74	1½ "

*Mix No. 1 — Concrete for heavy wearing surface.
 Mix No. 2 — Concrete exposed to weak acid or alkali solutions.
 Mix No. 3 — Concrete for watertight floors, foundations, swimming pools, septic tanks and for structural reinforced concrete, beams, columns, slabs, residence floors, etc.
 Mix No. 4 — Concrete for foundations and walls not subjected to weather or water pressure.

Table 8–2A shows the quantities of cement, sand and stone for one cubic yard of various concrete mixes for one to twenty-nine cubic yards. (See page 120.)

Typical concrete shapes are shown in cross section in Figure 8–3. To obtain the quantity of concrete, the *square foot* area of the cross section is computed and this is multiplied by the length to get the number of cubic feet. Divide the cubic feet by 27 to obtain the number of cubic yards. Shortcuts are shown in the following section.

Shortcuts to Estimating Quantities

Rather than referring to lengthy tables to determine the number of cubic yards of concrete in footings, walls, slabs and columns, there is an easier method. It is also much simpler than multiplying out the cubic feet and dividing by 27 to obtain the cubic yards.

TABLE 8–2A

Quantities of Cement, Sand and Stone for 1 to 29 Cubic Yards of
Various Concrete Mixes

Cu Yds Concrete	MIX 1 1:1:1 3/4 AGGREGATE			MIX 2 1:2:2 1/4 AGGREGATE			MIX 3 1:2.25:3 AGGREGATE			MIX 4 1:3:4 AGGREGATE		
	Sacks Cement	Yards Sand	Yards Stone	Sacks Cement	Yards Sand	Yards Stone	Sacks Cement	Yards Sand	Yards Stone	Sacks Cement	Yards Sand	Yards Stone
1	10	.37	.63	7.75	.56	.65	6.25	.52	.70	5.0	.56	.74
2	20	.74	1.26	15.50	1.12	1.30	12.50	1.04	1.40	10.0	1.12	1.48
3	30	1.11	1.89	23.25	1.68	1.95	18.75	1.56	2.10	15.0	1.68	2.22
4	40	1.48	2.52	31.00	2.24	2.60	25.00	2.08	2.80	20.0	2.24	2.96
5	50	1.85	3.15	38.75	2.80	3.25	31.25	2.60	3.50	25.0	2.80	3.70
6	60	2.22	3.78	46.50	3.36	3.90	37.50	3.12	4.20	30.0	3.36	4.44
7	70	2.59	4.41	54.25	3.92	4.55	43.75	3.64	4.90	35.0	3.92	5.18
8	80	2.96	5.04	62.00	4.48	5.20	50.00	4.16	5.60	40.0	4.48	5.92
9	90	3.33	5.67	69.75	5.04	5.85	56.25	4.68	6.30	45.0	5.04	6.66
10	100	3.70	6.30	77.50	5.60	6.50	62.50	5.20	7.00	50.0	5.60	7.40
11	110	4.07	6.93	85.25	6.16	7.15	68.75	5.72	7.70	55.0	6.16	8.14
12	120	4.44	7.56	93.00	6.72	7.80	75.00	6.24	8.40	60.0	6.72	8.88
13	130	4.82	8.20	100.76	7.28	8.46	81.26	6.76	9.10	65.0	7.28	9.62
14	140	5.18	8.82	108.50	7.84	9.10	87.50	7.28	9.80	70.0	7.84	10.36
15	150	5.56	9.46	116.26	8.40	9.76	93.76	7.80	10.50	75.0	8.40	11.10
16	160	5.92	10.08	124.00	8.96	10.40	100.00	8.32	11.20	80.0	8.96	11.84
17	170	6.30	10.72	131.76	9.52	11.06	106.26	8.84	11.90	85.0	9.52	12.58
18	180	6.66	11.34	139.50	10.08	11.70	112.50	9.36	12.60	90.0	10.08	13.32
19	190	7.04	11.98	147.26	10.64	12.36	118.76	9.84	13.30	95.0	10.64	14.06
20	200	7.40	12.60	155.00	11.20	13.00	125.00	10.40	14.00	100.0	11.20	14.80
21	210	7.77	13.23	162.75	11.76	13.65	131.25	10.92	14.70	105.0	11.76	15.54
22	220	8.14	13.86	170.05	12.32	14.30	137.50	11.44	15.40	110.0	12.32	16.28
23	230	8.51	14.49	178.25	12.88	14.95	143.75	11.96	16.10	115.0	12.88	17.02
24	240	8.88	15.12	186.00	13.44	15.60	150.00	12.48	16.80	120.0	13.44	17.76
25	250	9.25	15.75	193.75	14.00	16.25	156.25	13.00	17.50	125.0	14.00	18.50
26	260	9.64	16.40	201.52	14.56	16.92	162.52	13.52	18.20	130.0	14.56	19.24
27	270	10.00	17.00	209.26	15.12	17.56	168.76	14.04	18.90	135.0	15.02	20.00
28	280	10.36	17.64	217.00	15.68	18.20	175.00	14.56	19.60	140.0	15.68	20.72
29	290	10.74	18.28	224.76	16.24	18.86	181.26	15.08	20.30	145.0	16.24	21.46

Figure 8-3
Typical concrete shapes.

slab on beams

sidewalk slab

foundation wall

footing

slab on fill

column

retaining wall

Footings:

Multiply the cross section in square inches by .000257 to get the cubic yards in 1-lineal foot of footing. Multiply that result by the length of the footing.

Illustrative Example

Find the number of cubic yards of concrete in a footing 24" wide, 10" deep and 40' long.

$$24" \times 10" \times .000257 \times 40' = 2.47 \text{ yards}$$

Wall:

Multiply the square foot surface area by the wall thickness in inches. Multiply that result by .0031.

Illustrative Example

Find the number of cubic yards of concrete in a foundation wall 8' high, 50' long and 12" thick.

$$8' \times 50' \times 12" \times .0031 = 14.88 \text{ yards}$$

Floor slabs and Concrete walks:

Multiply the floor or walk area in square feet times the thickness in inches times .0031.

Illustrative Example

Find the number of cubic yards of concrete in a garage floor 24' wide, 25' long and 4" thick.

$$24' \times 25' \times 4" \times .0031 = 7.44 \text{ yards}$$

Columns:

Compute the cubic yards in a column as though it were a footing. Multiply the cross section in square inches by .000257. Multiply that result by the height of the column.

Illustrative Example

Find the number of cubic yards of concrete in a column 12" by 12" by 12' high.

$$12" \times 12" \times .000257 \times 12' = .44 \text{ cu yds}$$

Table 8–2B shows the cubic yards of concrete per lineal foot of foundation walls from one foot to twelve feet high.

For intermediate wall heights and thicknesses, interpolate between factors in the table.

To obtain the number of cubic feet of concrete, multiply the factors above by 27, or multiply the cubic yards determined by 27.

Window and door openings may be deducted by using the formula shown below:

Where:
A = **total sq ft area of opening**
t = **thickness of foundation wall** *in inches*
A $\times$ **t"** $\times$ **.0031** = **cu yds of concrete to be deducted**

TABLE 8–2B

Cubic Yards of Concrete per Lineal Foot of Foundation Wall

Wall Height In Feet	Thickness of Wall in Inches						
	6"	7"	8"	9"	10"	11"	12"
1	.0185	.0217	.0248	.0279	.0310	.0341	.0372
2	.0360	.0434	.0496	.0558	.0620	.0682	.0744
3	.0556	.0651	.0744	.0837	.0930	.1023	.1116
4	.0741	.0868	.0992	.1116	.1240	.1364	.1488
5	.0926	.1085	.1240	.1395	.1550	.1705	.1860
6	.1111	.1302	.1488	.1674	.1860	.2046	.2232
7	.1296	.1519	.1736	.1953	.2170	.2387	.2606
8	.1482	.1736	.1984	.2232	.2480	.2728	.2976
9	.1667	.1953	.2232	.2511	.2790	.3069	.3348
10	.1852	.2170	.2480	.2790	.3100	.3410	.3720
11	.2046	.2387	.2728	.3069	.3410	.3751	.4092
12	.2232	.2604	.2976	.3348	.3720	.4092	.4464

Example:
     Opening is 2' − 6" × 4' − 0" (area 10 sq ft)
     Thickness of wall is 10"
     10 × 10" × .0031 = .31 cu yds to be deducted

Table for Slabs and Walks

Sometimes the estimator wants to know how many square feet of floor or slab of a given thickness will be covered by one cubic yard of concrete. Table 8–3 shows this for slabs 2 inches to 12 inches thick.

TABLE 8–3

Square Foot Area That One Cubic Yard Will Cover

Thickness In Inches	Number Square Feet	Thickness In Inches	Number Square Feet
2	162	6	54
3	108	7	46
3½	93	8	40
4	81	9	36
5	65	10	32
5½	59	12	27

The number of square feet that a cubic yard of concrete will cover in floor slabs, driveways and walks, is easily determined. Simply divide the constant 324, by the thickness in inches.

Illustrative Example

How many square feet will one cubic yard of concrete cover in a floor slab 4.5" thick?

$$\frac{324}{4.5} = 72 \text{ sq ft}$$

Table 8–3A shows the cubic yards of concrete, gravel base and cement topping for concrete slabs, walks, patios, driveways, etc., for large areas of 200 to 3000 sq ft and 1″ to 12″ thick.

Table 8–3B shows the same data for small areas from 20 – 500 sq ft.

Estimating Material Costs

The cost of the materials for a cubic yard of concrete for a mix of 1:2 1/4:3 is readily obtained as follows, *when the local cost of cement, sand and stone (or gravel) has been established*:

TABLE 8–3A

Cubic Yards of Concrete Base and Topping in Slabs, Walks and Patios
with Areas 200 to 3,000 square feet, 1″ to 12″ Thick

| Ground Floor Area | CUBIC YARDS OF GRAVEL BASE; CONCRETE OR TOPPING FOR THICKNESS SHOWN | | | | | | | | | | | |
	1″	2″	3″	4″	5″	6″	7″	8″	9″	10″	11″	12″
200	0.62	1.24	1.86	2.48	3.10	3.72	4.34	4.96	5.58	6.20	6.82	7.44
300	0.93	1.86	2.79	3.72	4.65	5.58	6.51	7.44	8.37	9.30	10.23	11.16
400	1.24	2.48	3.72	4.96	6.20	7.44	8.68	9.92	11.16	12.40	13.64	14.88
500	1.55	3.10	4.65	6.30	7.75	9.30	10.85	12.40	13.95	15.50	17.05	18.60
600	1.86	3.72	5.58	7.44	9.30	11.16	13.02	14.88	16.74	18.60	20.46	22.32
700	2.17	4.34	6.51	8.68	10.85	13.02	15.19	17.36	19.53	21.70	23.87	26.04
800	2.48	4.96	7.44	9.92	12.40	14.88	17.36	19.84	22.32	24.80	27.28	29.76
900	2.79	5.58	8.37	11.16	13.95	16.74	19.53	22.32	25.11	27.90	30.69	33.48
1000	3.10	6.20	9.30	12.40	15.50	18.60	21.70	24.80	27.90	31.00	34.10	37.20
1100	3.41	6.82	10.23	13.64	17.05	20.46	23.87	27.28	30.69	34.10	37.51	40.92
1200	3.72	7.44	11.16	14.88	18.60	22.32	26.04	29.76	33.48	37.20	40.92	44.64
1300	4.03	8.06	12.09	16.12	20.15	24.18	28.21	32.24	36.27	40.30	44.33	48.36
1400	4.34	8.68	13.02	17.36	21.70	26.04	30.38	34.72	39.06	43.40	47.74	52.08
1500	4.65	9.30	13.95	18.60	23.25	27.90	32.55	37.20	41.85	46.50	51.15	55.80
1600	4.96	9.92	14.88	19.84	24.80	29.76	34.72	39.68	44.64	49.60	54.56	59.52
1700	5.27	10.54	15.81	21.08	26.35	31.62	36.89	42.16	47.43	52.70	57.97	63.24
1800	5.58	11.16	16.74	22.32	27.90	33.88	39.06	44.64	50.22	55.80	61.38	66.96
1900	5.89	11.78	17.67	23.56	29.45	35.34	41.23	47.12	53.01	58.90	64.79	70.68
2000	6.20	12.40	18.60	24.80	31.00	37.20	43.40	49.60	55.80	62.00	68.20	74.40
2100	6.51	13.02	19.53	26.04	32.55	39.06	45.57	52.08	58.59	65.10	71.61	78.12
2200	6.82	13.64	20.46	27.28	34.10	40.92	47.74	54.56	61.38	68.20	75.02	81.84
2300	7.13	14.26	21.39	28.52	35.65	42.78	49.91	57.04	64.17	71.30	78.43	85.56
2400	7.44	14.88	22.32	29.76	37.20	44.64	52.08	59.52	66.96	74.40	81.84	89.28
2500	7.75	15.50	23.25	31.00	38.75	46.50	54.26	62.00	69.75	77.50	85.25	93.00
2600	8.06	16.12	24.18	32.24	40.30	48.36	56.42	64.48	72.54	80.60	88.66	96.72
2700	8.37	16.74	25.11	33.48	41.85	50.22	58.59	66.96	75.33	83.70	92.07	100.44
2800	8.68	17.36	26.04	34.72	43.40	52.08	60.76	69.44	78.12	86.80	95.48	104.16
2900	8.99	17.98	26.97	35.96	44.95	53.94	62.93	71.92	80.91	89.90	98.89	107.88
3000	9.30	18.60	27.90	37.20	46.50	55.80	65.10	74.40	83.70	93.00	102.30	111.60

Illustrative Example

6¼ sacks cement	@ $4.00 =	$25.00
.52 cu yds sand	@ 7.00 =	3.64
.70 cu yds stone	@ 6.00 =	4.20
	Total cost	$32.84

TABLE 8-3B

Cubic Yards of Concrete, Base and Topping in Slabs, Walks and Patios
with Areas 20 to 500 Square Feet, 1″ to 12″ Thick

Ground Floor Area	CUBIC YARDS OF GRAVEL BASE; CONCRETE OR TOPPING FOR THICKNESS SHOWN											
	1″	2″	3″	4″	5″	6″	7″	8″	9″	10″	11″	12″
20	.06	.12	.19	.25	.31	.37	.43	.50	.56	.62	.68	.74
30	.09	.18	.28	.37	.47	.56	.65	.75	.84	.93	1.02	1.12
40	.12	.24	.37	.50	.62	.74	.87	.99	1.12	1.24	1.36	1.49
50	.16	.32	.47	.62	.78	.93	1.09	1.24	1.40	1.55	1.71	1.86
60	.19	.38	.56	.74	.93	1.12	1.30	1.49	1.67	1.86	2.05	2.23
70	.22	.43	.65	.87	1.09	1.30	1.52	1.74	1.95	2.17	2.39	2.60
80	.25	.50	.74	.99	1.24	1.49	1.74	1.98	2.23	2.48	2.73	2.98
90	.28	.56	.84	1.12	1.40	1.67	1.95	2.23	2.51	2.79	3.07	3.35
100	.31	.62	.93	1.24	1.55	1.86	2.17	2.48	2.79	3.10	3.41	3.72
120	.37	.74	1.12	1.49	1.86	2.23	2.60	2.98	3.35	3.72	4.09	4.46
140	.43	.87	1.30	1.74	2.17	2.60	3.04	3.47	3.91	4.34	4.77	5.21
160	.49	.99	1.49	1.98	2.48	2.98	3.47	3.97	4.46	4.96	5.46	5.95
180	.56	1.12	1.67	2.23	2.79	3.35	3.91	4.46	5.02	5.58	6.14	6.70
200	.62	1.24	1.86	2.48	3.10	3.72	4.34	4.96	5.58	6.20	6.82	7.44
220	.68	1.36	2.05	2.73	3.41	4.09	4.77	5.46	6.14	6.82	7.50	8.18
240	.74	1.49	2.23	2.98	3.72	4.46	5.21	5.95	6.70	7.44	8.18	8.93
260	.81	1.61	2.42	3.22	4.03	4.84	5.64	6.45	7.25	8.06	8.87	9.67
280	.87	1.74	2.60	3.47	4.34	5.21	6.08	6.94	7.81	8.68	9.55	10.42
300	.93	1.86	2.79	3.72	4.65	5.58	6.51	7.44	8.37	9.30	10.23	11.16
320	.99	1.98	2.98	3.97	4.96	5.95	6.94	7.94	8.93	9.92	10.91	11.90
340	1.05	2.11	3.16	4.22	5.27	6.32	7.38	8.43	9.49	10.54	11.59	12.65
360	1.12	2.23	3.35	4.46	5.58	6.70	7.81	8.93	10.04	11.16	12.28	13.39
380	1.18	2.36	3.53	4.71	5.89	7.07	8.25	9.42	10.60	11.78	12.96	14.14
400	1.24	2.48	3.72	4.96	6.20	7.44	8.68	9.92	11.16	12.40	13.64	14.88
420	1.30	2.60	3.91	5.21	6.51	7.81	9.11	10.42	11.72	13.02	14.32	15.62
440	1.36	2.73	4.09	5.46	6.82	8.18	9.55	10.91	12.28	13.64	15.00	16.37
460	1.43	2.85	4.28	5.70	7.13	8.56	9.98	11.41	12.83	14.26	15.69	17.11
480	1.45	2.98	4.46	5.95	7.44	8.93	10.42	11.90	13.39	14.48	16.37	17.86
500	1.55	3.10	4.65	6.20	7.75	9.30	10.85	12.40	13.95	15.50	17.05	18.60

Estimating Labor Costs

Where small quantities of concrete are needed, or where either a mixing machine or ready-mix concrete is not available, it is mixed by hand. Most builders however are equipped with a concrete mixing machine, or more than one with different capacities. The machines may also be rented. The rental cost varies with the capacity of the mixer and the length of time it is kept.

Ready-mix concrete is usually more economical than job-mixing, since it eliminates the use of a mixer and also the labor of handling materials to be mixed. In many situations ready-mix concrete reduces the labor cost of wheeling and pouring where the truck can back in close to the position where the concrete is to be placed. The concrete can be distributed to the proper spot with chutes attached to the rear of the truck.

Before an estimate of labor costs to mix and place concrete is made, the project should be carefully studied to determine the best procedure that will give proper consideration to all factors affecting the labor required. The following checklist of major items that the estimator should give particular attention to will serve as a guide:

1. Kind of work.
     Footings or foundations
     Walls, beams, girders, columns
     Floor slabs, slabs on ground
2. Method of mixing concrete.
     By hand
     By machine on the job
     Ready-Mix delivered
3. How is concrete to be placed?
     Directly from Ready-Mix truck
     By wheelbarrow from site of mixing
     What distance must it be wheeled?
4. Are plank runways to be built?
5. Is hoisting necessary?
     By hand
     By power hoist
6. Will weather conditions affect labor?
     Winter or summer
     Protection against rain, snow, or frost

Mixing by hand is slower than by machine. Ready-mix concrete does away with the labor of mixing on the job, but the cost of the ready-mix should be checked against on-the-job mixing. The labor required to wheel concrete long distances or hoist it up or down to the forms must be estimated in hours per cubic yard over and above actual mixing.

The labor cost to set up and shift runways should be added as a separate item rather than be included in the unit cost per yard.

Mixing and placing concrete in the winter requires great care. The aggregate has to be heated or some type of anti-freeze must be mixed with the batch. In some cases the poured concrete is protected by covering with straw, salt hay or tarpaulins to prevent frost action or actual freezing.

Table 8–4 shows the average number of hours of labor required for mixing and placing concrete for different kinds of work.

TABLE 8–4*

Approximate Hours of Labor Required for Mixing and Placing Concrete

Kind of Work. (Forms 25 Feet or Less From Mixer)	Hours Per Cubic Yard
Mixing by hand (small jobs)	2
Mixing by machine	1
Placing in footings and foundations	2
Placing in columns and 12 inch walls	4
Placing in walls 12 to 24 inches thick	3
Placing in floors and walks	2
Placing in stairs	4
Add for heating aggregate	½
Add for each additional 25 feet of wheeling	½

Finishing Concrete	Hours Per 100 Sq ft
Wood float finish on slabs	2
Steel trowel finish on slabs	3
Machine finishing	1–2
Cement wash on walls	3

*The labor shown in Table 8–4 is the total man-hours. The estimator should base his costs on the wages of the labor available. Mixing and placing of concrete is principally done by common labor. Cement finishing is done by masons. Under certain conditions and in non-union areas it may be necessary to figure a mason's time for the entire job.

REINFORCEMENT

Reinforcing in concrete should be figured only when the size, quantity, and shape of bars are known to the person preparing the estimate; otherwise someone familiar with reinforced concrete design should be called upon to submit a sub-bid.

Reinforcing is estimated by the *pound* or by the *ton* for both material cost and the labor to place it. The material cost is computed by multiplying the *lineal feet* of reinforcing times the weight per foot times the price per pound. Table 8–5 shows the weight per foot, diameter and area of various sizes of bars. (See page 128.)

The labor to bend and place reinforcing depends on the weights of the bars and whether they must be shaped and wired in place. Where the bars are less than 3/4-inch, a worker should be able to place them in position without tying at the rate of approximately 20 hours per ton. Where they are to be tied in place, the rate can be figured at about 24 hours per ton. These rates do not include unloading or hoisting to floors above grade. On jobs where bars over 3/4-inch are used, the rate for bending, placing and tying is approximately 18 hours per ton. Subcontractors substantially reduce these rates.

Wire Mesh Reinforcing

Wire mesh, or welded steel fabric reinforcing for slabs, driveways and walks comes in square or rectangular mesh and is sold by the roll usually 5 feet wide and 150 feet long. It is also available in flat sheets.

TABLE 8–5

Weights of Reinforcing Bars

No.	Nominal Diameter Inches	Area in Square Inches	Weight in Pounds Per Lineal Foot
*2	.250	0.05	0.167
3	.375	0.11	0.376
4	.500	0.20	0.668
5	.625	0.31	1.043
6	.750	0.44	1.502
7	.875	0.60	2.044
8	1.000	0.79	2.670
9	1.128	1.00	3.400
10	1.270	1.27	4.303
11	1.410	1.56	5.313

*No. 2 bar is a plain round bar.

A popular size used in residential work is 6 x 6 - 10/10. This indicates the wires are 6 inches on center each way forming 6 inch squares and that the wire is No. 10 gauge.

When laid, the mesh is lapped a full square.

The labor putting wire mesh reinforcing in floors and driveways varies with the job. In straight slab work like garage floors a worker should handle 200 square feet per hour, or a roll 5 feet by 150 feet in 4 hours.

FINISHING CONCRETE

After concrete slabs have had an initial set, the surface is given a float finish or a smooth troweled finish. The labor for doing this operation is added to the cost of mixing and placing the concrete.

When forms have been removed from concrete foundations, walls, ceilings, and so forth, the surface is sometimes given a cement wash to provide a more attractive finish. Table 8–4 shows the number of hours per 100 square feet of surface for finishing concrete. For cement plastering of surfaces, refer to Chapter 16.

TOPPING FOR FLOORS AND WALKS

Sometimes the top inch of a floor slab or walk consists of cement mortar which is laid over the concrete base. The mixture is generally sand and cement but occasionally a coarse aggregate not exceeding 3/8 inch is used. When a 1-inch topping is to be laid, the thickness of the concrete base is reduced 1 inch. On a slab, for example, 4 inches thick, the base would be 3 inches with a 1-inch topping.

The mixture of the mortar may be figured at 1 part cement to 2 parts sand. A cubic yard would require 9.5 sacks cement and 1.1 cu yds of sand.

To estimate the number of cubic yards required, the square-foot area is multiplied by 1/12 (1 inch is 1/12 of 1 foot) and the product is divided by 27 (1 cubic yard = 27 cubic feet).

Illustrative Example

To obtain the cubic yards of 1″ topping required on a slab that measures 20′ x 30′.

$$20' \times 30' = 600 \text{ sq ft}$$
$$\frac{600}{12''} = 50 \text{ cu ft}$$
$$\frac{50}{27} = 1.85 \text{ cu yds}$$

(See page 123 for shortcut estimating of slabs. Also see Table 8–3 for square feet one cubic yard covers.)

The labor to mix and place the topping may be estimated on the same basis as for concrete shown in Table 8–4. The labor for mixing is figured separately from the placing. Finishing by hand or machine should be added by using Table 8–4 as a guide.

ESTIMATING COSTS

The cost of material and labor for a specific concrete job may be figured either at a unit cost per cubic yard or the total material and total labor may be estimated. For example, a foundation wall 12″ thick 9′ high and 60′ long would contain (1′ x 9′ x 60′) 540 cubic feet or 20 cubic yards of concrete. The cost, then, of the material may be estimated by obtaining the cost of 1 cubic yard and multiplying by 20. The cost of the labor to mix and pour the concrete may be obtained from Table 8–4.

Illustrative Example

Cost of 1 Cubic Yard of Concrete, 1:2 1/4:3 Mix
(Assumed Prices and Wages)

6 1/4 sacks cement	@ $4.00	=	$25.00
.52 cu yds sand	@ 7.00	=	3.64
.70 ″ ″ stone	@ 6.00	=	4.20
	Total		$32.84

Cost of 20 cu yds, 20 × $32.84 = $656.80

Labor to Mix and Pour

2 cu yds per hour (see Table 8–4)
Labor for 20 cu yds, 2 × 20 = 40 hours
40 hours laborer's time @ $9.00 = $360.00

Total Cost of Job

Material	$ 656.80
Labor	360.00
Total	$1,016.80

If this same job was to be estimated on a unit cost per cubic yard:

Material for 1 cu yd (as above)	=	$32.84
Labor 2 hours (Table 8–4) @ $9.00	=	18.00
Unit cost for 1 cu yd	=	$50.84
Total cost for 20 cu yds 20 × $50.84	=	$1,016.80

Unit Costs

The following basic formula may be used to obtain unit costs per cubic yard of concrete by inserting local labor wages and material prices.

The mix of 1:2 1/4:3 is recommended in estimating for all but concrete that requires a special mixture. It does not include additives for quick setting, hardening, antifreeze or coloring.

Basic Formula

6¼ sacks cement	@	$	= $
.52 cu yds sand	@		=
.70 cu yds stone	@		= ____
Cost of material for 1 cu yd			= $
Hours labor to mix concrete	@		=
Hours labor to place concrete	@		= ____
Unit cost for 1 cu yd			= $

The number of hours necessary to mix and place the concrete should be selected from Table 8–4, depending upon the method of mixing, the kind of work, and on the distance from the site of mixing to the forms. The cost of building runways, hoisting or other operations should be added to the job cost as a separate item.

Illustrative Examples

Assuming the following material cost and wages, and using a mix of 1:2¼:3, compute a unit cost per cubic yard for machine mixed concrete to be placed in a foundation form.

6 ¼ sacks cement	@ $4.00 -	$25.00
.52 cu yds sand	@ 7.00 -	3.64
.70 cu yds stone	@ 6.00 -	4.20
	Material cost per cu yd -	$32.84
1 hour labor mixing by machine	@ $9.00 -	9.00
2 hours placing concrete	@ 9.00 -	18.00
	Unit cost per cu yd -	$59.84

Application of Unit Costs

The following illustrative examples are shown for applying unit costs to obtain the total cost of concrete work. Material prices and labor wage rates are the same as those assumed in the foregoing illustrations of unit costs.

Illustrative Example

A concrete walk 4 inches thick, 3 feet wide and 40 feet long is to be laid using a mix of 1:2¼:3. The concrete is machine mixed, and a float finish is given the top by a mason.

Quantity of concrete $\frac{4}{12} \times 3' \times 40' = 40$ cu ft

$$\frac{40}{27} = 1.48 \text{ cu yds}$$

Unit cost for 1 cu yd $59.84

Cost for 1.48 cu yds	= $	88.56
Surface to be finished 3′ × 40′ = 120 sq ft		
(Table 8–4) 2.4 hours mason's time @ $12	=	28.80
	$	117.36

The cost per square foot may be obtained by dividing $117.36 by 120 sq ft = $.978

Illustrative Example

To find the total cost of concrete to be machine mixed and placed in a retaining wall 12 inches thick, 6 feet high and 20 feet long, using a mix of 1:2¼:3. Develop a unit cost (page 130), and the total cost

$$1' \times 6' \times 20' = 120 \text{ cu ft}$$
$$\frac{120}{27} = 4.44 \text{ cu yds}$$

Cost of wall = 4.44 × $59.84 = $ 265.69

DAMAGEABILITY OF CONCRETE

Concrete has demonstrated its fire-resistive qualities in innumerable extreme fires. Its high value as a protection for steel lies in the similarity of its temperature coefficient to that of steel. Generally 2 inches of concrete protection is considered adequate against temperatures in a fire that would cause steel to buckle and warp.

The rate of heat conductivity of concrete is relatively slow, due in part to its porosity. As the surface of the concrete is dehydrated, the porosity increases, improving the insulating qualities with the result that, except in long-burning hot fires, the main damage is to the outer surface. Spalling of the surface (popping off of thin shale-like fragments) occurs frequently in hot fires, and this condition may be aggravated by alternate heating by the fire and cooling by water. The degree of spalling of concrete varies with the kind of aggregate used. Quartz, gravel, and chert tend to spall quicker than aggregates composed of traprock or limestone.

Fire damage to concrete floor slabs such as garage and basement floors, is seldom of a serious nature. There are a number of reasons for this. The burning of the structure or its contents is above the level of the floor and the heat rises. Water from fire hoses or sprinkler systems tends to keep the floor cooled down. Also contents that have not burned and falling debris from above frequently will cover portions of the floor surface affording a protection.

Soilage, caused by smoke, heat, tar or asphalt drippings from melted roofing, or stains from various contents materials during a fire, is not uncommon.

Cracks in concrete due to shrinkage, may develop during a fire. They may be merely surface cracks, or may penetrate completely through a wall. Concrete walls and floors may also develop serious cracking as a result of settlement, earthquake, or from hydrostatic pressure during flooding conditions which take place during excessive rains and overflow of rivers and streams.

Methods of Repair

Concrete, either plain or reinforced, in most cases can be readily repaired. Foundations, floor slabs, columns, beams, girders and walls which have been seriously injured can be repaired by removing the affected portions, placing new reinforcement where required, erecting forms and placing new concrete to be bonded to the existing structure. Surfaces of floor slabs that have been damaged can be chipped off, cleaned, and a new 1 to 1 1/2-inch topping applied over the old floor.

Each situation must be treated according to the particular conditions that are encountered. When preparing the details of a repair estimate, the estimator should consider whether or not it will be necessary to:

1. Cut away the damaged concrete.
2. Build, erect and strip necessary forms.
3. Place new reinforcing bars or mesh.
4. Prepare old concrete surface to be bonded to the new concrete.
5. Mix and place the concrete.
6. Finish concrete surface as required.

(1) All building or contents debris in the working area should be removed, and any undamaged machinery or fixtures that are in the way of the operation will have to be moved or shored up. The concrete that is to be replaced should be cut away with power chisels, or by hand. Reinforcing bars that have been affected by heat should be cut off, allowing enough of the old steel to protrude in order to tie or weld it to the new.

(2) The building, erecting, and stripping of forms have already been discussed in this chapter.

(3) For estimating reinforcing materials and labor refer to that section in this chapter.

(4) The preparation of the old surface for bonding to the new concrete is probably the most important requirement for successful repair work. After the surface has been hacked and chipped, it must be thoroughly cleaned with stiff wire brushes and washed with a caustic solution to remove all dirt and grease. In some instances steam hosing is necessary, or a solution of muriatic acid is applied followed by washing with clean water.

There are several reliable bonding products available which can either be mixed in the concrete batch, or can be made into a grout by mixing with cement and water and brushing on to the old surface.

(5) The method of estimating materials and labor for mixing and placing concrete has been previously discussed in this chapter.

(6) Where concrete structures have been repaired, if appearance is important, it may be necessary to chip or machine grind the joints where the new meets the old. In addition to smoothing the joining surfaces, a cement wash coat may have to be brushed over the entire old and new concrete to provide a uniform color.

Repairs to Concrete Floors

Floor surfaces that are spalled, chipped, or otherwise injured will have to be cut away an inch or so deep to expose the old aggregate and provide a rough bonding surface for the new concrete.

Estimating the labor required to cut away old concrete, or to scarify and chip the surface for bonding, is so variable that each job must be studied and judged according to the hardness of the old concrete, whether power hammers are available, or hand tools will have to be used.

The surface must be carefully cleaned and prepared for bonding as outlined under "Methods of Repair." A new topping of cement mortar 1 to 1 1/2 inches thick is then placed and finished (see Topping for Floors and Walks).

Concrete floors that are broken through or heaved will have to be taken up and replaced with a new section bonded to the adjacent old section.

Repairs to Cracks

Cracks in structural concrete can be repaired by chiseling out the cracks, removing all dirt and loose material, and then filling them with cement mortar of a 1:3 mixture. Repairs should be estimated on the basis of time and materials.

Repairing Spalled Walls

Spalled walls, not injured structurally, can be repaired by applying a coat of cement plaster to the surface. All loose material should be chipped away. After cleaning the surface a wash coat is usually applied; it consists of cement, a small amount of sand, and one of the patented bonding ad-mixtures. The cement plaster or stucco is then put on. (For estimating, see Chapter 16.)

Repairs to Soiled Walls

Soilage of concrete from smoke, roof tar drippings, and stains from various contents, can be removed by one of the following methods:

(1) Wash with caustic solution.
(2) Steam clean.
(3) Sand blast.
(4) Cement wash.
(5) Paint.

The estimator must determine from actual conditions what steps are necessary, and what method is most likely to restore the appearance of the surface.

CHAPTER 9

Masonry

Masonry is work done by a mason who lays up either stone found in nature or units which have been manufactured such as brick, tile or cement block. Mortar is used between the joints to bind the masonry units together. It may be either cement mortar or lime mortar, or a combination of both.

There are many different types of manufactured masonry units used in building construction. There is a variety of material used in making bricks, and they come in a number of sizes, shapes and finishes. Hollow tile and concrete, cinder or gypsum blocks are made in different sizes and shapes.

This chapter deals with the more commonly encountered kinds of masonry. The principles outlined, however, are applicable to all of them. With careful attention given to the cost of materials, and with proper allowances for labor variations on any specific type of masonry, little difficulty should be found in estimating costs.

MORTAR

The desirable physical properties of mortar are that it shall be easily worked by the mason, will not lose its water too rapidly to the masonry units, will be sufficiently strong and durable for the purpose used, and that it will have good bonding qualities. These properties are dependent on the proportions of cement, sand, water, and lime (if used), and also on the quality of those materials.

Materials

Structurally the proper proportioning of the ingredients of mortar is very important. In estimating the *cost* of masonry, it is relatively unimportant. The cost of the materials for mortar when compared to the overall cost of the masonry, is small. Any difference between the cost of mortar materials as a result of varying the proportions will make an insignificant difference in the total estimate.

The Portland Cement Association recommends for different types of mortar service the proportions shown in Table 9–1.

TABLE 9–1

Recommended Mortar Mixtures—Proportions by Volume

Type of Service	Cement	Hydrated lime or lime putty	Mortar sand in damp loose condition.
For ordinary service	1 masonry cement* ____ or ____ 1 portland cement	. . . . ½ to 1¼	2¼ to 3 4½ to 6
Subject to extremely heavy loads, violent winds, earthquakes or severe frost action. Isolated piers	1 masonry cement* plus 1 portland cement ____ or ____ 1 portland cement	. . . . 0 to ¼	4½ to 6 2¼ to 3

* ASTM Specifications C91, Type II.

The addition of hydrated lime to mortar makes it easier for the mason to handle; otherwise it would be stiff and difficult. Furthermore the addition of lime in quantities up to 10 or 15 per cent makes the cement adhere better to the sand. Quantities in excess of that amount reduce both the tensile strength of the mortar and its adhesive, or bonding qualities.

The *volumetric* proportions shown in Table 9–1 can be converted to *weight* proportions by multiplying the unit volumes shown by the weight per cubic foot of the materials, which may be assumed to be as follows:

Masonry cement . . . . 　Weight printed on bag
Portland cement . . . . 　94 lbs
Hydrated lime . . . . 　40 lbs
Mortar sand, damp and loose . . . 　85 lbs (approximately)

Illustrative Example

Using a volume mix of 1 portland cement, ¼ hydrated lime, and 3 of sand, the proportions by weight would be:

1 cement　　@　　94 lbs = 94 lbs
¼ lime　　　@　　40 lbs = 10 lbs
3 sand　　　@　　85 lbs = 255 lbs

Using 1 sack of cement, the mixture would require 10 pounds of lime and 255 pounds of sand.

To convert this to a yardage basis, assuming that 1 yard of sand weighs approximately 2,550 pounds, it would require 1 cubic yard of sand, 10 sacks of cement and 2½ sacks of lime (100 pounds).

For convenient and practical estimating purposes, since the amount of lime is subject to judgment, the following quantities of materials are recommended for estimating the cost of *1 cubic yard* of mortar.

Cement	10 sacks
Sand	1 cubic yard
Hydrated lime	2 sacks

Any admixtures that are to be combined with the mortar to produce hardness or color, or to prevent freezing, should be included in the cost of the materials.

Mixing

Mortar is usually mixed by laborers, or masons' helpers, while they are tending masons by carrying materials to be placed within their reach, and shifting scaffold planks or horses. On smaller jobs the mixing is done by hand; otherwise a portable mixer is used. The labor to mix the mortar is included in the over-all labor allowance made for masons' helpers. No special or separate computation is necessary.

Estimating Mortar Costs

By using the previously recommended quantities of materials for a cubic yard of mortar, and by applying local prices, a cost per cubic yard of mortar can be obtained.

Illustrative Example

(Assume Local Material Prices As Shown)

10 sacks cement	@ $4.00	= $40.00
1 cu yd sand	@ 7.00	= 7.00
2 sacks hyd. lime	@ 2.00	= 4.00
Cost per cu yd		= $51.00

Mortar Required for Masonry

The amount of mortar needed to lay up masonry units varies with the type and size of unit involved and also with the *thickness* of the joints between the units. The larger the unit and the narrower the joint between, the less mortar will be required.

When estimating losses, very little attention is given to such refinements as the joint thickness unless an unusually large quantity of masonry is being considered. As a general rule for estimating, using a 3/8 inch joint, Table 9–2 shows the quantities of mortar to be used for the types of masonry shown. An allowance of 10 per cent waste is included.

Where some of the cells of the concrete block, or the hollow tile are to be filled, an allowance for more mortar must be made accordingly.

CONCRETE BLOCK MASONRY

Concrete block is manufactured in two weights. One is made of lightweight aggregate such as cinders and is frequently referred to as cinder block. A unit actual size 7 5/8″ × 7 5/8″ × 15 5/8″ but called 8″ × 8″ × 16″ weighs about 30 pounds. Its

TABLE 9–2

Approximate Quantities of Mortar Required
for Various Types of Masonry Units

Type of Masonry 3/8" Joints	Nominal size of Unit in Inches	Cubic Yards of Mortar		Cubic Feet of Mortar	
		Per 1,000 Units	Per 100 Units	Per 1,000 Units	Per 100 Units
4" brick masonry	2¼ × 3¾ × 8	.33	.033	9.0	.9
8" brick masonry	2¼ × 3¾ × 8	.50	.05	13.5	1.35
12" brick masonry	2¼ × 3¾ × 8	.60	.06	16.2	1.62
Hollow terra cotta wall tile*	6 × 12 × 12	1.0	.10	27	2.7
(Laid cells horizontal)	8 × 12 × 12	1.3	.13	35	3.5
(Laid cells horizontal)	10 × 12 × 12	1.6	.16	42	4.2
(Laid cells horizontal)	12 × 12 × 12	2.0	.20	54	5.4
8" Concrete block masonry	8 × 8 × 16	1.5	.15	40	4.0
12" Concrete block masonry	12 × 8 × 16	1.7	.17	46	4.6
Gypsum partition tile	3 × 12 × 30	. . .	. . .	. . .	5.0
Gypsum partition tile	4 × 12 × 30	. . .	. . .	. . .	6.0
Gypsum partition tile	6 × 12 × 30	. . .	. . .	. . .	9.0

*For hollow terra cotta tile laid on end, add 20 per cent more mortar to figures shown

light weight reduces the cost of handling. It has better insulating qualities, and also provides a good nailing base for furring strips. The heavier block is made with aggregates of pebbles, crushed stone or slag. A nominal unit 8" × 8" × 16" weighs about 45 pounds. Both units may be used interchangeably in construction.

Concrete block is an economical type of construction. It is used in dwellings, and may be finished by painting or by application of brick veneer, cement plaster or stucco. The lightweight cinder block provides an excellent bonding surface for plaster both inside and outside. Because of its fire-resistive qualities it is used frequently in the construction of farm buildings, garages, and one-story mercantile and manufacturing buildings. It is also used in retaining walls and fences. Figures 9–1 and 9–2 show the common shapes and sizes of concrete block. (See pages 138–139.)

Bond Beams

In some jurisdictions concrete block walls must, by regulation, be stabilized at the top with a continuous concrete beam or cap to which is anchored the timber wall plate. The width of the bond beam conforms with that of the wall, and the depth may range from 8 to 10 inches. It is reinforced with rods. Side forms are placed on top of the wall and concrete is placed in the form with proper reinforcing.

(a) Steel sash units

Regular   Corner   Pier between windows

(b) Wood sash (jamb) units

Regular   Corner

(c) Bullnose units (radius varies 1 to 3-in.)

Single   Double   Single with sash corner

(d) Capping units

W = nominal 6", 8", 10"   12" grade (4" shelf)
and 12"

(h) Return or corner angle units

6"   10"   12"   6" Bull Nose

(i) Joist unit

(j) 4-in. high units

Regular stretcher   Single corner   Regular sash   Double corner   Corner sash

Figure 9–1

Common shapes and sizes of concrete block. Courtesy Portland Cement Association, Skokie, IL 60076

Regular stretcher

One plain end
(single corner)

Both ends plain
(double corner or pier)

(a) Two-core 8x8x16-in. units

Regular stretcher

One plain end
(single corner)

Both ends plain
(double corner or pier)

(b) Three-core 8x8x16-in. units

(c) Cross sections

Figure 9-2

Reinforced Concrete Block

In areas subject to frequent high winds, or earthquake shocks, concrete block walls must be reinforced both vertically and laterally and tied in with the footing. If the type of reinforcing used cannot be determined by inspection, it is recommended that the local building code be consulted for proper specifications.

Waterproofing

The earth side of the wall should have two coats of portland cement plaster (1 cement to 2 1/2 parts of fine sand). Each coat should be about 1/4 inch thick. The first coat is applied over dampened block and scratched to make a bond for the finish coat which is applied 24 hours later.

Usually a coat of hot asphalt or tar is applied over the plaster to seal the wall against moisture. The entire waterproofing operation should extend down and over the edge of the footing.

Estimating Quantities

Concrete blocks are made in a variety of sizes, although the most popular is the 8″ × 8″ × 16″. Its actual measurements are 7 5/8″ × 7 5/8″ × 15 5/8″, the difference being an allowance made in manufacturing for a 3/8″ joint. Units of special shape are available for door jambs, and for corners.

Illustrative Example

Gross wall area	60′ × 16′	= 960 sq ft
Deduct openings: window .	4′ × 6′ = 24 sq ft	
Door	6′ × 8′ = 48 sq ft	= -72 sq ft
	Wall area less openings	888 sq ft
	Plus 1/8	111 sq ft
	Number of units required	999

Figure 9–3

A unit 8″ × 8″ × 16″ contains (8″ × 16″) 128 square inches of surface area which is 128/144 or 8/9ths of a square foot. To estimate the number of 8″ × 8″ × 16″

blocks needed in a wall, compute the square foot area of the wall, and *deduct all openings in full*. Add 1/8 to the number of square feet obtained to get the required number of concrete blocks, or multiply the square foot area by 1.125.

    Corners should be deducted, making certain that it is done only once. A building 20 feet by 30 feet has a perimeter of 100 lineal feet. In computing the number of 8-inch block, two of the walls should be shortened by 16 inches. The lineal feet of wall to be used in estimating are:

$$
\begin{array}{rl}
2(30') & = 60' - 0'' \\
2(18'\text{-}8'') & = \underline{37' - 4''} \\
& \;\; 97' - 4''
\end{array}
$$

Figure 9–4

TABLE 9–2A

Number of 8″ × 8″ × 16″ Concrete Blocks Per Lineal Foot
of Foundation Wall

Wall Height In Inches	Wall Height In Feet		Number of Blocks Per Lineal Foot of Foundation
8	.67	(0′–8″)	.754
16	1.33	(1′–4″)	1.500
24	2.00	(2′–0″)	2.250
32	2.67	(2′–8″)	3.004
40	3.33	(3′–4″)	3.746
48	4.00	(4′–0″)	4.500
56	4.67	(4′–8″)	5.254
64	5.33	(5′–4″)	6.000
72	6.00	(6′–0″)	6.750
80	6.67	(6′–8″)	7.504
88	7.33	(7′–4″)	8.246
96	8.00	(8′–0″)	9.000
104	8.67	(8′–8″)	9.754
112	9.93	(9′–4″)	10.500
120	10.00	(10′–0″)	11.250

    Window and door openings may be deducted by multiplying the surface area of such openings by 1.125 to determine the number of blocks to be deducted.

If the height of the wall is 12 feet, the area is 12 feet by 97 1/3 feet = 1,168 sq ft. To obtain the number of 8″ × 8″ × 16″ concrete block, assuming there are no openings to be deducted, one-eighth is added.

$$\begin{array}{rl} \text{Wall surface area} = & 1,168 \text{ sq ft} \\ \text{Plus } 1/8 & \underline{146} \\ \text{Number of block} & 1,314 \end{array}$$

Gable ends of walls are sometimes constructed by carrying the concrete blocks up to the underside of the roof boards. The blocks are cut to the roof slope. A few extra blocks should be allowed for cutting waste in such cases, but no additional labor need be made as the laying rate is not sufficiently affected.

Estimating Labor

A mason and a laborer work together in laying concrete block. The laborer mixes and carries mortar and blocks to the mason. The laborer also shifts, or assists in shifting, scaffold plank and wood horses when the mason moves to a new area of operation.

The unit of measurement used in estimating concrete block masonry is 100 blocks. The number of hours required for a mason and a laborer to lay 100 blocks varies with the kind of work and the size and weight of block being handled. The rate of laying block decreases where the work involves pilastered walls, numerous window or door openings, piers, or breaks and corners. The lightweight cinder blocks lay up faster than the heavyweight units. The 12-inch-thick blocks lay up slower than those 8 inches or less in thickness.

Table 9–3 shows the approximate number of hours required for a mason and a laborer to lay 100 units of lightweight concrete block under average conditions. The cost of structural scaffolding to be erected by carpenters and the cost of hoisting should be added as a separate item whenever either is required.

TABLE 9–3

Approximate Hours of Labor Required to Lay 100 Lightweight Concrete Masonry Units of Various Sizes

Nominal Size of Unit in Inches	Kind of Work	Mason	Laborer	Blocks per Hour
8 × 8 × 16	Foundations and piers	5.5	5.5	18
8 × 12 × 16	Foundations and piers	6.5	6.5	15
4 × 8 × 12	Superstructures	3.5	3.5	˙29
4 × 8 × 16	Superstructures	4	4	25
8 × 8 × 12	Superstructures	4.5	4.5	22
8 × 8 × 16	Superstructures	5	5	20
8 × 12 × 16	Superstructures	6	6	17

For heavyweight concrete blocks add 10 percent to the hours shown for lightweight units. These hours include pointing and cleaning.

Estimating Costs

The cost of concrete block masonry is the combined cost of the blocks, the mortar required to lay them up, and the cost of labor. If 1 1/2 yards of mortar will lay 1,000 8-inch blocks, it will take .15 yards for each 100 blocks. The cost of a yard of mortar should be computed and multiplied by .15 to obtain the mortar cost for each 100 blocks.

The cost of labor is determined by multiplying the rate of wages for a mason and a laborer times the number of hours shown in Table 9–3.

Illustrative Example

10 sacks cement	@ $ 4.00 =	$ 40.00
1 cu yd sand	@ 7.00 =	7.00
2 sacks hydrated lime	@ 2.00 =	2.00
Cost of 1 cu yd of mo		$ 51.00
(Table 9–2) Mortar for 100 concrete block		.15
		7.65
100 (8″ × 8″ ×16″) concrete block	@ $ .50 =	50.00
(Table 9–3) 5 Hours-mason	@ 12.00 =	60.00
(Table 9–3) 5 Hours-laborer	@ 9.00 =	45.00
Unit cost for 100 blocks	=	$162.65
Unit cost per block $162.65	=	$1.63
100		
Unit cost per sq ft, $1.63 + 1/8 =		$1.83

Unit Costs

The basic formula for the unit cost of concrete block may be further simplified as follows:

	100 concrete block	@	**$**	**= $**
(Table 9–2)	**Cu yds mortar**	@		**=**
(Table 9–3)	**Hours—mason**	@		**=**
(Table 9–3)	**Hours—laborer**	@		**=** _____
	Unit Cost For 100 Block			**= $**
	Unit Cost Per Block =		**$ ____**	**= $**
			100	

CONCRETE BLOCK CHIMNEYS

Manufacturers of concrete block make a lightweight unit 16 by 16 by 8 inches (in height). A single 8 by 8-inch tile flue lining can be set inside the unit. The blocks laid one on top of the other with mortar joints form a chimney.

Mortar is filled in between the flue lining and the concrete block. A cubic yard of mortar will lay approximately 100 lineal feet of chimney. To find the cost of mortar required, multiply the cost of 1 cubic yard by the length of the chimney divided by 100.

Illustrative Example

If the cost of 1 cubic yard of mortar is computed at $50.00 and the chimney is 30 feet high, the cost of the mortar required would be:

$$\frac{30}{100} \times \$50.00 = \$15.00$$

Since the blocks are 8 inches thick, the number required is obtained by multiplying the height of the chimney by 1.5. A chimney 30 feet high would require $1.5 \times 30 = 45$ blocks.

A mason and a laborer working together can lay on the average 4 feet per hour including setting of the flue lining.

Flue lining $8'' \times 8''$ comes in 2-foot lengths.

The total cost of building a 30-foot concrete block chimney can be obtained in the following manner, assuming that the flue lining begins 5 feet from the base. Prices and wages are assumed.

45 concrete blocks 16 $''$ × 16 $''$ × 8 $''$	@	$2.50 =	$112.50
13 - 2 ft lengths 8 $''$ × 8 $''$ TC flue	@	1.50 =	19.50
Mortar $\frac{30}{100}$ × $50.00		=	15.00
Mason $\frac{30}{4}$ = 7 1/2 hrs	@	12.00 =	90.00
Helper 7 1/2 hrs,	@	9.00 =	67.50
		Total cost =	$304.50

DAMAGEABILITY OF CONCRETE BLOCKS

In general the damageability of concrete as discussed in Chapter 8 applies to concrete blocks. Whether composed of aggregate of slag, gravel, or cinders, the 8-inch and thicker concrete blocks provide excellent insulation against heat up to temperatures similar to those common in chimneys, 600° to 1000° F. In moderate fires they have been found to stand up well. There is usually no spalling such as is characteristic of concrete, but the blocks will tend to crumble after long exposure to extreme heat. Because of the thinness of the face shell, which is approximately 1 1/2 inches thick, the small amount of water used in their manufacture, and the air cells, concrete blocks are less effective as an insulation against the heat of long burning fires than a solid concrete wall of equal thickness.

Cracks may develop in concrete block walls from intense heat, subsidence, vibration, blasting, earthquake, impact or pressure from an external force such as earthslide, and hydrostatic pressure. If the forces are of sufficient magnitude, the walls will bulge and finally collapse. Numerous cases are on record of *unbraced* concrete block walls of buildings under construction having been blown over during strong windstorms.

Concrete block masonry is subject to soilage by smoke, tar or asphalt roof drippings, and also from contact with contents materials. Concrete block walls, unless reinforced by steel or pilasters, offer somewhat less resistance to lateral pressures caused by explosion, impact, or by expanding water-soaked contents than similar walls of solid concrete or brick masonry.

METHODS OF REPAIR

Sections of concrete block walls that have been damaged are readily cut out and new sections can be installed. The blocks are removed at mortar joints stepping back the sides. Old mortar is removed where the new blocks are to be set in place. When the patching-in of new sections is being estimated, special consideration should be given to the labor necessary to lay the stepped-back blocks where the new wall joins the old. Unless a very large area of wall is involved, it is better to prepare the estimate on a time and material basis rather than trying to approximate a unit cost. The cost of demolishing the damaged wall in preparation for laying the new block should be treated as a *separate labor item*. In some cases the walls adjacent to the damaged section will require propping or bracing to keep them in line and plumb during the operation. If the area being replaced extends very far vertically down the surface of the wall, leaving the sides unsupported, they should be checked for plumbness and held in place by timber supports until the new wall has thoroughly set. Where a roof has been destroyed, has collapsed into the interior, or is so badly damaged that its replacement is necessary, the side walls generally need stabilizing if any extensive repairs are contemplated. Each situation should be carefully examined and analyzed to determine the extent and method of repairs that will be undertaken.

Soilage of concrete block masonry is estimated in the same manner as soilage of concrete masonry, which is discussed in Chapter 8. The basis of estimating cleaning depends on the kind of stain or soilage. Scraping, steam cleaning, and sandblasting may have to be resorted to in order to remove such deposits as asphalt tar dripping that comes from roofing. In certain instances it will be necessary to provide a uniform color to the surface by whitewashing, painting, cement washing, and in extreme cases by cement plastering.

Concrete block walls sometimes escape structural damage but the surface that was exposed to heat may have chipped, pitted, crumbled, or been otherwise scarred in various ways. Rather than replacing an otherwise sound wall, it may be satisfactory to apply portland cement plaster to the entire surface. The specifications for such restoration are a matter of negotiation and agreement between the parties concerned.

Fire surface cracks in concrete block may be pointed with mortar and if necessary a coating of cement wash or paint can be applied. Larger cracks caused by heat, settlement, blasting or other perils can be pointed or grouted with cement mortar as long as no structural damage is evident that would require replacing the units of masonry.

BRICK MASONRY

Brick masonry is a term applied to masonry employing solid units made of burned clay, the "Standard Brick" size being 2 1/4 $\times$ 3 3/4 $\times$ 8 inches long. There are many sizes, kinds and qualities of brick depending upon the type of clay, the molding process, and the manner of firing.

NOTE: While the coring types shown are typical for solid units, they do not necessarily apply to the specific types of units with which they are shown above. They will vary with the manufacturer.

Figure 9-5

Sizes of modular and non-modular brick. Courtesy Brick Institute of America, McLean, VA 22101

Figure 9–5 shows the sizes of the various Modular and Non-Modular brick.

Common brick are machine molded of ordinary clay. The position in the kiln affects the shape and hardness of the brick, and to some extent the finished color. Overburned brick may be darker in color and some of them are irregular in shape. In certain localities, brick are sold *kiln-run*, while in others, those of uniform color and shape are selected and sold at a slightly higher price.

Face brick are usually wire-cut, and have a uniform color and finish. They are used as face or veneer brick. Pressed brick are also known as face brick.

Glazed brick have one surface that is glazed during firing by being coated with mineral salts. They are mainly used where sanitary wall surfaces are specified.

Firebrick, made from fire clay to stand high temperatures, are used in lining furnace fire boxes, and also fire places.

The use of brick for structural purposes is almost unlimited. Although cement block has supplanted them in foundations, brick are used extensively in exterior and interior walls of mercantile and manufacturing buildings, dwelling walls, porches and steps, chimneys, and columns. They are also used in sidewalks, patio floors, and garden walls.

Usually when brick are used as a veneer in a solid brick wall, the veneer is bonded to the main wall by making every sixth course a header course. When brick are used as a veneer on the exterior of a frame building, or one of concrete block or hollow tile, metal clips are used, preferably at every second or third course. In the case of a frame building the metal ties are nailed to the sheathing and imbedded in the mortar joints of the veneer masonry. When brick are being used as veneering against concrete block or hollow tile construction, the clips also may be nailed or imbedded in the backing joints.

Estimating Quantities—Walls

The unit of measurement for brick masonry is 1,000 brick. A "Standard Brick" is approximately 3 3/4 inches thick. A wall one brick thick is normally referred to as a 4-inch wall; one 2 bricks thick is an 8-inch wall; one 3 bricks thick a 12-inch wall, and so on, in multiples of 4 inches. The actual measured thickness of the wall in inches may be 3 3/4 or 4 inches, 8 or 9 inches, or 12 or 13 inches. A cubic foot of brick wall with joints of about 1/4 to 3/8 inches contains approximately 21 brick. For estimating the number of Standard Brick per square foot in walls 4 inches to 24 inches thick, with mortar joints 1/4 to 3/4 inches, use Table 9–3A.

TABLE 9–3A

Number of Standard Brick in Walls 4″ to 24″ Thick and Mortar Joints Shown

Thickness of Wall In Inches	Number of Brick Per Square Foot			
	1/4″ to 3/8″	1/2″	5/8″	3/4″
4	7	6.3	6.09	5.81
8	14	12.6	12.18	11.62
12	21	18.9	18.27	17.43
16	28	25.2	24.36	23.24
20	35	31.5	30.45	29.05
24	42	37.8	36.54	34.86

Illustrative Example

Gross wall area 12″ × 30″		360 sq ft
Deduct openings:		
Window	4′ × 5′ = 20 sq ft	
Door	7′6″ × 3′ 6″ = 26 sq ft	−46
Wall area less openings		314 sq ft
(21 brick for a 12″ wall with 3/8″ joint)		×21
Number of brick required		6,594

(If this wall were 8 inches thick, the number of brick would be 314 × 14 = 4,396)

Pilasters

Where pilasters are part of a wall, the total area of the wall is first obtained, and the projecting pilasters ignored. The area of the pilasters alone is then computed and the number of brick is determined by the thickness of the pilaster.

Figure 9–6

Illustrative Example

Wall area excluding pilasters 10 × 40 ft	400	sq ft
Times 14 brick for 8 inch wall	×14	
Number of bricks per square foot in the 8 inch wall		5,600
Area of pilasters 2 × 1⅓' × 10'	26 ⅔ sq ft	
Times 7 brick for a 4 inch wall, ⅜" joint	×7	
Number of brick in pilasters		187
Total number of brick		5,787

Estimating Quantities—Piers

The number of brick in piers and columns can be figured either by taking the surface area of one side and multiplying by the number of brick according to thickness, or simply by determining the number of cubic feet in the pier or column and multiplying by 21 brick per cubic feet.

Illustrative Example

Method by Surface Area
(16 in = 1⅓' ft)   1⅓' × 9'   = 12 sq ft
Times 28 brick for a 16-inch wall, ⅜" joint   ×28

Total number of brick   336

Method 2 by Cubic Feet
1⅓' × 1⅓' × 9'   = 16 cu ft
Times 21 brick per cubic ft, ⅜" joint   ×21

Total number of brick   336

Figure 9–7

Estimating Quantities—Brick Chimneys

Brick chimneys are built with different sizes and numbers of flues. Nominal flue sizes are 8" × 8" × 12" and 12" × 12" × 12". Interior chimney walls may be 4 inches thick when lined with terra cotta flue, but exterior lined chimney walls should be 8 inches thick. When the exterior wall of a building forms one side of a chimney, care should be taken not to include it in computing the number of brick in the chimney.

To figure the number of brick in a chimney, draw a sketch (as in Figure 9–8) showing the number of brick in each course. Multiply the number of brick *per course* by the number of courses in a lineal foot of chimney. The height of the chimney in feet multiplied by the number of brick per lineal foot will give the total brick required.

Illustrative Example

An interior chimney with 4-inch walls has an 8" × 8" and an 8" × 12" flue. Using Figure 9–8 we find there are 11 brick per course or 50 brick per lineal foot of height. If the chimney is 30 feet high, it will require 1,500 brick.

Mortar for chimney brick may be obtained from Table 9–2. A mason and a laborer work as a team building a chimney. Table 9–4 shows the approximate hours of labor required per 1,000 brick. Flue lining is set in place by the mason erecting the chimney and is included in the hours of labor shown. (See page 151.)

NUMBER OF BRICKS PER LINEAL FOOT FOR CHIMNEYS
WITH FLUE SIZES SHOWN

	WALL THICKNESS	BRICK PER COURSE	BRICK PER LINEAL FOOT
	4"	6	28
	4"	7	32
	4"	11	50
	8"	16	73
	8"	18	82
	8"	20	91
	8"	29½	135

(BASED ON 3/8" MORTAR JOINTS.)

Figure 9–8

Estimating Quantities—Corners, Sills, and Waste

In figuring the quantity of brick in building walls, it should be remembered that corners are to be deducted.

Waste in brick masonry is a small factor for consideration and it is not customarily added in estimating losses unless large quantities of brick are involved. Generally between 2 and 5 percent is adequate for normal breakage in handling and cutting if it is desired to add waste.

Window sills and window or door lintels are treated in the same manner discussed under concrete block masonry. Where *brick* window sills are laid the number of brick may be obtained by dividing by 2.5 the length of the sill in inches.

Estimating Labor

In laying brick, as in concrete block masonry, a mason and a laborer work together. It is customary, therefore, in most brick masonry to figure the labor on the basis of an equal number of hours for both a mason and a laborer. The mason is attended by the laborer who mixes the mortar, arranges the plank and horse scaffolding, and supplies materials.

The unit of measurement in brick masonry is 1,000 brick. The number of hours required to lay 1,000 brick depends on the kind of work. Straight walls without openings take less time per thousand than walls with many openings, or pilasters, where proper plumbing of jambs and corners requires extra time.

Table 9–4 shows the approximate number of hours required for a mason and a laborer to lay 1,000 brick under normal working conditions for the kind of work indicated. The cost of scaffolding to be erected by carpenters and any necessary hoisting should be added as a separate item to the estimate.

The number of hours it will take for a mason to lay 1,000 brick varies generally with those factors that affect all labor production as outlined in Chapter 2. The time also varies specifically with the type of structure involved, that is, whether it is a solid wall, a chimney, piers, 4-inch veneer, a fireplace, and so forth. Table 9–4 covers the more common types of work.

TABLE 9–4

Approximate Hours of Labor Required to Lay 1,000 Brick

Kind of Work	Mason	Laborer
8″ or 12″ foundation walls	10	10
8″ common brick walls, normal openings	12	12
8″ common brick walls, large openings, pilasters	14	14
12″ common brick walls, normal openings	10	10
12″ common brick walls, large openings, pilasters	12	12
4″ veneer on frame or masonry—ordinary grade	16	16
4″ veneer on frame or masonry—high grade	20	20
Masonry piers and columns	14	14
Common brick chimneys	15	15
Ordinary brick fireplaces	20	20
Brick veneering fireplaces	24	24
Washing down brick masonry (per 100 sq ft)	1	1

Estimating Costs

The factors to be included in estimating the cost of brick masonry are the brick, the mortar, and the labor.

Table 9–2 shows the approximate amount of mortar required for 1,000 brick in masonry walls of various thicknesses. Multiply the cost of 1 cubic yard of mortar by the percentage of a yard indicated.

The cost of labor for laying 1,000 brick may be obtained by multiplying the prevailing rate of wages for a mason and a laborer by the number of hours shown in Table 9–4 for the kind of work being done.

Illustrative Example

Assume that it has been determined that 28,000 common brick are required in the 8-inch exterior walls of a ranch type dwelling. Prices and wages are for illustration only.
Cost of 1 cu yd of mortar, (see page 136, mortar for concrete block $51.00)
Cost of mortar for 1,000 brick (Table 9–2) .5 × $51.00 = $25.50

Labor to lay 1,000 brick (Table 9–4)

12 mason's hours	@ $12.00 =	$ 144.00
12 laborer's hours	@ 9.00 =	108.00
Total labor cost for 1,000	=	$ 252.00

Application of material and labor cost per 1,000 brick

28,000 common brick	@ $100.00 per M =	$	2800.00
Mortar for 28,000 brick	@ 25.50 " " =		714.00
Labor for 28,000 brick	@ 252.00 " " =		7056.00
	Total		$10,570.00
Cost per 1,000 brick	$10,570	= $	377.50
	28		

Unit Costs

Using the same material and labor wages shown in the previous illustrative example, the *unit cost* for laying 1,000 common brick would be:

1,000 common brick	@	$100.00 =	$100.00
½ cubic yard mortar	@	51.00 =	25.50
12 hours mason labor	@	12.00 =	144.00
12 hours laborer	@	9.00 =	108.00
Unit cost per 1,000 brick =			$377.50

Having obtained a unit cost per 1,000 brick, the cost of 28,000 brick would be:

28 × $377.50 = $10,570.00

The unit cost of veneer brick, glazed brick, and other kinds may be obtained in the same way. The *local price* of the materials, and the *prevailing wage rates* should be used in each instance. The approximate hours of labor per 1,000 brick are shown in Table 9–4.

Illustrative Examples Using Assumed Prices and Wages
Illustrative Example No. 1

Unit Cost for Ordinary Grade 4-inch Veneer on Frame Wall.

	1,000 common brick	@	$100.00 =	$100.00
(Table 9–2)	.33 cu yds mortar	@	51.00 =	16.83
(Table 9–4)	16 hours mason	@	12.00 =	192.00
(Table 9–4)	16 hours laborer	@	9.00 =	144.00
			Unit cost for 1,000 brick =	$452.83

Illustrative Example No. 2

*Unit Cost for 12-inch Brick in Wall Having Large Window
and Door Openings and Pilasters.*

	1,000 common brick	@	$100.00 =	$100.00
(Table 9–2)	.60 cu yds mortar	@	51.00 =	30.60
(Table 9–4)	12 hours mason	@	12.00 =	144.00
(Table 9–4)	12 hours laborer	@	9.00 =	108.00
			Unit cost for 1,000 brick =	$382.60

Illustrative Example No. 3

Unit Cost for Brick Piers and Columns 12 inches by 12 inches

	1,000 common brick	@	$100.00 =	$100.00
(Table 9–2)	.60 cu yds mortar	@	51.00 =	30.60
(Table 9–4)	14 hours mason	@	12.00 =	168.00
(Table 9–4)	14 hours laborer	@	9.00 =	126.00
			Unit cost for 1,000 brick =	$424.60

DAMAGEABILITY OF BRICK MASONRY

Damage by Fire

Bricks are made principally of clay products, and during the manufacturing process they are fired at temperatures ranging as high as 2,000°F. It is therefore understandable that good quality brick masonry is able to withstand exposure to very hot fires without suffering serious structural damage. As is true with other kinds of masonry, the first injury is soilage or discoloration by smoke. As the fire progresses and the heat intensifies, the exposed face of the brick may crack or spall. Alternate heating and then cooling with water from the fire hoses tend to aggravate this condition. Soft, underburned brick will show the effect of fire more rapidly than hardburned brick under similar circumstances. Thin masonry walls suffer damage more readily than thicker walls. Old brick masonry, especially when laid with sand-lime mortar, reacts to severe fire less favorably than new brickwork laid with good portland cement mortar. The reason is mainly due to the effect of extreme heat and the water on the bond which has already been weakened by old age. It is not uncommon to have walls of this kind fall during a fire, particularly under pressure of dropping timbers or debris, or from side thrusts of collapsing floor and roof structures. Masonry of this type may also present a difficult problem when local authorities, for public safety, condemn entire walls even though on the surface they show only a slight damage resulting from the fire.

Brickwork usually does not disintegrate and crumble in extremely hot fires of long duration as does concrete block masonry, or certain types of inferior concrete. There are, however, instances of brick actually melting in temperatures in excess of 2,500°F. when exposed over a long period of time.

Good brick masonry has a modulus of elasticity approximating that of good concrete, but it does not have steel reinforcing imbedded in it to care for the tensile stresses induced during expansion and contraction. There is, therefore, a greater danger of cracks developing in brick masonry. They may run the full height of a wall or show up at the corners, near the window sills, or at the window or door heads. The cracks may follow the joints in the masonry or pass completely through the brick itself. Thin walls are more inclined to develop cracks than thick walls.

Damage by Water

Good quality brick masonry is not readily damaged by water, and it can withstand submersion over long periods of time. Poor quality mortar may be washed out of the joints by the strong action of a fire hose. It may also be washed out by the action of successive heavy rains or during submersion, as in the case of flooding or wave action. The brick itself will absorb water up to approximately 10 to 18 per cent by weight when submerged for 24 hours. Although the absorption of water slightly reduces the compressive strength of brick, they dry out readily to show no permanent damage. Freezing of water absorbed in the brick very seldom has any ill effects because the pores are only partially filled, leaving room for expansion during the frost action.

Soft brick weathers more rapidly than a hard-burned good quality brick. It is not uncommon to see actual erosion of soft brick in old buildings. This is especially noticeable at corbeled cornices, belt courses, window sills, and other parts of a structure where the brick projects out beyond the face.

Efflorescence, a white deposit on brickwork, is the crystallization of salts on the surface of the brick. It appears on new brickwork frequently, and is attributed to the presence of certain salts within the brick itself or sometimes in the mortar used in laying them up. It is inherent in the masonry. Linseed oil brushed over the surface removes the deposit but the treatment has to be repeated every few years as the oil is washed off.

Damage by Cracking

Cracks in brick masonry may be caused by heat, explosion, or by disturbance of the foundation supporting it. Blasting operations, earthquake, and, in rare cases, vibration caused by trucks or trains passing in the immediate vicinity will cause cracking in brickwork. Settlement due to soil subsidence, and the action of frost may also be a factor. Poorly constructed foundations and footings and inferior materials or workmanship in the masonry increase the possibilities of cracking. Brick walls that have been struck by vehicles, or subjected to either impact or pressure such as may occur during a collapse of a portion of the structure or during violent tornado-like windstorms, are subject to cracking in addition to other damage.

One of the tasks confronting the person making the survey and estimate of cracked masonry is to distinguish the cracks that may be directly attributed to the immediate casualty from the old unrelated cracks.

METHODS OF REPAIRING BRICK MASONRY

Brickwork that has been damaged can be stripped down to where it is undamaged, and rebuilt. The upper sections of multistoried brick buildings involved in a fire may have several courses of brick removed to reach a sound base for rebuilding or, when indicated, whole stories may be demolished and new walls erected. Most times it is preferable to engage a structural or consulting engineer to advise on the extent of repairs that are necessary. When faced with condemnation by local authorities it is good procedure to consult a well-qualified and reputable engineer as to whether any doubt exists concerning the justification of condemning the masonry.

Sectional Repairs

When a portion of the top of a brick wall must be removed and rebuilt, or a hole in a wall must be repaired, the primary factors to be considered are:

1. The labor to remove the damaged brick
2. The scaffolding necessary
3. Hoisting equipment
4. Shoring and sidewalk canopy
5. Rebuilding the new section

Laborers, as a general rule, demolish the damaged brickwork under the supervision of a mason or the contractor. The work on upper walls is usually done from the inside so that outside scaffolding is not required. If an exterior scaffold is needed, the cost must be added. The means of getting new materials up and debris down from upper stories should be considered according to the circumstances of each individual job. A building freight elevator may be available or it might be necessary to erect a hand- or machine-operated hoist. The site of the job, the height of the work above ground, and amount of material to be hoisted will have to be given consideration.

Shoring of beams, girders, rafters, or other structural supports bearing on the brick masonry should be added to the cost. When the need of a sidewalk canopy to protect pedestrians is indicated, or when required by local ordinance, the cost should be figured and included.

Rebuilding the new section should be figured as any other new job except when estimating the labor, adequate time should be allowed for the masons to tooth-in or step-in the new brick with the old where they join.

Unit costs may be used where there is a sufficient quantity of brick to be laid to approximate a rate of hours per thousand. Otherwise the repairs can be more accurately estimated on a material and time basis.

Repairing Spalled Brickwork

When the exterior face of a brick wall has been chipped and spalled with no structural injury, the damaged layer of brick can be cut out with power drills and hammers or chiseled out by hand. New matching brick can be laid. This method is recommended for veneer brick. When a large area of wall has been spalled, it is usually possible, with the consent of the owner, to clean it off and apply portland cement stucco over the entire surface.

On secondary structures, if the spalling is not serious, cement washing or oil painting may be satisfactory to all parties. Much depends on the individual situation.

Spalled interior walls that are to be furred and lined with wall board or plaster, will usually require little more than a coat of cement mortar, cement wash, or whitewash to remove smoke odor. In some instances nothing is done inasmuch as they are permanently concealed by the wall treatment. If the walls are exposed, cement mortar plaster, cement washing, or painting are generally acceptable as methods of repair.

Repairs to Cracks

The repairing of cracks in brick masonry can usually be done by a tuck-pointer, or mason. Where the crack is not large, and it follows the mortar joint, grouting and filling the joint is generally satisfactory. The old mortar is chiseled out as deeply as possible and, when filled, the new joint is struck.

Large cracks sometimes present a more serious problem, depending on the importance of appearances after the repairs are made. If grouting and filling are not acceptable as a method of repair, it may be necessary to cut out the face brick and replace them with new ones. In difficult situations, because skillful repairs to brick masonry are highly specialized, an expert mason should be consulted and asked to make the estimate.

Repairs to Soiled Walls

Brick masonry that has been soiled by smoke, roof tar drippings, or stains from contact with contents can usually be cleaned. (See Repairs to Soiled Walls, Chapter 8 on "Concrete.") There are, in most cities, firms that make a specialty of cleaning exteriors of brick, stone, or concrete structures. The method used will depend on the type of stain. Most stains can be removed by washing down with caustic solutions or brushing and steam cleaning. On secondary buildings and some older buildings it is often satisfactory to paint over old brick work, thereby covering up the stain.

HOLLOW TERRA COTTA TILE MASONRY

Hollow terra cotta tile is used extensively in building construction because it is relatively light in weight (an 8" × 12" × 12" block weighs about 25 pounds) is sufficiently strong, and has excellent fire-retardant qualities. The name *terra cotta* comes from the Latin words meaning "baked earth," which aptly names the process by which it is made. The hard-burned units are dense and strong. The semi-porous units are the more commonly used. They are exposed to less firing in the kiln but are strong enough for all but heavy load-bearing walls. The outer wall of a tile block is called the "shell." The inner "cells" are divided by thin partitions called "webs."

Hollow tile is used for the walls of dwellings and for small manufacturing and mercantile structures. It is also used for curtain and nonbearing walls in steel frame buildings, for fire-protecting steel columns, beams and girders, and for flat-arch type floors. The units are made in a variety of shapes and sizes depending on their use in the structure. They are made for corners, end blocks, and window and door jambs or lintels. The standard wall units are 12" × 12" giving them a surface area of 1 square foot; the thickness ranges from about 2 to 12 inches. The most commonly used are the 8-, 10- and 12-inch thick unit.

Estimating Quantities

The unit of measurement of hollow terra cotta tile is usually 100 blocks. Since a standard wall tile is 12" × 12", each unit covers 1 square foot. To obtain the number of units required, compute the surface area in square feet, being certain *to deduct all window and door openings in full*. The number of hollow tile will be equal to the number of square feet. Generally waste is not added unless large quantities are involved, or unless an unusual amount of cutting is to be done. When waste is added, 2 to 5 percent is considered adequate.

Corners should be deducted, but one must be certain not to deduct them twice. Lintels and sills should be estimated as outlined under concrete block masonry.

The materials and quantity of mortar required are discussed under "Mortar" in this chapter.

Estimating Labor

A mason and a laborer work together in laying hollow tile block as with concrete block and brick masonry. The rate at which the units can be laid depends on the thickness of the tile blocks and on the kind of work being done. Table 9–5 shows the approximate number of hours required to lay 100 hollow tile.

TABLE 9–5
Approximate Hours of Labor Required to Lay 12" × 12" Hollow Wall Tile

Size of Unit	Mason	Laborer
6" × 12" × 12"	5	5
8" × 12" × 12"	6	6
10" × 12" × 12"	7	7
12" × 12" × 12"	8	8

Estimating Costs

The cost of laying hollow terra cotta wall tile is obtained by determining the cost of the tile, the mortar, and the labor required to lay them.

From Table 9–2 the approximate quantity of mortar for 100 units may be found for the thickness of the wall being constructed. Multiply the percentage indicated by the cost of 1 cubic yard. Table 9–5 shows the approximate number of hours required to lay 100 wall tile of different thicknesses. Multiply the number of mason and laborer hours indicated by the prevailing rate of wages.

Illustrative Example

Cost of Cubic Yard Mortar (page 136)			$51.00
Mortar required for 100 tile (Table 9–2)			× .13
			$ 6.63
100 T.C. tile 8″ × 12″ × 12″	@	$ .50 =	50.00
(Table 9–5) 6 hours mason	@	12.00 =	72.00
(Table 9–5) 6 hours laborer	@	9.00 =	54.00
	Unit cost per 100 tile =		$182.63
	Unit cost per tile =		1.83

LINTELS

In concrete block construction a reinforced concrete lintel is usually provided for each door and window opening. The size of the lintel and the size and the number of reinforcing rods used are determined by the load above the arch. In some buildings a steel angle lintel is used. The type of construction employed can be observed by inspection. Where a reinforced concrete lintel exists, the cross section generally conforms with the cross section of the block, 8 by 8 inches or 8 by 12 inches, and the lintel extends approximately 8 inches beyond each side of the opening. Estimating the cost of cast-on-the-job lintels consists in figuring the wood forms, the concrete and reinforcing, plus the labor of mixing and placing the materials (see Chapter 8 on Concrete).

In brick veneer or solid brick wall construction, the lintel for windows and doors consists of a steel angle iron for veneer and two, back to back, for solid walls. The size and weight of the angle irons depend on the span and weight carried above. The exact size and shape can be determined in the field. (See Structural Steel, Chapter 21.)

Stone lintels were frequently used in older buildings and reinforced concrete lintels are still used in some solid masonry wall construction.

SILLS

Windows and doors in masonry walls are provided with sills. In commercial structures, particularly, precast concrete sills are used or they are cast in place on the job. Their cost should be estimated the same as for lintels.

Limestone sills are frequently used in better residential construction, and in offices and public buildings. Measurements can usually be obtained in the field and prices should be obtained from local supply sources. The price is governed by the shape, dimensions, number of sills required and shipping costs.

COPING

Masonry walls that project above the roof are customarily finished by capping with stone, concrete, or terra cotta tile coping. The main purpose of coping is to seal the top of the wall against the eroding effects of snow, water, and freezing. A secondary purpose is to provide an architectural finish to the wall.

Glazed terra cotta wall coping is available in widths of 9, 13, and 18 inches. Each section is 2 feet long. Also available are starter, end, angle, and corner coping.

Cut flagstone coping is occasionally used. Lengths vary from 18 to 30 inches and widths are adequate to overlap each side of the wall an inch or two.

Dressed limestone coping is cut according to specifications as to shape, width, thickness and length.

When estimating, the proper sizes of coping units may be obtained at the place of the job. Price should be determined locally.

All coping is set in regular mortar and laid by a mason and a laborer. A mason and a laborer can lay approximately 100 lineal feet of terra cotta or flagstone coping a day. Cut stone coping takes longer and should be estimated in accordance with the circumstances of each job.

The cost of hoisting, or otherwise getting the coping up to where it is to be laid must be added to the estimate as a separate item.

CHAPTER 10

Rough Carpentry

Carpentry is work done by a carpenter, who builds with wood and with certain composition materials. About 80 percent of all building construction involves a substantial amount of carpentry work. Some knowledge of this trade is important for anyone who makes or checks building loss estimates. It is a major factor in the construction of dwellings, barns and other farm buildings, and also plays a large part in commercial structures which have masonry walls, wood floors, partitions and roofs. Some of these are public garages, apartment houses, warehouses, motels, and older office and loft buildings, mills, schools, and hotels. There are very few buildings that do not contain some form of carpentry. Carpenters also build and erect concrete forms, scaffolding, fences, and the various types of wooden yard fixtures. The carpentry trade is very broadly divided into two classifications: *rough carpentry* and *finish carpentry* or *millwork*.

TYPES OF ROUGH CARPENTRY

While the line of separation between rough and finish carpentry cannot always be sharply defined, the following work and materials are generally considered as applying to rough carpentry.

Sleepers.
Forms.
Framing.
   Sills, plates, ridge pieces,
   Studding, fire stops, etc.
   Joists, bridging and bracing,
   Rafters, wood trusses, and purlins.
Sheathing, roof boards and rough floors.
Furring and grounds.
Window and door frames and bucks.
Temporary and rough cellar stairways.
Fences, trellises, arbors and other yard fixtures.
Rough hardware.
Cutting for other trades.
Scaffolding.
Debris chutes, canopies, etc.

TYPES OF BUILDING CONSTRUCTION
EMPLOYING CARPENTRY

There are four common classifications of building construction which require a major or a substantial amount of carpentry. They are customarily referred to as *frame* (including frame and stucco), *brick veneer, ordinary brick,* and *mill or slow-burning.*

Frame Construction

Buildings of frame construction, except for the foundations, interior wall treatment, or stucco finish when it exists, are constructed almost entirely of wood. The principle for erecting the basic framework or skeleton is the same in all frame buildings. A lateral bottom member called a *sill* is laid on the ground (*a mudsill*), on piers, or on top of a masonry type foundation. *Studding* (called *scantlings* in some sections of the country) are fastened vertically to the sill and capped with a member similar to the sill but called a *plate.* This assembly provides the general framework for exterior walls. *Rafters,* which are members that support the roof, are set on top of the upper plates. In gable roofs the rafters are either butted together at the ridge or joined to a *ridge pole.* Floor *joists,* frequently called floor *beams,* are laid horizontally across the building and supported on the exterior plates, and on girders carrying partitions. Solid or cross-*bridging* is placed between the joists at intervals to stiffen and strengthen them.

Interior frame partitions, erected to separate the area into rooms, closets, and other service quarters, are constructed in a manner similar to the exterior walls. The partition studs may be secured at the bottom directly to doubled joists or to a *shoe* (interior sill) which is laid across the joists, or on the *subflooring.*

Wood or composition *sheathing* is applied to the exterior walls. Roof *sheathing* (called *roof decking* in some areas) is laid over the rafters, and rough or *subflooring* is put down over the floor joists.

A weather surface of roofing material is laid over the roof decking, and the exterior is appropriately finished with wood, metal, or composition *siding.* In a *frame* and *stucco* building the exterior finish is composed of wire lath and portland cement stucco. *Windows* and *exterior doors* are installed. *Cornices* and other exterior *wood trim* are put into place.

Interior finishes are then applied. These may be *lath and plaster, wood paneling,* or *composition board* such as gypsum board, Celotex, or Masonite. *Finish floors* are laid, *interior doors, trim, cupboards, stairways,* and other *millwork* is installed.

During the construction, and before covering the walls and ceilings, the *electrical wiring, plumbing* and *heating* facilities are roughed in. The final operations are generally the installation of *fixtures* for the service equipment, and the *decorating.*

There are many modifications of all types of construction depending upon the purpose for which the building is to be used, its climatic location, architecture, local custom and practice, and the amount of money to be spent.

Brick Veneer

A building which in most respects is constructed as that described for frame construction but which has a masonry veneer (usually stone or brick) is classified as *brick veneer* construction.

Ordinary Brick

A building which is constructed similar to a frame building, but which has solid 8-inch masonry exterior walls, is known as *ordinary brick* construction. While the exterior walls are usually brick, they may also be cement block, or some other type of masonry. In multiple-story structures such as apartment houses, mercantile or manufacturing buildings, the exterior walls are thicker varying with the height and local building code requirements.

Mill, or Slow-Burning

This type of construction employs heavy timber framing. Exterior walls are masonry, being either brick, stone or concrete. Wood *girders* of large dimensions are set on iron, steel, or wood *columns* or *posts*, the ends bearing in or on the walls. *Beams* not less than 10 inches deep and spaced 8 to 10 feet on center are placed between the girders on *beam hangers*. The floors are usually 3- to 4-inch tongue and groove or *splined* plank flooring spiked to the beams. The roof construction is relatively the same as the floor construction but somewhat lighter. It is surfaced with roll type or built-up asphalt roofing.

ESTIMATING MATERIALS

The material used most in carpentry is building lumber which comes from a dozen or more commercially useful trees. The *conifers* or *needle-leaved* trees comprise the group that supplies the majority of woods used for structural purposes. They are the so-called *evergreens* such as fir, spruce, hemlock, pine and cedar. The *broad-leaved* trees generally supply wood for interior floors, trim and cabinet work. These are the hardwoods such as oak, maple, birch, ash, walnut, and white wood (basswood). In some sections of the country, depending on choice and local supply, the different types of lumber are used interchangeably.

BOARD MEASURE

Nearly all lumber, from the time it is surveyed in the forest until it is sawed from the logs into timbers and boards, is measured and sold by the unit of the *board foot*. Special millwork like moldings, window and door trim, lattice, handrail, balusters, closet pole, shelf cleats, some types of baseboard, molded exterior trim, wood gutters, and fence pickets are sold by the lineal foot. Plywood, particle board, flakeboard etc., as distinguished from lumber is sold by the square foot.

A board foot is one square foot of wood one inch thick. The symbol for board feet is FBM (feet board measure). See Figure 10–1.

The number of board feet in a piece of wood is obtained by multiplying the width times the thickness in inches, times the length in feet, and dividing by 12.

Figure 10–1
Each a board foot.

Table 10–1 shows the number of board feet in various sizes of lumber of different lengths. Table 10–2 gives the factor to multiply a piece of lumber by to determine the number of board feet it contains. (See page 164.)

$$\frac{\text{Width in inches} \times \text{Thickness in inches} \times \text{Length in feet}}{12}$$

Illustrative Example

A 2″ × 6″ × 10′ contains 10 FBM

$$\frac{2'' \times 6'' \times 10'}{12} = 10 \text{ FBM}$$

The total number of FBM in a list of framing lumber would be obtained as follows:

$$6 \text{ pieces } 2'' \times 4'' \times 8' = \frac{\cancel{6} \times \cancel{2} \times 4 \times 8}{\underset{\cancel{2}}{\cancel{12}}} = \quad 32 \text{ FBM}$$

$$10 \text{ pieces } 2'' \times 6'' \times 12' = \frac{10 \times 2 \times 6 \times \cancel{12}}{\cancel{12}} = \quad 120 \text{ FBM}$$

$$4 \text{ pieces } 2'' \times 10'' \times 10' = \frac{\cancel{4} \times 2 \times 10 \times 10}{\underset{3}{\cancel{12}}} = \quad 66\tfrac{2}{3} \text{ FBM}$$

$$8 \text{ pieces } 4'' \times 4'' \times 18' = \frac{\overset{3}{8} \times 4 \times 4 \times \overset{6}{\cancel{18}}}{\underset{\cancel{3}}{\cancel{12}}} = \quad \underline{192 \text{ FBM}}$$

$$\text{Total } = \quad 410\tfrac{2}{3} \text{ FBM}$$

TABLE 10–1

Number of Board Feet in Various Sizes of Lumber
Length of Piece

Size Inches	8′	10′	12′	14′	16′	18′	20′	22′
2×3	4	5	6	7	8	9	10	11
2×4	5⅓	6⅔	8	9⅓	10⅔	12	13⅓	14⅔
2×6	8	10	12	14	16	18	20	22
2×8	10⅔	13⅓	16	18⅔	21⅓	24	26⅔	29⅓
2×12	16	20	24	28	32	36	40	44
3×4	8	10	12	14	16	18	20	22
3×6	12	15	18	21	24	27	30	33
3×8	16	20	24	28	32	36	40	44
3×10	20	25	30	35	40	45	50	55
3×12	24	30	36	42	48	54	60	66
4×4	10⅔	13⅓	16	18⅔	21⅓	24	26⅔	29⅓
4×6	16	20	24	28	32	36	40	44
4×8	21⅓	26⅔	32	37⅓	42⅔	48	53⅓	58⅔
6×6	24	30	36	42	48	54	60	66
6×8	32	40	48	56	64	72	80	88
6×10	40	50	60	70	80	90	100	110
8×8	42⅔	53⅓	64	74⅔	85⅓	96	106⅔	117⅓
8×10	53⅓	66⅔	80	93⅓	106⅔	120	133⅓	146⅔
8×12	64	80	96	112	128	144	160	176
10×10	66⅔	83⅓	100	116⅔	133⅓	150	166⅔	183⅓
10×12	80	100	120	140	160	180	200	220
12×14	112	140	168	196	224	252	280	308

TABLE 10–2

Multipliers to Compute the Number of Board Feet
in any Length of Dimension Lumber

Nominal Size In Inches	Multiply Length By	Nominal Size In Inches	Multiply Length By
2 × 2	0.333	4 × 4	1.333
2 × 3	0.500	4 × 6	2.000
2 × 4	0.667	4 × 8	2.667
2 × 6	1.000	4 × 10	3.333
2 × 8	1.333	4 × 12	4.000
2 × 10	1.667		
2 × 12	2.000	6 × 6	3.000
		6 × 8	4.000
3 × 3	0.750	6 × 10	5.000
3 × 4	1.000	6 × 12	6.000
3 × 6	1.500		
3 × 8	2.000	8 × 8	5.333
3 × 10	2.500	8 × 10	6.667
3 × 12	3.000	8 × 12	8.000

Illustrative Example

1- 2″ × 8″ × 14′ joist contains:
1.333 × 14′ = 18.7 FBM
(or)
6- 2″ × 10″ × 16′ joists contain:
6 × 1.667 × 16′ = 160 FBM

TABLE 10–2A

Standard Sizes of Lumber
(American Lumber Standard 1970)

Type of Lumber	Nominal Size in Inches		Actual Size in Inches	
	Thickness	Width	Thickness	Width
Dimension	2	4	1.5	3.5
Plank	2	6	1.5	5.5
and	2	8	1.5	7.25
Joist	2	10	1.5	9.25
	2	12	1.5	11.25
S 4 S	4	4	3.5	3.5
	4	6	3.5	5.5
	4	8	3.5	7.25
	4	10	3.5	9.25
Timbers	6	6	5.5	5.5
	6	8	5.5	7.5
	6	10	5.5	9.5
S 4 S	8	8	7.5	7.5
	8	10	7.5	9.5
Common	1	4	¾	3.5
Boards		6		5.5
S 4 S		8		7.25
		10		9.25
		12		11.25

			Width	
			Overall	Face
	1	6	5½	5 1/8
Shiplap		8	7¼	6 7/8
Boards		10	9¼	8 7/8
		12	11¼	10 7/8
	1	4	3 3/8	3 1/8
Tongue		6	5 3/8	5 1/8
and		8	7 1/8	6 7/8
Grooved		10	9 1/8	8 7/8
Boards		12	11 1/8	10 7/8

MILLING WASTE

Matched boards are those which have the edges milled in a *tongue and groove*, or a *lap joint* (shiplap), to join them tightly together. Stock of this kind is used as surface lumber, as distinguished from framing lumber, and consists of such types as rough and finish flooring, exterior sheathing, siding, roof boards, paneling, and ceiling boards.

Most sawed lumber is planed (dressed) at the lumber mill before shipping. The result is that the nominal size, for example a 2″ × 4″ stud, turns out to measure 1 1/2″ × 3 1/2″. Prior to September 1, 1970, when the new Lumber Standards were accepted by the government, the same 2″ × 4″ stud measured 1 5/8″ × 3 5/8″. And before that the same 2″ × 4″ actually measured 1 3/4″ × 3 3/4″. The difference (in the case of the 2″ × 4″) was termed *milling waste.* That term is no longer accurate because the present 1 1/2″ × 3 1/2″ stud is *not* cut from a piece 2″ × 4″ but rather something less in size. See Table 10–2A for normal and actual sizes.

Milling waste is not significant in estimating where planks, timbers, joists, and similar dimension lumber are involved. It does become important, however, when estimating quantities of surface lumber such as common boards, shiplap, tongue and grooved boards, and bevel siding. The supplier will compute his charges on the basis of the *nominal size* and if a factor is not added for the so-called "milling waste," when placing the order, the quantity delivered will fall short of what is needed.

Table 10–3 shows the factor by which the area to be covered is multiplied to obtain the exact amount needed. This factor is obtained for *any piece of surface lumber* by dividing the *nominal* size by the *actual* face size. For example, the factor for a 1″ × 6″ square edge board (actual face size 5.5″) is 6 divided by 5.5 or 1.09.

This formula, $\dfrac{\text{nominal size}}{\text{actual face size}}$ = factor, is useful when there is no Table to refer to. *Normal cutting and fitting waste is to be added.*

Use of Table 10–3 to Compute Material Needed

To obtain from Table 10–3 the exact quantity of material to cover an area, multiply the measured square foot area by the factor in the Table for the particular type of material to be used.

Illustrative Example

1. The measured area in which 1" × 8" shiplap is to be applied to the exterior walls of a building is 100' × 9' or 900 sq ft. The Table gives a factor of 1.16 for 1" × 8" shiplap.

900 sq ft × 1.16 = 1,044 FBM needed to cover the area.

2. Tongue and grooved rough flooring, 1" × 6", is to be laid in a room 15' × 30' or 450 sq ft. The Table gives a factor of 1.17 for 1" × 6" T&G boards.

450 sq ft × 1.17 = 527 FBM needed to cover the area.

3. The area of the exterior walls of a building to be covered with 1" × 8" bevel siding is 1,200 sq ft. The Table gives a factor of 1.28 for 1" × 8" bevel siding.

1,200 sq ft × 1.28 = 1,536 FBM needed to cover the area.

TABLE 10–3

Factor by Which Area to be Covered is Multiplied to Determine Exact Amount of Surface Material Needed (From Western Wood Products Association)

Item	Nominal Size	Width Overall	Face	Area Factor
Shiplap	1"×6"	5½"	5 1/8"	1.17
	1×8	7¼	6 7/8	1.16
	1×10	9¼	8 7/8	1.13
	1×12	11¼	10 7/8	1.10
Tongue and Grooved	1×4	3 3/8	3 1/8	1.28
	1×6	5 3/8	5 1/8	1.17
	1×8	7 1/8	6 7/8	1.16
	1×10	9 1/8	8 7/8	1.13
	1×12	11 1/8	10 7/8	1.10
S 4 S	1×4	3½	3½	1.14
	1×6	5½	5½	1.09
	1×8	7¼	7¼	1.10
	1×10	9¼	9¼	1.08
	1×12	11¼	11¼	1.07
Solid Paneling	1×6	5 7/16	5 7/16	1.19
	1×8	7 1/8	6¾	1.19
	1×10	9 1/8	8¾	1.14
	1×12	11 1/8	10¾	1.12
*Bevel Siding	1×4	3½	3½	1.60
	1×6	5½	5½	1.33
	1×8	7¼	7¼	1.28
	1×10	9¼	9¼	1.21
	1×12	11¼	11¼	1.17

Note: This area factor is strictly so-called milling waste. The cutting and fitting waste must be added.

* 1-inch lap

CUTTING WASTE

In addition to the so-called *milling waste*, an allowance should be made for material wasted in cutting and fitting it into place. This will average 3 to 5 percent in framing lumber, which includes a certain amount used on jobs for such things as saw-horses or minor scaffolding. On surface materials such as rough floor and roof boards, also board sheathing, paneling, etc. the waste will average 5 to 10 percent depending on whether the material is applied horizontally or diagonally.

ESTIMATING FRAMING MATERIALS

When estimating quantities of framing lumber, it should be remembered that the lengths needed for plates, sills, joists, and rafters will have to be cut from the nearest length sold at the lumber yard. A rafter or joist measuring 11 feet 6 inches must be cut from a 12-foot length.

Where studs can be obtained *pre-cut* at the lumber yard, that may be the most economical way to buy them rather than cutting them on the job and wasting part. This will be governed by the length of studs needed.

Sills and Plates

Sills and plates are taken off by the lineal foot. The sill in a shed 10 feet wide and 20 feet long would have 60 lineal feet which is the equivalent of the perimeter of the building. If the plate on top of the studs carries around the entire building, as it would if the roof were flat or slightly pitched, the number of lineal feet of plate would, for all practical purposes be the same as the lineal feet of sill. Assuming that the sill was 2 by 6 inch stock, and that the plate was 4 by 4 inch stock, then the material would be listed in this manner.

$$\text{Sill—60 lin ft } 2'' \times 6'' = 60 \text{ FBM}$$
$$\text{Plate—60 lin ft } 4'' \times 4'' = 80 \text{ FBM}$$

Exterior Studding

Exterior studding is estimated by determining the number of studs required, their size, and their length. In house framing, studding is usually spaced 16 inches center to center. Many rural and secondary buildings, and some dwelling-type structures located outside of the jurisdiction of building code requirements, will be found to have studding on centers of 18, 20 and sometimes 24 inches. Except in barn structures, ice houses, refrigerated warehouses or buildings requiring unusual wall thickness and strength, studding is customarily 2″ × 4″ stock. Non-bearing curtain partitions frequently have 2″ × 3″ studs.

However, 2″ × 6″ studding 2′ on center and fully insulated in between is becoming popular for exterior walls in dwellings. The number of 2″ × 6″ studs needed in walls, for spacings 12″ to 36″ is shown in Table 10–4. The number of FBM of 2″ × 6″ studding in walls for spacings 16″, 18″, 24″ and 30″ may be obtained by simple formulas shown on page 171.

Some estimators prefer to take off the exact number of studs, and window and door framing, and add a percentage for cutting waste. This may be done by counting and measuring the members if they have not been destroyed. When they cannot be physically counted, a sketch should be drawn to be used for the take-off.

Where the exact number of studs is desired, the lineal feet of wall is divided by the spacing in feet. *One stud is added for the end.* See also Tables 10–4 and 10–5.

Illustrative example

A partition is 40 feet long and the studding is 16 inches on center.

$$40 \div \frac{16}{12} = 30 + 1 \text{ for the end} = 31 \text{ studs}$$

TABLE 10–4

Number of Studs, Joists and Rafters Required in Walls, Floors and Roofs

Length of Wall, Floor or Roof	Spacing on Center					
	12"	16"	20"	24"	30"	36"
8'	9	7	6	5	4	4
9	10	8	6	6	5	4
10	11	9	7	6	5	4
11	12	9	8	7	5	5
12	13	10	8	7	6	5
13	14	11	9	8	6	5
14	15	12	9	8	7	6
15	16	12	10	9	7	6
16	17	13	11	9	7	6
17	18	14	11	10	8	7
18	19	15	12	10	8	7
19	20	15	12	11	9	7
20	21	16	13	11	9	8
21	22	17	14	12	9	8
22	23	18	14	12	10	8
23	24	18	15	13	10	9
24	25	19	15	13	11	9
25	26	20	16	14	11	9
26	27	21	17	14	11	10
27	28	21	17	15	12	10
28	29	22	18	15	12	10
29	30	23	18	16	13	11
30	31	24	19	16	13	11
32	33	25	20	17	14	12
34	35	27	22	18	15	12
36	37	28	23	19	15	13

To the above there should be added members for doubling studs at windows, doors, corners; doubling joists under partitions, at stairwells and around chimneys; doubling rafters at roof openings around chimneys, etc.

Table 10–4A shows the board feet in 2" × 4" stud walls and partitions 8' to 28' long and 8', 10', and 12' high; 12", 16" and 20" centers. Formulas for obtaining the board feet of 2" × 4" and 2" × 6" studs in walls and partitions, 12", 16", 20" and 24" on center are shown on page 171.

TABLE 10–4A

Number of Board Feet Needed in 2" × 4" Stud Walls and Partitions 8' to 28' Long, End Studs Included. Add Plate and Shoe from End Columns, Add Window and Door Framing.

(For studding 24" on center, use one-half the FBM shown for studding 12" on center. Minor differences are due to end studs.)

Lin Ft of Partition or Wall	12" Spacing Height of Wall			16" Spacing Height of Wall			20" Spacing Height of Wall			Plate and Shoe Single 2"x4"	Double 2"x4"
	8'	10'	12'	8'	10'	12'	8'	10'	12'		
8	48	60	72	38	47	56	32	40	48	5	11
9	54	67	80	43	54	64	34	43	52	6	12
10	59	74	88	48	60	72	37	47	56	7	14
11	64	80	96	48	63	76	43	53	64	8	15
12	70	87	104	54	67	80	46	56	68	8	16
13	75	94	112	59	74	88	48	60	72	9	18
14	80	100	120	64	80	96	51	63	76	10	19
15	86	107	128	64	80	98	54	67	80	10	20
16	91	114	136	70	87	104	59	73	88	11	22
17	96	120	144	75	94	112	61	76	92	12	23
18	1 02	127	152	80	100	120	64	80	96	12	24
19	1 07	134	160	80	103	124	67	83	99	13	26
20	1 12	140	168	86	107	128	70	87	104	14	27
21	118	147	176	91	114	136	74	93	112	14	28
22	123	154	184	96	120	144	75	97	116	15	30
23	128	160	192	96	123	148	80	100	120	16	31
24	134	167	200	101	127	152	82	102	122	16	32
25	139	174	208	107	134	160	85	107	128	17	34
26	144	180	216	112	140	168	91	113	136	18	35
27	150	187	224	112	143	172	93	116	140	18	36
28	155	194	232	118	147	176	96	120	144	19	38

FORMULAS FOR BOARD FEET OF 2″ × 4″
STUDDING IN WALLS AND PARTITIONS
INCLUDING END STUDS*

Where: L = length of wall

H = height of wall

Studding 12″ on Center: (L′ + 1) × .67 × H′ = FBM

Example: A wall is 20′ long and 8′ high

(20′ + 1) × .67 × 8′ = 112.56 FBM

Studding 16″ on Center: (L′ × .75 + 1) × .67 × H′ = FBM

Example: A wall is 28′ long and 10′ high

(28′ × .75 + 1) × .67 × 10′ = 147.4 FBM

Studding is 20″ on Center: (L′ × .6 + 1) × .67 × H′ = FBM

Example: A wall is 24′ long and 8′ high

(24′ × .6 + 1) × .67 × 8′ = 82.5 FBM

Studding is 24″ on Center: (L′ × .5 + 1) × .67 × H′ = FBM

Example: A wall is 12′ long and 8′ high

(12′ × .5 + 1) × .67 × 8′ = 37.5 FBM

FORMULAS FOR BOARD FEET OF 2″ × 6″
STUDDING IN WALLS AND PARTITIONS
INCLUDING END STUDS

Where: L = length of wall

H = height of wall

Studding is 16″ on Center: (L′ × .75 + 1) × H′ = FBM

Studding is 18″ on Center: (L′ × .67 + 1) × H′ = FBM

Studding is 24″ on Center: (L′ × .5　+ 1) × H′ = FBM

Studding is 30″ on Center: (L′ × .4　+ 1) × H′ = FBM

TABLE 10–5

Multipliers for Obtaining the Number of
Studs, Joists and Rafters in the Length
of Floors, Walls or Roofs

Spacing In Inches	Factor to Multiply the Width of Room, Length of Wall or Roof	
12	1.00	
16	.75	
18	.67	Add one for
20	.60	end in each
24	.50	case

---

*The results, using these formulas, may be slightly different in some instances from the figures shown in Table 10–4A which have been rounded out to whole numbers. Plates and shoes, and framing around openings should be added.

Examples

A room 16' wide requires joists 16" o.c.
Ans. 16' × .75 + 1 = 13 joists.
A wall 30' long requires studs 18" o.c.
Ans. 30' × .67 + 1 = 21 studs.
A gable roof 40' long requires rafters 24" o.c.
Ans. 40' × .50 + 1 = 21 rafters.

Partition Studding

Studs for partitions are determined in the same way as exterior studs. Members may be counted individually taking into consideration extra studs at door openings and corners such as at closets, or where joining other partitions or exterior walls.

The use of either Table 10–4 or Table 10–5 is a quick method for determining the number required.

Pre-cut Studs

Where pre-cut studs are available in the size desired, rather than cutting them on the job, the number required may be taken from Table 10–4 (page 169), or computed using Table 10–5 (page 171).

Pre-cut studs come in lengths 92 5/8", 93", 94 1/2" and 96".

Floor Joists

Joists in a dwelling are generally 2-inch nominal stock and their depth ranges from 6 to 12 inches. In other types of buildings the thickness may be 2, 3, or even 4 inches depending on the floor load and the span. The depth of the large timbers may be as much as 16 inches. Table 10–7 shows the maximum safe span for joists under various live loads with either plastered or unplastered ceilings below.

Where double or sometimes triple joists are needed under load bearing partitions, or at stairwells, they must be added in the material list.

The number of joists may be counted directly at the premises if still visible, or may be computed from Table 10–4 or Table 10–5

The number of board feet of joists and ceiling beams of various sizes, 16, 20 and 24 inches on center may be read directly from Table 10–5B for floor areas of 200 to 3,000 square feet. Quantities include on allowance for one end joist and cutting and fitting waste.

Girder, Door, and Window Headers

A *girder* is a horizontal beam of iron or wood, or a combination, used to support vertical loads. Where an I-beam is used as a girder in frame construction, a 2" × 4" or 2" × 6" is bolted to the top and the joists are nailed to that. There is less shrinkage and less sagging or deflection in this arrangement than in girders composed solely of doubled or tripled joists. A recommended type of basement girder is made up of 2-2" × 12" members with a 12" steel plate in between and the three members bolted firmly together. Sometimes, in place of a steel plate, a 1/2" piece of plywood is glued and nailed between members. This works well as it brings the header flush with the 2" × 4" studding. (Illustrations of girders in Figure 10–2)

Window and door headers can have 2-2" × 4"s *on edge* up to spans of 2 feet. Larger spans require stonger framing. As in girders, where 2" × 8" to 2" × 12" members are used a piece of 1/4" to 1/2" plywood is glued and nailed between two members. Table 10–6 shows header sizes and maximum spans recommended.

Table 10–5A shows simple formulas to obtain the number of board feet in floor and ceiling joists of various sizes and on centers of 16", 20", and 24".

TABLE 10–5A

Formulas for Estimating the FBM of Joists of Various Sizes Per Floor

(A = Area of The Floor or the Ground-floor Area of the Dwelling) Quantities include an allowance for an end joist and cutting and fitting waste.

Size of Joist	Inches Center to Center	Formula
2" × 4"	16	.55A
	20	.44A
	24	.37A
2" × 6"	16	.85A
	20	.68A
	24	.56A
2" × 8"	16	1.10A
	20	.88A
	24	.73A
2" × 10"	16	1.35A
	20	1.08A
	24	.90A
2" × 12"	16	1.65A
	20	1.32A
	24	1.10A

Illustrative Example

A 2-story dwelling, 32' × 70', has a *ground-floor* area of 2,240 sq ft. The first floor joists are 2" × 12", the second floor are 2" × 10", and the second floor ceiling joists are 2" × 6". All are on 16" centers.

First floor joists,	2" × 12"	2,240 × 1.65 =	3,696 FBM
Second "	2" × 10"	2,240 × 1.35 =	3,024 "
Ceiling joists,	2" × 6"	2,240 × .85 =	1,904 "
		Total	8,624 "

The number of board feet may be read directly from: Table 10–5B for floor areas of 200 to 3,000 square feet.

TABLE 10–5B

Board Feet of Joists, 2" × 4" to 2" × 12" for Areas 200 to 3,000
Square Feet (16", 20" and 24" on center)

Quantities include allowance for end joist, cutting and fitting waste.

Ground Floor Area	2" × 4" On Center			2" × 6" On Center			2" × 8" On Center			2" × 10" On Center			2" × 12" On Center		
	16"	20"	24"	16"	20"	24"	16"	20"	24"	16"	20"	24"	16"	20"	24"
200	110	88	73	170	136	114	220	176	147	270	216	180	330	264	220
300	165	132	110	255	204	170	330	264	220	405	324	270	495	396	330
400	220	176	147	340	272	227	440	352	293	540	432	360	660	528	440
500	275	220	183	425	340	283	550	440	367	695	540	450	825	660	550
600	330	264	220	510	408	340	660	528	440	810	648	540	990	792	660
700	385	308	257	595	476	397	770	616	514	945	756	630	1155	924	770
800	440	352	293	680	544	454	880	704	587	1080	864	720	1320	1056	880
900	495	396	330	765	612	510	990	792	660	1215	972	810	1485	1188	990
1000	550	440	367	850	680	567	1100	880	734	1350	1080	900	1650	1320	1100
1100	605	484	404	935	748	624	1210	968	807	1485	1188	990	1815	1415	1210
1200	660	528	440	1020	816	680	1320	1056	880	1620	1296	1080	1980	1584	1320
1300	715	572	477	1105	884	737	1430	1144	954	1755	1404	1170	2145	1716	1430
1400	770	616	514	1190	952	794	1540	1232	1027	1890	1512	1260	2310	1848	1540
1500	825	660	550	1275	1020	850	1650	1320	1100	2025	1620	1350	2475	1980	1650
1600	880	704	587	1360	1088	907	1760	1408	1177	2160	1728	1440	2640	2112	1760
1700	935	748	624	1445	1156	964	1870	1496	1247	2295	1836	1530	2805	2244	1870
1800	990	792	660	1530	1224	1021	1980	1584	1320	2430	1944	1620	2970	2376	1980
1900	1045	836	697	1615	1292	1077	2090	1672	1394	2565	2052	1710	3135	2508	2090
2000	1100	880	734	1700	1360	1134	2200	1760	1467	2700	2160	1800	3300	2640	2200
2100	1155	924	770	1785	1428	1190	2310	1848	1540	2835	2268	1890	3465	2772	2310
2200	1210	968	807	1870	1496	1247	2420	1936	1614	2970	2376	1980	3630	2904	2420
2300	1265	1012	844	1955	1564	1304	2530	2024	1688	3105	2484	2070	3795	3036	2530
2400	1320	1056	880	2040	1632	1360	2640	2112	1760	3240	2592	2160	3960	3168	2640
2500	1375	1100	917	2125	1700	1417	2750	2200	1834	3375	2700	2250	4125	3300	2750
2600	1430	1144	953	2210	1768	1474	2860	2288	1908	3510	2808	2340	4290	3432	2860
2700	1485	1188	990	2295	1836	1530	2970	2376	1980	3645	2916	2430	4455	3564	2970
2800	1540	1232	1027	2380	1904	1587	3080	2464	2054	3780	3024	2520	4620	3696	3080
2900	1595	1276	1064	2465	1972	1644	3190	2552	2128	3915	3132	2610	4785	3828	3190
3000	1650	1320	1100	2550	2040	1700	3300	2640	2200	4050	3240	2700	4950	3960	3300

Figure 10–2

Girder construction and framed openings.

TABLE 10–6

Window and Door Opening
Maximum Spans for Various
Headers

Header Sizes	Maximum Span
2-2″ × 4″	2′
2-2 × 6	4′
2-2 × 8	5′
2-2 × 10	8′
2-2 × 12	10′

Note: The use of plywood interlay is recommended. (See Illustrations in Figure 10–2)

TABLE 10-7

Span Calculations provide for carrying the live loads shown and the additional weight of the joists and double flooring.

SPAN OF JOISTS		20# L.L.	30# LIVE LOAD		40# LIVE LOAD		50# LIVE LOAD		60# LIVE LOAD	
SIZE	SPACING	Plaster Clg.	Plaster Clg.	No Plaster	Plaster Clg.	No Plaster	Plaster Clg.	No Plaster	Plaster Clg.	No Plaster
2 x 4	12"	8' - 8"								
	16"	7' - 11"								
	24"	6' - 11"								
2 x 6	12"	13' - 3"	11' - 6"	14' - 10"	10' - 8"	13' - 2"	10' - 0"	12' - 0"	9' - 6"	11' - 1"
	16"	12' - 1"	10' - 6"	12' - 11"	9' - 8"	11' - 6"	9' - 1"	10' - 5"	8' - 7"	9' - 8"
	24"	10' - 8"	9' - 3"	10' - 8"	8' - 6"	9' - 6"	8' - 0"	8' - 7"	7' - 7"	7' - 10"
2 x 8	12"	17' - 6"	15' - 3"	19' - 7"	14' - 1"	17' - 5"	13' - 3"	15' - 10"	12' - 7"	14' - 8"
	16"	16' - 0"	13' - 11"	17' - 1"	12' - 11"	15' - 3"	12' - 1"	13' - 10"	11' - 5"	12' - 9"
	24"	14' - 2"	12' - 3"	14' - 2"	11' - 4"	12' - 6"	10' - 7"	11' - 4"	10' - 1"	10' - 6"
2 x 10	12"	21' - 11"	19' - 2"	24' - 6"	17' - 9"	21' - 10"	16' - 8"	19' - 11"	15' - 10"	18' - 5"
	16"	20' - 2"	17' - 6"	21' - 6"	16' - 3"	19' - 2"	15' - 3"	17' - 5"	14' - 6"	16' - 1"
	24"	17' - 10"	15' - 6"	17' - 10"	14' - 3"	15' - 10"	13' - 5"	14' - 4"	12' - 8"	13' - 3"
2 x 12	12"	26' - 3"	23' - 0"	29' - 4"	21' - 4"	26' - 3"	20' - 1"	24' - 0"	19' - 1"	22' - 2"
	16"	24' - 3"	21' - 1"	25' - 10"	19' - 7"	23' - 0"	18' - 5"	21' - 0"	17' - 5"	19' - 5"
	24"	21' - 6"	18' - 8"	21' - 5"	17' - 3"	19' - 1"	16' - 2"	17' - 4"	15' - 4"	16' - 0"
3 x 8	12"	20' - 0"	17' - 7"	24' - 3"	16' - 4"	21' - 8"	15' - 4"	19' - 10"	14' - 7"	18' - 4"
	16"	18' - 6"	16' - 1"	21' - 4"	14' - 11"	19' - 1"	14' - 1"	17' - 4"	13' - 4"	16' - 0"
	24"	16' - 5"	14' - 3"	17' - 9"	13' - 2"	15' - 9"	12' - 4"	14' - 4"	11' - 9"	13' - 3"
3 x 10	12"	25' - 0"	22' - 0"	30' - 2"	20' - 6"	27' - 1"	19' - 3"	24' - 10"	18' - 4"	23' - 0"
	16"	23' - 2"	20' - 3"	26' - 8"	18' - 10"	23' - 10"	17' - 8"	21' - 9"	16' - 10"	20' - 2"
	24"	20' - 7"	17' - 11"	22' - 3"	16' - 7"	19' - 10"	15' - 7"	18' - 1"	14' - 10"	16' - 8"

* Weyerhaeuser Company, Lumber and Plywood Division, St. Paul, Minn.

TABLE 10-8

SPAN OF RAFTERS

Span Calculations provide for carrying the live loads shown and the additional weight of the rafters, sheathing and wood shingles

SIZE	SPACING	15# L.L.		20# L.L.		30# L.L.		40# L.L.	
		No Plaster	Plaster	No Plaster	Plaster	No Plaster	Plaster	No Plaster	Plaster
2 x 4	12"	12'-7"	8'-11"	11'-4"	8'-4"	9'-8"	7'-6"	8'-7"	6'-11"
	16"	11'-0"	8'-2"	9'-11"	7'-7"	8'-5"	6'-10"	7'-5"	6'-3"
	24"	9'-1"	7'-2"	8'-2"	6'-8"	6'-11"	6'-0"	6'-1"	5'-6"
2 x 6	12"	19'-2"	13'-8"	17'-4"	12'-9"	14'-10"	11'-6"	13'-2"	10'-8"
	16"	16'-10"	12'-6"	15'-2"	11'-8"	12'-11"	10'-6"	11'-6"	9'-8"
	24"	13'-11"	11'-0"	13'-0"	10'-3"	10'-8"	9'-3"	9'-6"	8'-6"
2 x 8	12"	25'-1"	17'-11"	22'-9"	16'-10"	19'-7"	15'-3"	17'-5"	14'-1"
	16"	22'-1"	16'-6"	20'-0"	15'-5"	17'-1"	13'-11"	15'-3"	12'-11"
	24"	18'-5"	14'-7"	16'-7"	13'-8"	14'-2"	12'-3"	12'-6"	11'-4"

* Weyerhaeuser Company, Lumber and Plywood Division, St. Paul, Minn.

Rafters

A rafter of 2″ × 6″ is very common in house construction, although on bungalows, porches, and many secondary buildings 2″ × 4″ rafters are used. In many of the earlier built homes, expecially those with large steeply pitched roofs, rafters are heavier stock. In many commercial type buildings with flat roofs the size of the rafters may be 3 or 4 inches thick and 8 to 12 inches deep. The dimensions of rafters depend to a large degree on the span, the weight of the roofing material, and the expected wind and snow load as shown in Table 10–8.

TABLE 10–9A

Framing for Gable and Hip Roofs for Pitch and Centers shown, and Using 2″ × 4″ Rafters, including Ridge Pole . . . Board Feet Required. Interpolate for Other Areas

Ground Floor Area	16″ CENTER TO CENTER							24″ CENTER TO CENTER					
	ROOF PITCH							ROOF PITCH					
	1/8	1/6	5/24	1/4	1/3	1/2		1/8	1/6	5/24	1/4	1/3	1/2
200	114	116	120	124	132	156		78	80	82	86	92	108
300	171	174	180	186	198	234		117	120	123	129	138	162
400	228	232	240	248	264	312		156	160	164	172	184	216
500	285	290	300	310	330	390		195	200	205	215	230	270
600	342	348	360	372	396	468		234	240	246	258	276	324
700	399	406	420	434	462	546		273	280	287	301	322	378
800	456	464	480	496	528	624		312	320	328	344	368	432
900	513	522	540	558	594	702		351	360	369	387	414	486
1000	570	580	600	620	660	780		390	400	410	430	460	540
1100	627	638	660	682	726	858		429	440	451	473	506	594
1200	684	696	720	744	792	936		468	480	492	516	552	648
1300	741	754	780	806	858	1014		507	520	533	559	598	702
1400	798	812	840	868	924	1092		546	560	574	602	644	756
1500	855	870	900	930	990	1170		585	600	615	645	690	810
1600	912	928	960	992	1056	1248		624	640	656	688	736	864
1700	969	986	1020	1054	1122	1326		663	680	697	731	782	918
1800	1026	1044	1080	1116	1188	1404		702	720	738	774	828	972
1900	1083	1102	1140	1178	1254	1482		741	760	779	817	874	1026
2000	1140	1160	1200	1240	1320	1560		780	800	820	860	920	1080
2100	1197	1218	1260	1302	1386	1638		819	840	861	903	966	1134
2200	1254	1276	1320	1363	1452	1716		858	880	902	946	1012	1188
2300	1311	1334	1380	1426	1518	1794		897	920	943	989	1058	1242
2400	1368	1392	1440	1488	1584	1872		936	960	984	1032	1104	1296
2500	1425	1450	1500	1550	1650	1950		975	1000	1025	1075	1150	1350
2600	1482	1508	1560	1612	1716	2028		1014	1040	1066	1118	1196	1404
2700	1539	1566	1620	1674	1782	2106		1053	1080	1107	1161	1242	1458
2800	1596	1624	1680	1736	1848	2184		1092	1120	1148	1204	1288	1512
2900	1653	1682	1740	1798	1914	2262		1131	1160	1189	1247	1334	1566
3000	1710	1740	1800	1860	1980	2340		1170	1200	1230	1290	1380	1620

Table 10–9A shows the number of board feet in gable and hip roofs framing using 2″ × 4″ rafters, 16″ and 24″ on center with roof pitches from 1/8 to 1/2.

Table 10–9B shows the number of board feet in gable and hip roofs framing using 2″ × 6″ rafters, 16″ and 24″ on center with roof pitches from 1/8 to 1/2.

TABLE 10–9B

Framing for Gable and Hip Roofs for Pitch Centers Shown, and Using 2″ × 6″ Rafters, including Ridge Pole . . . Board Feet Required. Interpolate for Other Areas.

Ground Floor Area	16″ CENTER TO CENTER							24″ CENTER TO CENTER					
	ROOF PITCH							ROOF PITCH					
	1/8	1/6	5/24	1/4	1/3	1/2		1/8	1/6	5/24	1/4	1/3	1/2
200	164	170	174	180	192	228		114	116	120	124	132	156
300	246	255	261	270	288	342		171	174	180	186	198	234
400	328	340	348	360	384	456		228	232	240	248	264	312
500	410	425	435	450	480	570		285	290	300	310	330	390
600	492	510	522	540	576	684		342	348	360	372	396	468
700	574	595	609	630	672	798		399	406	420	434	462	546
800	656	680	696	720	768	912		456	464	480	496	528	624
900	739	765	783	810	864	1026		513	522	540	558	594	702
1000	820	850	870	900	960	1140		570	580	600	620	660	780
1100	902	935	957	990	1056	1254		627	638	660	682	726	858
1200	984	1020	1044	1080	1152	1368		684	696	720	744	792	936
1300	1066	1105	1131	1170	1248	1482		741	754	780	806	858	1014
1400	1148	1190	1218	1260	1344	1596		798	812	840	868	924	1092
1500	1230	1274	1305	1350	1440	1710		855	870	900	930	990	1170
1600	1312	1360	1392	1440	1536	1824		912	928	960	992	1056	1248
1700	1394	1445	1479	1530	1632	1938		969	986	1020	1054	1122	1326
1800	1476	1530	1566	1620	1728	2052		1026	1044	1080	1116	1188	1404
1900	1558	1615	1653	1710	1824	2166		1083	1102	1140	1178	1254	1482
2000	1640	1700	1740	1800	1920	2280		1140	1160	1200	1240	1320	1560
2100	1722	1785	1827	1890	2016	2394		1197	1218	1260	1302	1386	1638
2200	1804	1870	1914	1980	2112	2508		1254	1276	1320	1364	1452	1716
2300	1886	1955	2001	2070	2208	2622		1311	1334	1380	1426	1518	1794
2400	1968	2040	2088	2160	2304	2736		1368	1392	1440	1488	1584	1872
2500	2050	2125	2175	2250	2400	2850		1425	1450	1500	1550	1650	1950
2600	2132	2210	2262	2340	2496	2964		1482	1508	1560	1612	1716	2028
2700	2214	2295	2349	2430	2592	3078		1539	1566	1620	1674	1782	2106
2800	2296	2380	2436	2520	2688	3192		1596	1624	1680	1736	1848	2184
2900	2378	2465	2523	2610	2784	3306		1653	1682	1740	1798	1914	2262
3000	2460	2550	2610	2700	2880	3420		1710	1740	1800	1860	1980	2340

SIMPLE FORMULAS FOR ESTIMATING FBM
OF RAFTER FRAMING IN GABLE AND HIP ROOFS

The formulas shown in Table 10–9C may be used to estimate the number of board feet (FBM) in any gable and hip roof in place of Tables 10–9A and 10–9B. Simply multiply the appropriate factor shown for the pitch and the size and spacing of the rafters. Overhang is not included in the formulas. Add 5% to the quantity developed for each 12" of *horizontal* overhang.

A = ground floor (horizontal) area:

TABLE 10–9C

Size of Rafter and Centers	Pitch of Roof					
	1/8	1/6	5/24	1/4	1/3	1/2
2" × 4" - 16" on center	.57×A	.58×A	.60×A	.62×A	.46×A	.78×A
" - 24" " "	.39×A	.40×A	.41×A	.43×A	.46×A	.54×A
2" × 6" - 16" " "	.82×A	.85×A	,87×A	.90×A	.96×A	1.14×A
" - 24" " "	.57×A	.58×A	.60×A	.62×A	.66×A	.78×A

Illustrative Example No. 1

A dwelling 40' × 50' has a *ground-floor* area of 2,000 sq ft. The roof pitch is 1/3 and the roof has 2" × 4" rafters, 24" on center. The total FBM for the rafters is:

2,000 × .46 =   920 FBM

If the rafters were 16" on center, the answer would be:

2,000 × .66 = 1,320 FBM

Where the roof extends over areas outside the perimeter of the dwelling, such as over a carport, breezeway, open porch or other non-dwelling area, use the horizontal area covered by the roof in place of the *ground-floor* area of the dwelling.

The overhang should be added to the computed length, and the next even foot length of rafter is used. If the overhang is 18 inches then the length of the rafter to be used would be 18 feet.

The number of rafters are obtained in the same way as the number of joists. The length of the roof in feet is divided by the rafter spacing in feet, and one rafter is added for the end.

Table 14–2 on page 271 shows the rafter length when the height and width are known. For example, if the height is 8' and the width is 30' the rafter length is 17'.

A *ridge pole* (ridge piece) is the horizontal framing member at the ridge of the roof to which the rafters are framed and attached. It is listed in the framing schedule by length and size, which should be nominally 2 inches deeper than the rafter. In other words where 2" × 6" rafters are used, the ridge pole should be 2" × 8".

Hip Roofs, Dormers, and Intersecting Roofs

The board feet (FBM) of rafter framing in dormers will be only slightly greater in most cases than the board feet that are eliminated from the main roof by installing the dormer. The board feet of rafter framing in a hip roof, *all sides with the same pitch*, will be approximately the same as a gable roof *of the same pitch*. There may be a slight increase if larger hip rafters are used.

The board feet of rafter framing for gable and hip roofs, including those that intersect, may be read directly from Tables 10–9A and 10–9B. These tables are based on the ground floor (horizontal) square foot area from 200 to 3,000 square feet and roof pitches from 1/8 to 1/2.

BRIDGING

Floor joists need to be reinforced so that the floor load is distributed over many joists and any deflection of the joists is uniform. To accomplish this, bridging is placed between the joists. There are three types in use: *wood cross bridging, wood solid bridging*, and *metal cross bridging*. (See Figure 10–3 on page 182.)

Wood cross bridging can be purchased at most lumber yards in 50 piece bundles or it can be cut on the job. The material used can be 1″ × 3″, 1″ × 4″, or 2″ × 2″. The length of the pieces (struts) will depend on the depth and spacing of the joists as will the angle of the cut, since the piece is nailed even with the upper edge of one joist and even with the lower edge of the next joist. Two pieces of bridging are placed between each joist, crossing one another. In small rooms one line of bridging is adequate but in larger rooms the lines of bridging should be spaced 6′ to 8′ feet apart. The bridging is placed before the floor or subfloor is laid.

Wood solid bridging as its name implies is composed of a line of solid pieces usually of the same stock as the joists. Obviously, solid bridging has to be staggered in line in order to nail it between joists.

Metal cross bridging are factory-made 18 ga. steel braces placed in the same manner as wood cross bridging. They are available in cartons of 200.

Bridging Costs

The labor to install bridging is shown in Table 10–15 (page 194) and varies with the type used. The following examples illustrate redicut and cut-on-the-job wood cross bridging, and metal bridging:

A room 20′ long has 2″ × 10″ × 12′ joists, 16″ on center. There are 15 spaces to be bridged.

Wood cross bridging cut on the job: Table 10–10 shows that 2.15 lin ft of bridging is needed for each foot of row. The row here is 20 ft. Therefore, 20 × 2.15 = 43 lin ft is needed plus 10% waste or 47 lin ft. (See page 183.)

WOOD CROSS BRIDGING

METAL CROSS BRIDGING

SOLID BRIDGING

Figure 10–3
Bridging.

Using 1″ × 3″ at 15¢ per foot and labor at $12 per hour, this bridging will cost:

$$47 \text{ lin ft } 1″ × 3″ @ 15¢ = \$ \ 7.05$$
(Table 10–15) 8 sets per hour = 1.875 hrs
$$@ \$12 \ = \ \underline{22.50}$$
Cost for 15 sets $29.55
Cost per set   1.97

Wood cross bridging, redicut: Assume the price of bridging, and carpenter wages as shown.

30 pcs or, 15 sets bridging = $ 6.00
(Labor Table 10–15) 1 hour @ $12.00 = $\underline{12.00}$

15 sets Total $18.00
Cost per set $ 1.20

Metal bridging: Metal bridging for the same job would cost, installed:

30 pcs, @ $.20 ea.	=	$ 6.00
(Labor Table 10–15) 20 sets per hour =		9.00
15 sets Total		$15.00
Cost per set		1.00

TABLE 10–10
Material Required for Wood Bridging Cut
on the Job

Size Joist In Inches	Spacing In Inches	Lineal Feet Per Set (2)	Lineal Feet Per-Foot-of-Row
2 × 6	16	2.57	1.92
2 × 8	16	2.70	2.02
2 × 10	16	2.87	2.15
2 × 12	16	3.06	2.30
2 × 8	20	3.31	2.00
2 × 10	20	3.45	2.07
2 × 12	20	3.61	2.17
2 × 8	24	3.94	2.00
2 × 10	24	4.05	1.97
2 × 12	24	4.19	2.10

Note: Add to the total lineal feet developed from the Table at least 10% cutting waste.

Example: A room 20 feet wide has two rows of bridging. The 2″ × 12″ joists are 16″ on center

2 × 20 lin ft × 2.30 ft per-foot of-row =	92.0	lin ft
Add 10% waste	9.2	″ ″
Total lin ft =	101.2	″ ″
Round out to	101	lin ft

(Per set method would be, 30 sets × 3.06 = 91.8 to which must be added 10% waste.)

ROOF-TRUSS CONSTRUCTION

Trusses are pre-cut and assembled on the ground, at the job site or at a prefabricating plant that is set up for mass production. Trusses are triangular frames for gable roofs and in residences can be designed to span 40 feet. Special trusses are available for hip roofs. It is estimated that the roofs in 60 to 70 percent of all new residential construction are or will be truss-construction. Figure 10–4 shows the construction of a typical truss.

The size of the members, that is, the bottom cord, rafter-cord and braces, depends on the span and load requirements. On residential roofs, where the spans are generally less than 32 feet, the entire truss can be fabricated from 2″ × 4″ 's. Local building codes and FHA requirements govern the use of these light framing materials, particularly since nominal 2″ × 4″ timbers are now actually 1 1/2″ × 3 1/2″. Much also depends on the wind and snow loads in the area. An acceptable and practical design employs 2″ × 6″ top cord with all other units 2″ × 4″ material.

There are numerous methods of connecting the members of a truss. Earlier designs used 1/2″ plywood "gussets" which were applied to both sides and nailed and glued. More recently, truss components are attached with one of the numerous types of metal connector plates. These range from plain zinc-coated steel, 22 to 14 gauge, with pre-drilled holes; to crimped, barbed, corrugated, and flared types. In truss assembly plants these are applied with presses or rollers, or pneumatic hammers.

Trusses in residential construction are usually placed 24″ on center. The lower cord serves as the ceiling joist and the roof boards or plywood sheathing are nailed to the rafter-cord.

Where trusses can be purchased from a local fabricator, the cost is considerably less than attempting to make them on the job with inexperienced labor. The following is an example of the material and estimated labor to build one truss on the job.

Figure 10–4
Metal-plate type truss.

Illustrative Example

To build a truss with 2″ × 6″ rafters, 2″ × 4″ cord and braces, using metal connectors. The span is 28 feet and the pitch is 1/6, 4″ rise per-foot-of-run. Assume first truss fabricated and templates in place.

Using the truss in Figure 9–4 the length of the members can be scaled directly. (A line drawing or sketch to scale is recommended.)

Top cord	2 - 2″ × 6″ × 18′ - 0″ =	36.0	FBM
Bottom cord	2 - 2″ × 4″ × 16′ - 0″ =	21.3	″
Center post	1 - 2″ × 4″ × 5′ - 0″ =	3.3	″
Vertical braces	2 - 2″ × 4″ × 2′ - 6″ =	3.3	″
Diagonal braces	2 - 2″ × 4″ × 7′ - 0″ =	9.3	″
		73.2	FBM

Assume cost @ 40¢	$	.40
	$	29.28
16 Metal truss plates @ 60¢		9.60
	$	38.88

2 hours labor @ $12   24.00
Total cost, labor and material $ 62.88

Note: Laying out and fabricating the first truss takes about 5 or 6 hours. With templates in place 2 hours is adequate.

Table 10–11 shows how to compute the board feet of material in trusses with spans from 20 feet to 30 feet, and pitches or 1/6, 5/24 and 1/4. For practice purposes the members in Table 10–11 may be used to figure either King post trusses or W-Trusses. Add any overhang to the top cord. Also add connectors, whether metal, or wood gussets glued and nailed.

TABLE 10–11

Board Feet of Material for Trusses with 2″ × 6″
Top Cord, Other Members 2″ × 4″. No Overhang
Contemplated

	SPAN IN FEET				
	20	24	26	28	30
Top Cord (2″ × 6″)					
Pitch					
1/6	2-12′	2-14′	2-16′	2-16′	2-18′
5/24	2-12′	2-14′	2-16′	2-18′	2-18′
1/4	2-14′	2-16′	2-16′	2-18′	2-18′
Bottom Cord (2″ × 4″)					
Pitch					
1/6	2-12′	2-14′	2-16′	2-16′	2-18′
5/24	2-12′	2-14′	2-16′	2-16′	2-18′
1/4	2-12′	2-14′	2-16′	2-16′	2-18′
Diagonals (2″ × 4″)					
Pitch					
1/6	2-8′	2-10′	2-10′	2-12′	2-12′
5/24	2-10′	2-12′	2-12′	2-12′	2-14′
1/4	2-10′	2-12′	2-14′	2-14′	2-16′

(1/6 pitch = 4 inch per-ft-run, 5/24 = 5 inch per-ft-run, ¼ = 6 inch per-ft-run).
Note : Add 2 or 3 ft of 1″ × 4″ to joint 2″ × 4″ bottom cord. Add metal connecting members, nails etc.

Table 10–11A shows the number of board feet of lumber in W-Type trusses, 24″ on center, that have 2″ × 4″ and 2″ × 6″ top cords or rafters and 2″ × 4″ bottom cords, braces and struts included. The Table covers ground floor areas 200 to 3,000 sq ft, and roof pitches 1/8″ to 1/2″.

TABLE 10–11A

Roof Trusses . . . 24″ on Center . . . Including Braces and Struts, Board
Feet Required for Rafter Size and Roof Pitch Shown (W-type trusses)

Ground Floor Area	2″ × 4″ RAFTER AND CORD							2″ × 6″ RAFTER . . . 2″ × 4″ CORD					
	ROOF PITCH							ROOF PITCH					
	1/8	1/6	5/24	1/4	1/3	1/2		1/8	1/6	5/24	1/4	1/3	1/2
200	210	212	214	218	224	240		246	248	252	256	264	288
300	315	318	321	327	336	360		369	372	378	384	396	432
400	420	424	428	436	448	480		492	496	504	512	528	576
500	525	530	535	545	560	600		615	620	630	640	660	720
600	630	636	642	654	672	720		738	744	756	768	792	864
700	735	742	749	763	784	840		861	868	882	896	924	1008
800	840	848	856	872	896	960		984	992	1008	1024	1056	1152
900	945	954	963	981	1008	1080		1107	1116	1134	1152	1188	1296
1000	1050	1060	1070	1090	1120	1200		1230	1240	1260	1280	1320	1440
1100	1155	1166	1177	1199	1232	1320		1353	1364	1386	1408	1452	1584
1200	1260	1272	1284	1308	1344	1440		1476	1488	1512	1536	1584	1728
1300	1365	1378	1391	1417	1456	1560		1599	1612	1638	1664	1716	1872
1400	1470	1484	1498	1526	1568	1680		1722	1736	1764	1792	1848	2016
1500	1575	1590	1605	1635	1680	1800		1845	1860	1890	1920	1980	2160
1600	1680	1696	1712	1744	1792	1920		1968	1984	2016	2048	2112	2304
1700	1785	1802	1819	1853	1904	2040		2091	2108	2142	2176	2244	2448
1800	1890	1908	1926	1962	2016	2160		2214	2232	2268	2304	2376	2592
1900	1995	2014	2033	2071	2128	2280		2337	2356	2394	2432	2508	2736
2000	2100	2120	2140	2180	2240	2400		2460	2480	2520	2560	2640	2880
2100	2205	2226	2247	2289	2352	2520		2583	2604	2646	2688	2772	3024
2200	2310	2332	2354	2398	2464	2640		2706	2728	2772	2816	2904	3168
2300	2415	2438	2461	2507	2576	2760		2829	2852	2898	2944	3036	3312
2400	2520	2544	2568	2616	2688	2880		2952	2976	3024	3072	3168	3456
2500	2625	2650	2675	2725	2800	3000		3075	3100	3150	3200	3300	3600
2600	2730	2756	2782	2834	2912	3120		3198	3224	3276	3328	3432	3744
2700	2835	2862	2889	2943	3024	3240		3321	3348	3402	3456	3564	3888
2800	2940	2968	2996	3052	3136	3360		3444	3472	3528	3584	3696	4032
2900	3045	3074	3103	3161	3248	3480		3567	3596	3654	3712	3828	4176
3000	3150	3180	3210	3270	3360	3600		3690	3720	3780	3840	3960	4320

BASIC FORMULA FOR ROOF TRUSSES

The following formula may be used to estimate the board feet of material in roof trusses for any dwelling with roof pitches shown. Two types of truss are shown. One with 2″ × 4″ rafters and cords: the other with 2″ × 6″ rafters and 2″ × 4″ cords, and struts and braces are included for each type. Multiply the factor shown by the ground-floor area of your dwelling.

TABLE 10–11B

Pitch of Roof

Rafter Size 24″ Center To Center	1/8	1/6	5/24	1/4	1/3	1/2
2″ × 4″	1.05	1.06	1.07	1.09	1.12	1.20
2″ × 6″	1.23	1.24	1.26	1.28	1.32	1.44

Illustrative Example

A dwelling 35′ × 80′ has a *ground-floor* area of 2,800 square feet. The roof pitch is 1/3 and the rafters are 2″ × 6″ with no overhang. The above formula shows a factor of 1.32 by which the area is to be multiplied to obtain the board feet of truss material.

$$1.32 \times 2,800 = 3,696 \text{ FBM}$$

For each 12″ of horizontal overhang add 3% to the above figure.

RAPID ESTIMATING OF FRAMING BY AREA
OF WALLS, FLOORS, OR ROOF

Sometimes it is necessary to rapidly estimate the framing in a given area of floor, wall or roof, knowing the size of the framing members and the center to center spacing. From Table 10–12 the framing in any part of a structure can be quickly computed, or the framing in an entire building can be determined. (See page 188.)

For example, assume without even seeing the building that the question arises as to the total amount of framing it has. The building submitted is 24′ × 40′ long and has a roof rise of 6′. The exterior walls average 10′ high and there are 160′ of partitioning in it. Total framing is determined from Table 10–12, page 188, as follows:

Exterior Walls - 2″ × 4″ studs 16″ o.c.
Area   2(24′ + 40′) × 10′   = 1280 sq ft × .50 = 640 FMB
Partitions   - 2″ × 4″ studs 16″ o.c.
Area   160′ × 10′   = 1600 sq ft × .50 = 800 ″
Floor joists   - 2″ × 10″ -   16″ o.c.
Area   24′ × 40′   = 960 sq ft × 1.25 = 1,200 ″
Ceiling Beams- 2″ × 6″ -   16″ o.c.
Area   24′ × 40′   = 960 sq ft × .75 = 720 ″
Rafters   - 2″ × 6″ -   16″ o.c.
(Rafter length from Table 14–2, page 271,
is 13.5 ft) 2 × 13.5′ = 27′ × 40′   = 1080 sq ft × .75 = 810 FMB
                                                                                4,170 FMB
                                                  Plus 10%   417 ″
                                                  Total   4,587 FMB

To the total framing developed from Table 10–12 a percentage should be added to cover miscellaneous framing such as bridging, door and window heads, doubling of beams and rafters where necessary and doubling studs at walls, openings, etc. In the example shown 10 percent is considered adequate.

TABLE 10–12

Board Feet of Framing Per Square Foot of Walls, Floors and Roofs

Size of Framing In Inches	Center to Center In Inches		
	12	16	20
1 × 2	.167	.125	.10
2 × 2	.333	.250	.20
2 × 3	.500	.375	.30
2 × 4	.667	.500	.40
2 × 6	1.00	.750	.60
2 × 8	1.33	1.00	.80
2 × 10	1.67	1.25	1.00
2 × 12	2.00	1.50	1.20
3 × 6	1.50	1.13	.90
3 × 8	2.00	1.50	1.20
3 × 10	2.50	1.88	1.50
3 × 12	3.00	2.25	1.80

Note: When spacing is 24 inches on center, use half the figure shown for 12 inches on center.

NAILS AND NAILING

Figure 10–5 shows 14 common nails with their gauge, length and "penny" (d) designation. Most of these sizes are also available in finishing nails that have a small head which can be set below the surface in finish carpentry. Up until a few decades past these nails, along with flooring cut nails, lath nails, and roofing nails, were, with some exceptions, the only types available and the ones commonly used in house construction.

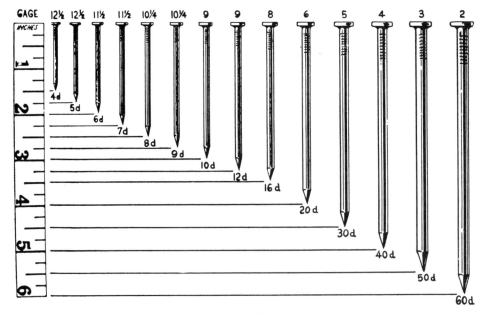

Figure 10–5

Common nail sizes.

Add shoe and plate, also framing around openings in walls and partitions. Add doubling of joists and rafters where required. This Table does not include end studs, rafters or joists.

There is a nail for almost every type of construction material, some with special heads, points or size. Some have screw threads, spiral, annular and knurled threads. There are nails for framing, roofing, drywall, underlayment, flooring, hardboard, trim, cedar shakes, asbestos siding, wood siding, and gypsum lath to name a few of the special purpose nails. Some are coated with rosin or cement for better holding power although the threaded nail, which can be driven, holds best.

Table 10–13 lists the penny system of nails, both common and finishing, and shows the length, gauge and approximate number to the pound.

TABLE 10–13

Penny Nail System

Size	Length in Inches	Gauge Number	Approximate Number to Pound		
			Common	Finishing	Casing
2d	1	15	850		
3d	1¼	14	550	640	
4d	1½	12½	350	456	
5d	1¾	12½	230	328	
6d	2	11½	180	273	228
7d	2¼	11	140	170	178
8d	2½	10¼	100	151	133
9d	2¾	9½	80	125	100
10d	3	9	65	107	96
12d	3¼	9	50		60
16d	3½	8	40		50
20d	4	6	31		
30d	4½	5	22		
40d	5	4	18		
50d	5½	3	14		
60d	6	2	12		

Special purpose nails

There are a number of special purpose nails manufactured with particularly designed heads, threads and points, depending upon their use. These three features are shown in Figure 10–6. Some of the more common special purpose nails are shown and described in Figures 10–7 and 10–8 by courtesy of Independent Nail, Inc., Bridgewater, Mass. 02324. (See page 190–192.)

Nails required

Builders are not always in agreement on the pounds of nails needed for specific items of rough carpentry. In fact, not all agree on the size or type to use, which has an important bearing on the amount used. Table 10–14 gives the approximate number of pounds and size of nail for various types of rough carpentry using the "penny" system. For special purpose nails the manufacturer or supplier should be consulted. (page 193.)

HEADS

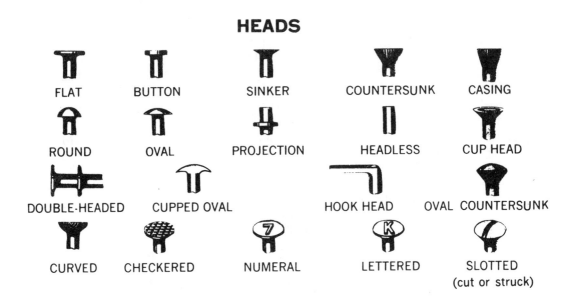

FLAT BUTTON SINKER COUNTERSUNK CASING

ROUND OVAL PROJECTION HEADLESS CUP HEAD

DOUBLE-HEADED CUPPED OVAL HOOK HEAD OVAL COUNTERSUNK

CURVED CHECKERED NUMERAL LETTERED SLOTTED (cut or struck)

THREADS

SCREW-TITE® SPIRAL   STRONGHOLD® SCREW THREAD

STRONGHOLD® ANNULAR

Combinations or variations of these threads are available.

POINTS

REGULAR DIAMOND LONG DIAMOND BLUNT CONICAL NEEDLE SHEARED BEVEL CHISEL SHEARED SQUARE

Figure 10–6

Nail heads, threads and points. Courtesy Independent Nail, Inc., Bridgewater, MA 02324

INTERIOR HARDBOARD NAILS

- **STRONGHOLD®** Annular Thread for tight, smooth decorative, prefinished plywood and interior hardboard paneling applications.
- Small, button head for flush driving helps make head inconspicuous.
- Color-matching eliminates need for nail-setting and puttying.
- Hardened Steel — heat treated, quenched, and tempered steel to make this slender shank nail easily driveable.

EXTERIOR HARDBOARD SIDING NAILS

- **SCREWTITE®** Spiral Thread for weathertight siding applications and ease of driving into hardboard siding.
- Stiff Stock Steel — high carbon, high tensile steel aids driving.
- Pilot point to help start the nail straight and ease a path for the thread.
- Flat, countersunk head for flush driving and best apearance of installed hardboard siding.
- Hot-dipped Galvanized for maximum corrosion resistance for a steel nail.

BLUNT-PQINT WOOD SIDING NAILS

- **STRONGHOLD®** Annular Thread for weathertight siding applications.
- Stiff Stock Steel — high carbon, high tensile steel aids driving.
- Slender shank minimizes splitting.
- Large, slightly countersunk head (checkered) for ease of flush driving.
- Blunt diamond point crushes wood fibers, effectively minimizing splitting in wood siding patters, or close to end or edge of wood siding patterns.
- Hot-dipped Galvanized for maximum corrosion resistance for a steel nail.

FLOORING NAILS

- **SCREWTITE®** Spiral Thread effectively minimizes "squeaks" in floor systems by pulling flooring tight. Turning action provided by thread minimizes splitting of hardwood flooring.
- Flat, countersunk head permits flush driving.
- Stiff Stock Steel — high carbon, high tensile steel for driving into less dense flooring.
- Hardened Steel — heat treated, quenched, tempered steel for ease of driving in dense hardwood flooring.

COPPER SLATING NAILS

- **STRONGHOLD®** Annular Thread for maximum holding power in wood or plywood.
- **SCREWTITE®** Spiral Thread for maximum shear, racking, lateral load resistance and high holding power.

ASBESTOS SIDING FACE NAILS

- **STRONGHOLD®** Annular Thre for weathertight siding applic tion.

Figure 10–7

Special purpose nails. Courtesy Independent Nail, Inc., Bridgewater, MA 02324

UNDERLAYMENT NAILS

- **STRONGHOLD®** Annular Thread effectively minimizes "squeaks" in floor systems by minimizing nail-popping, which can also mar floor coverings.
- Flat, slightly countersunk head permits flush driving for smooth underlayment surface.
- Regular steel for driving into less dense underlayment and subflooring.
- Hardened Steel — heat treated, quenched and tempered steel for ease of driving into denser underlayment and subflooring.

TRUSSED RAFTER POLE BARN NAILS

- **STRONGHOLD®** Annular Thread for maximum withdrawal resistance in engineered wood construction.
- **SCREWTITE®** Spiral Thread for maximum shear resistance in engineered wood construction.
- The generally smaller wire diameters than those of comparable common nails minimize splitting of drier, thinner lumber, framing, and decking.
- Stiff Stock Steel — high carbon, high tensile steel for driving into medium density wood species.
- Hardened Steel — heat treated, quenched, and tempered steel for driving into high density, dry wood species and for maximum rigidity.

ROOFING NAILS

- **STRONGHOLD®** Annular Thread for best holding power when fastening asphalt shingles and roll roofing over wood or plywood decking or sheathing.
- Large, flat head for maximum bearing area.
- Hot-dipped Galvanized for maximum corrosion resistance for a steel nail.

ASBESTOS SIDING FACE NAILS

- **STRONGHOLD®** Annular Thread for weathertight siding application.
- Hot-dipped Galvanized for maximum corrosion resistance for a steel nail.

CEDAR SHAKE NAILS

- **STRONGHOLD®** Annular Thread for weathertight cedar shingle and shake applications.
- Slender shank minimizes shake and shingle splitting.
- Small, checkered, button head blends with "textured" shingle and shake surfaces.
- Hot-dipped Galvanized for maximum corrosion resistance for a steel nail.

DRYWALL NAILS (type 500)

- **STRONGHOLD®** Annular Thread effectively minimizes nail-popping that can mar finished drywall appearance.
- Long diamond point for the "thumb" start.
- Flat, slightly countersunk, thin-rimmed head aids flush driving with minimum disruption of drywall surfaces.

Figure 10–8

More special purpose nails. Courtesy Independent Nail, Inc., Bridgewater, MA 02324

TABLE 10–14

Common Nails Required for Rough Carpentry
(Based on Framing Members 16″ o.c.)

Type Material	Kind of Nail	Pounds
Studding 16″ o.c.	10d, 16d	10 per 1,000 FBM
Joists and Rafters		
2″ × 6″	16d	9 per 1,000″
2″ × 8″	″	8 per 1,000″
2″ × 10″	″	7 per 1,000″
2″ × 12″	″	6 per 1,000″
Average house framing	8d, 10d, 16d, 20d	10–15 per 1,000″
Wood bridging 1″ × 3″	8d	1 per 12 sets
Shiplap and boards for walls, floors and roof		
1″ × 6″	8d	25–30 per 1,000 FBM
1″ × 8″	″	20–25 per 1,000″
1″ × 10″	″	15–20 per 1,000″
Plywood (Plyscord) sheathing and subfloor		
5/16″, 3/8″, 1/2″	6d	10 per 1,000 Sq ft
5/8″, ¾″	8d	20 per 1,000 ″ ″
Insulating board (Composition)	2d galv.	10 per 1,000 ″ ″
Furring on masonry	8d masonry	½ per 100 lin ft
Furring on studs	8d common	½ per 100 ″ ″

LISTING FRAMING IN AN ESTIMATE

A systematic, orderly listing in an estimate of the rough carpentry materials or operations, is most essential for accurate results. The location of the work should be clearly identified as to floor, room or portion of the structure involved. The number of framing pieces, their dimensions, and size, should be properly noted. The kind of work should be indicated either immediately before, or after, each item listed, showing that it is *sheathing, rough flooring, roof boards, studding, rafters, joists, plates, blocking, girders, bridging*, and so forth. In this way an estimator can easily recheck the take-off, and the board feet may be readily extended and totaled. Each type of rough carpentry in the listing can also be examined to establish an adequate labor charge.

ESTIMATING FRAMING COSTS

Table 10–15 shows the average number of hours required for carpenters to install various kinds of framing under normal working conditions. These rates may be used for developing unit costs for material and labor, or they may be used when it is desired to show the labor cost separately from the material cost. There may also be occasions when an hourly rate per 1,000 FBM is not applicable to a specific job operation. In that case the cost of the material should be estimated, and the number of hours of la-

TABLE 10-15

Average Hours to Install Rough Carpentry

Kind of Work	Hours Per 1,000 FBM	FBM Per Hour
Sills and plates, bolted and grouted	20	50
Studding, 2" × 4", 16" o.c.	25	40
Joists, 2" × 6" and 2" × 8"	22	45
2" × 10" and 2" × 12"	20	50
Girders, built-up from 2" stock	20	50
Rafters, 2" × 6" and 2" × 8", plain gable roof	30	34
" 2" × 6" and 2" × 8", plain hip roof	33	31
" 2" × 6" and 8" flat roof	26	39
Average for completely framing house	20-30	34-50
Exterior wall sheathing 1" × 6", and 1" × 8"	16	63
Subflooring 1" × 4"	18	56
" 1" × 6", and 1" × 8"	15	67
Roof boards, flat roof 1" × 6", and 1" × 8"	15	67
" gable roof 1" × 6", and 1" × 8"	20	50
" hip and cut up 1" × 6", and 1" × 8"	22	45

	Per 1,000 Sq ft	Sq Ft Per Hour
Plywood subfloor ....5/8"	12	83
" subfloor ....¾ "	16	63
" sheathing .... 5/16"	14	71
" sheathing ....3/8"	16	63
" roof decking (flat roof) ....½"	12	83
" roof decking (gable roof) ....5/8"	16	63
" roof decking (cut-up) ....5/8"	18	56
Insulating (composition) sheathing ....½"	12-20	50-83
" " roof decking ....1½"	20-30	33-50

Bridging - wood 1" × 3", cut on job	1-hour per 8 sets
" wood 1" × 3", redi cut	1-hour per 15 sets
" metal	1-hour per 20 sets
" wood-solid	1-hour per 10 pcs
Furring on masonry incl shimming	4-hours per 100 lin ft
" on studding including shimming	3-hours per 100 lin ft

Note: Adjust hourly rates to unusal difficult work conditions and where boards are laid diagonally. Matched boards take a little longer than square edged.

---

bor may be determined by judgment. This judgment rate is based on the estimator's best approximation of the length of time it will take one, two, or more carpenters to perform the particular work. Care should be taken not to use this method on any extensive amount of carpentry that includes numerous kinds of framing, as it then becomes an overall guess rather than a thoughtful analysis.

Unit cost for framing

Unit cost for framing is developed from a unit of 1,000 FBM of lumber. The total cost is the sum of the cost of the lumber, the nails, and the labor to install the material. The material cost is obtained in the locality where the loss occurred. The number of hours of labor may be taken from Table 10–15 and multiplied by local wage rates. The following basic formula is used to develop a unit cost for framing.

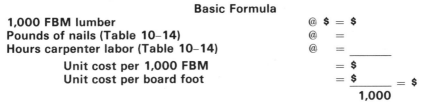

Basic Formula

1,000 FBM lumber	@ **$** = **$**	
Pounds of nails (Table 10–14)	@ =	
Hours carpenter labor (Table 10–14)	@ =	
Unit cost per 1,000 FBM	= **$**	
Unit cost per board foot	= **$** / **1,000** = **$**

The following illustrative examples show the method of developing various unit costs by using *assumed material prices and wage rate* for carpenters. The unit costs have been carried out to the nearest whole cent per board foot.

Illustrative Examples

Sills and Plates

1,000 FBM lumber	@ $ 400.00 =	$ 400.00
8 lb nails	@ .60 =	4.80
20 hours carpenter labor	@ 12.00 =	240.00
Unit cost per 1,000 FBM	=	$ 644.80
Unit cost per board foot	=	$ .64

Studding

1,000 FBM lumber	@ $ 400.00 =	$ 400.00
10 lb nails	@ .60 =	6.00
25 hours carpenter labor	@ 12.00 =	300.00
Unit cost per 1,000 FMB		$ 706.00
Unit cost per board foot		$ .71

2'× 10" Floor joists

1,000 FBM lumber	@ $ 400.00 =	$ 400.00
6 lb nails	@ .60 =	3.60
20 hours carpenter labor	@ 12.00 =	240.00
Unit cost per 1,000 FBM		$ 643.60
Unit cost per board foot		$ .64

2" × 6" Rafters— Gable Roof

1,000 FBM lumber	@ $ 400.00 =	$ 400.00
8 lb nails	@ .60 =	4.80
30 hours carpenter labor	@ 12.00 =	360.00
Unit cost per 1,000 FBM		$ 764.80
Unit cost per board foot		$ .76

2" × 10" Rafters—Flat Roof

1,000 FBM lumber	@ $ 400.00 =	$ 400.00
6 lb nails	@ .60 =	3.60
24 hours carpenter labor	@ 12.00 =	288.00
Unit cost per 1,000 FBM		$ 691.60
Unit cost per board foot		$ .69

SPECIAL CONSIDERATIONS IN ESTIMATING
ROUGH CARPENTRY LABOR

The principal factors that affect labor in general are discussed under "Estimating Labor" in Chapter 2. Each of these should be reviewed and carefully considered when estimating carpentry labor specifically.

In most instances it is not good practice to include in carpentry items, the labor cost of tearing out and removing the damaged materials. This work, sometimes done by carpenters, sometimes by *common labor*, should be treated as a separate cost for the entire job under a "Wrecking and Debris" item, or similar classification. Any attempt to intermingle the time to tear out and cart debris with the installing of new materials, particularly when done by labor at a different wage rate, is apt to result in inaccuracies. There are exceptions where a carpenter who is repairing a particular part of a structure will have to tear out damaged materials in repairing or replacing the damaged section. Such situations should be recognized when encountered. The estimator should not be influenced or confused by the existence of burned or otherwise damaged materials when determining the time required to install *new* material. One should visualize the area surrounding the operation as being completely cleared of all damaged material, and consider that the workmen generally will be unhampered by it.

For this reason, carpenters can cut, fit, or install materials at approximately the same hourly rate when repairing damaged buildings as they can when working on new construction, provided the working conditions are *reasonably* comparable. This is especially true where a large quantity of framing, sheathing, siding, and rough flooring is being replaced. When new work joins the old, as in the case of applying new siding, laying new flooring, or roof boards that have to be notched or stepped back to fit in with the undamaged part, more time will have to be given for that kind of an operation. The estimator should anticipate job conditions that will slow down the production rate of a carpenter. Working areas that restrict free movement or require awkward handling of materials affect the labor factor accordingly. Difficult patch jobs that need care and study on the part of a carpenter while doing the actual work should be thought out intelligently before the probable hours of labor are decided upon.

The unit of material for lumber is 1,000 FBM. The rate of production for a carpenter, working with lumber, is based on the average number of hours it will take to erect, install, or apply, 1,000 FBM. The rate varies with the kind of work being done, and is affected by conditions on the job which require more or less handling. Small-dimension framing members generally require more hours per 1,000 FBM than larger timbers. For example, it takes about 33 hours per 1,000 FBM to frame 2" × 4" rafters whereas 2" × 6" rafters can be framed in the same roof at an average of 30 hours per 1,000 FBM. This is readily understandable because the cutting, and fitting of a 2" × 6" does not take much longer than a 2" × 4" rafter but the former has 50 per cent more board feet per rafter; therefore, the rate per 1,000 FBM is less.

When it is necessary to hoist material mechanically or by hand to upper floors to get it to the place in a building where it is to be installed, the labor and the cost of using the equipment must be included. It may be added to the hours per 1,000 FBM if done by hand, or if it is a major operation, it can be treated as a specific item.

Figures 10–9 and 10–10 show carpenters framing in a building using automatic nailing machines. Where a sizeable amount of work is to be done these machines save considerable time. (See pages 198 and 199.)

ESTIMATING ROUGH SURFACE MATERIALS
(SHEATHING, SUBFLOORING, ROOF DECKING)

There is a wide difference of opinion among builders, contractors, and others who estimate as to whether openings should be deducted when measuring areas for sheathing, subflooring and roof boards. Some believe that openings should be deducted in full, while others contend that the surfaces should be figured as solids. Yet there are those who figure that a percentage (usually one half) of each opening or that openings with an area in excess of 10 square feet should be deducted. The principal argument in favor of including all or part of the openings in the area measurement is that its inclusion takes care of the additional labor required to cut and fit around them. However, this does not dispose of the question of what becomes of the excess material for the area of the opening. If no waste for cutting and fitting is added, then the practice of including openings has some justification, but usually the waste is added.

A more accurate method is to deduct the areas of all openings, add the normal waste for cutting and fitting, and make allowances of extra labor for the openings encountered in each instance.

Board sheathing, subflooring and roof decking

These items of rough carpentry are nominal 1-inch boards that are applied to the surface area of walls, floors, and roofs. Except where rough sawed lumber is used, the materials are dressed, and may be either square-edged or matched lumber. The boards are available in widths that range from 4 inches to 12 inches, but the 4-, 6- and 8-inch are most commonly used.

Milling and cutting waste in surface lumber is explained on pages 166 and 168. Nominal sizes versus actual sizes delivered are shown on page 167 and in Table 10–3.

Table 10–16 gives the combined milling and cutting waste that should be added to the area to be covered by the particular material in order to have the correct quantity to do the job. (See page 200.)

Composition Sheathing

The most commonly used composition board sheathing is the type known as insulating sheathing. It is made of ground cane or wood fibers impregnated with waterproofing compounds. The outside, and sometimes the inside, is coated with asphalt. The material is manufactured in 1/2-inch and 25/32-inch thicknesses. The 1/2-inch is available in widths of 4 feet and lengths of 8, 9, and 12 feet. The 25/32-inch sheathing comes in standard sizes of 8, 9 and 12 feet. The edges are matched with tongue and groove, or lap joint. Larger sizes are also available.

Quantity measurements are obtained by determining the area to be covered and adding about 5 per cent for cutting waste. For estimating purposes about 1 pound of 3d nails is required per 100 square feet of 1/2-inch insulating board, and 1 1/2 pounds of 4d nails for 25/32-inch insulating board.

Figure 10–9

Pneumatic nailing tool being used in framing a building. Courtesy Senco Products, Inc., Cincinnati, OH 45244

Figure 10–10

Pneumatic nailing tools being used to install bridging. Courtesy Senco Products, Inc., Cincinnati, OH 45244

TABLE 10–16

Approximate Milling and Cutting Waste* in 1-inch
Boards, to be Added to Area.

Item	Nominal Size, Inches	Laid Horizontal	Laid Diagonal
Shiplap	6	22%	27%
	8	21	26
	10	18	24
	12	15	21
Tongue & Groove	4	33	38
	6	22	27
	8	20	26
	10	18	23
	12	15	20
Square edge	4	19	24
	6	14	20
	8	15	20
	10	13	18
	12	12	17
Matched Solid Paneling Patterns	6	24	29
	8	24	29
	10	19	24
	12	17	22

*This includes 5 percent cutting and fitting waste which
should be adjusted to the particular operation being considered.

Plywood Sheathing, Subflooring and Roof Decking

Plywood is made in two types—Exterior and Interior. The former is moisture-resistant being made with a waterproof glue, and panels of "C" or better grade.

Interior plywood is made with a water-resistant glue and sometimes with a waterproof glue. Panel backs and interior plies are of lower grade than used in exterior types.

Plywood sheathing, called *Plyscord* when the panels are scored every 16 inches for ease of nailing, comes in sheets 3 and 4 feet wide and 6 to 10 feet long in multiples of 1 foot. Thicknesses range from 5/16, 3/8, 1/2, 5/8 to 3/4 inches. Cutting and fitting waste will range from 5 percent for run-of-the-mill straight work to as much as 20 percent for difficult, cut-up roof, floor and wall areas. Recommended thicknesses, as follows:

Exterior Walls	Minimum Thicknesses
2" × 4" Studs 16" o.c.	5/16"
2" × 4" Studs 24" o.c.	3/8"
Subflooring	
2" × 10" Joists 16" o.c.	5/8"
2" × 10" Joists 24" o.c.	3/4"
Roof Decking	
2" × 6" Rafters 16" o.c.	1/2
2" × 6" Rafters 24" o.c.	5/8"

Nails used for 5/16" to 1/2" are 6d, and for 5/8" to 3/4" are 8d. Sheets are nailed about 6 to 8 inches along the edges and 10 to 12 inches otherwise. Table 10–14 shows the quantity needed. Page 197 outlines how to estimate quantities, which is the same as for composition boards. In other words, compute the area and add about 5 percent for cutting and fitting depending on the job circumstances.

The hours of labor to apply plywood sheathing, subflooring and roof decking is shown in Table 10–15 on page 194.

Tables 10–17 through Table 10–22 show the quantity of various roof decking materials for roofs with pitches from 1/8 to 1/2 and for ground floor (horizontal) areas from 200 to 3,000 square feet. The quantities may be read directly from the Tables. Milling and cutting waste is included. (See pages 202–208.)

LISTING SHEATHING, SUBFLOORING
AND ROOF DECKING

A systematic listing, as discussed under framing, should be used for items of sheathing, subflooring, and roof decking. Most estimators list these items along with those of framing rather than putting them in a separate schedule. In this way they appear in their proper sequence with respect to actual repair operations. Roof decking follows rafter framing, sheathing follows exterior studs, and subflooring follows joist framing.

The measurements of the area to be covered, the extension of square feet and the addition of waste should be shown.

Illustrative Example

North Side of Roof
Rafters	15 pcs	2" × 6" × 16'	240 FBM
Ridge	20 lin ft	2" × 8"	28 FBM
Roof boards (1" × 6") 16' × 23' = 368 sq ft			
Waste 20%			74
			442 FBM

North Exterior Wall
Wall studs	20 pcs	2" × 4" × 10'	134 FBM
Repair sill	12 lin ft	2" × 6"	12 FBM
Repair plate	12 lin ft	4" × 4"	16 FBM
Sheathing (1" × 6") 14' × 20' = 280 sq ft			
Waste 20%			56 FBM
			336 FBM

Second Floor Bedroom
Floor joists	3 pcs	2" × 10" × 12'	60 FBM
Bridging	18 lin ft	1" × 3"	5 FBM
Subflooring (1" × 6") 6' × 12' = 72 sq ft			
Waste 20%			15
			87 FBM

TABLE 10–17

Roof Decking for Gable and Hip Roofs, Including Intersecting Roofs, with Pitch of 1/8 (milling and cutting waste is included)

Ground Floor Area Sq Ft	Roof Surface Area	Shiplap 1" by		Tongue & Groove Boards 1" by			Square Edge Boards 1" by		Plywood, Insulation & Particleboard
	Sq Ft	6" & 8" Board Ft	10" Board Ft	4" Board Ft	6" & 8" Board Ft	10" Board Ft	4" Board Ft	6" & 8" Board Ft	Square Feet
200	206	252	244	274	252	244	246	236	226
300	309	378	366	411	378	366	369	354	339
400	412	504	488	548	504	488	492	472	452
500	515	630	610	685	630	610	615	590	565
600	618	756	632	822	756	732	738	708	678
700	721	882	854	959	882	854	861	826	791
800	824	1008	976	1096	1008	976	984	944	904
900	927	1134	1098	1233	1134	1098	1107	1062	1017
1000	1030	1260	1220	1370	1260	1220	1230	1180	1130
1100	1133	1386	1342	1507	1386	1342	1353	1298	1243
1200	1236	1512	1464	1644	1512	1464	1476	1416	1356
1300	1339	1638	1586	1781	1638	1586	1600	1534	1469
1400	1442	1764	1708	1918	1764	1708	1722	1652	1582
1500	1545	1890	1830	2055	1890	1830	1845	1770	1695
1600	1648	2016	1952	2192	2016	1952	1968	1888	1808
1700	1751	2142	2074	2329	2142	2074	2091	2006	1921
1800	1854	2268	2196	2466	2268	2196	2214	2124	2034
1900	1957	2394	2318	2603	2394	2318	2337	2242	2147
2000	2060	2520	2440	2740	2520	2440	2460	2360	2260
2100	2163	2646	2562	2877	2646	2562	2583	2478	2373
2200	2266	2772	2684	3014	2772	2684	2706	2596	2486
2300	2369	2898	2806	3151	2898	2806	2829	2714	2600
2400	2472	3024	2928	3288	3024	2928	2953	2832	2712
2500	2575	3150	3050	3425	3150	3050	3075	2950	2825
2600	2678	3276	3172	3562	3276	3172	3198	3068	2938
2700	2781	3402	3294	3699	3402	3294	3321	3186	3051
2800	2884	3528	3416	3836	3528	3416	3440	3304	3164
2900	2987	3654	3538	3973	3654	3538	3567	3422	3277
3000	3090	3780	3660	4110	3780	3660	3690	3540	3390

Note: For types of roof decking not shown, use Roof Surface Area column. Add the appropriate milling and cutting waste.

TABLE 10–18

Roof Decking for Gable and Hip Roofs, Including Intersecting Roofs, with Pitch of 1/6 (milling and cutting waste is included)

Ground Floor Area Sq Ft	Root Surface Area Sq Ft	Shiplap 1" by 6" & 8" Board Ft	Shiplap 1" by 10" Board Ft	Tongue & Groove Boards 1" by 4" Board Ft	Tongue & Groove Boards 1" by 6" & 8" Board Ft	Tongue & Groove Boards 1" by 10" Board Ft	Square Edge Boards 1" by 4" Board Ft	Square Edge Boards 1" by 6" & 8" Board Ft	Plywood, Insulation & Particleboard Square Feet
200	212	258	250	282	258	250	252	244	234
300	318	387	375	423	387	375	378	366	351
400	424	516	500	564	516	500	504	488	468
500	530	645	625	705	645	625	630	610	585
600	636	774	750	846	774	750	756	732	702
700	742	900	875	987	900	875	882	854	819
800	848	1030	1000	1128	1030	1000	1008	976	936
900	954	1160	1125	1269	1160	1125	1134	1098	1053
1000	1060	1290	1250	1410	1290	1250	1260	1220	1170
1100	1166	1420	1375	1551	1420	1375	1386	1342	1287
1200	1272	1551	1500	1692	1550	1500	1512	1464	1404
1300	1378	1680	1625	1833	1680	1625	1638	1586	1521
1400	1484	1810	1750	1974	1810	1750	1764	1708	1638
1500	1590	1935	1875	2115	1935	1875	1890	1830	1755
1600	1696	2064	2000	2256	2064	2000	2016	1952	1872
1700	1802	2195	2125	2397	2195	2125	2142	2074	1989
1800	1908	2322	2250	2538	2322	2250	2268	2196	2106
1900	2014	2450	2375	2679	2450	2375	2394	2318	2223
2000	2120	2580	2500	2820	2580	2500	2520	2440	2340
2100	2226	2710	2625	2961	2710	2625	2646	2562	2457
2200	2332	2840	2750	3102	2840	2750	2772	2684	2574
2300	2438	2970	2875	3243	2970	2875	2898	2806	2691
2400	2544	3100	3000	3384	3100	3000	3024	2928	2808
2500	2650	3225	3125	3525	3225	3125	3150	3050	2925
2600	2756	3354	3250	3666	3354	3250	3276	3172	3042
2700	2862	3483	3375	3807	3483	3375	3402	3294	3159
2800	2968	3610	3500	3948	3610	3500	3528	3416	3276
2900	3074	3740	3625	4089	3740	3625	3654	3528	3393
3000	3180	3870	3750	4230	3870	3750	3780	3660	3510

Note: For types of roof decking not shown, use Roof Surface Area column. Add the appropriate milling and cutting waste.

TABLE 10–19

Roof Decking for Gable and Roofs, Including Intersecting Roofs, with
Pitch of 5/24 (milling and cutting waste is included)

Ground Floor Area Sq Ft	Roof Surface Area Sq Ft	Shiplap 1" by		Tongue & Groove Boards 1" by			Square Edge Boards 1" by		Plywood, Insulation & Particleboard
		6" & 8"	10"	4"	6" & 8"	10"	4"	6" & 8"	
	Sq Ft	Board Ft	Board Ft	Board Ft	Board Ft	Board Ft	Board Ft	Board Ft	Square Feet
200	218	266	258	290	266	258	260	250	240
300	327	400	387	435	400	387	390	375	360
400	426	532	516	580	532	516	520	500	480
500	546	665	645	725	665	645	650	625	600
600	654	800	774	870	800	774	780	750	720
700	763	930	900	1015	930	900	910	875	840
800	872	1064	1030	1160	1064	1030	1040	1000	960
900	981	1200	1160	1305	1200	1160	1170	1125	1080
1000	1090	1330	1290	1450	1330	1290	1300	1250	1200
1100	1199	1463	1420	1595	1463	1420	1430	1375	1320
1200	1308	1600	1550	1740	1600	1550	1560	1500	1440
1300	1417	1730	1680	1885	1730	1680	1690	1625	1560
1400	1526	1860	1810	2030	1860	1810	1820	1750	1680
1500	1635	2000	1935	2175	2000	1935	1950	1875	1800
1600	1744	2130	2064	2320	2130	2064	2080	2000	1920
1700	1853	2260	2195	2465	2260	2195	2210	2125	2040
1800	1962	2400	2322	2610	2400	2322	2340	2250	2160
1900	2071	2530	2450	2755	2530	2450	2470	2375	2280
2000	2180	2660	2580	2900	2660	2580	2600	2500	2400
2100	2289	2800	2710	3045	2800	2710	2730	2625	2520
2200	2398	2930	2840	3190	2930	2840	2860	2750	2640
2300	2507	2060	2970	3335	3060	2970	2990	2875	2760
2400	2616	3200	3100	3480	3200	3100	3120	3000	2880
2500	2725	3325	3225	3625	3325	3225	3250	3125	3000
2600	2834	3460	3354	3770	3460	3354	3380	3250	3120
2700	2943	3590	3483	3915	3590	3483	3510	3375	3240
2800	3052	3724	3610	4060	3724	3610	3640	3500	3360
2900	3161	3860	3740	4205	3860	3740	3770	3625	3480
3000	3270	3990	3870	4350	3990	3870	3900	3750	3600

Note: For types of roof decking not shown, use Roof Surface Area column.
Add the appropriate milling and cutting waste.

TABLE 10–20

Roof Decking for Gable and Hip Roofs, Including Intersecting Roofs, with
Pitch of 1/4 (milling and cutting waste is included)

Ground Floor Area Sq Ft	Roof Surface Area	Shiplap 1" by		Tongue & Groove Boards 1" by			Square Edge Boards 1" by		Plywood Insulation & Particleboard
	Sq Ft	6" & 8" Board Ft	10" Board Ft	4" Board Ft	6" & 8" Board Ft	10" Board Ft	4" Board Ft	6" & 8" Board Ft	Square Feet
200	224	274	264	298	274	264	266	258	246
300	336	411	396	447	411	396	400	387	369
400	448	548	528	600	548	528	532	516	492
500	560	685	660	745	685	660	665	645	615
600	672	822	792	894	822	792	800	774	738
700	784	959	924	1043	959	924	930	900	861
800	896	1096	1056	1192	1096	1056	1064	1030	984
900	1008	1233	1188	1340	1233	1188	1200	1160	1107
1000	1120	1370	1320	1490	1370	1320	1330	1290	1230
1100	1232	1507	1452	1639	1507	1452	1463	1420	1353
1200	1344	1644	1584	1788	1644	1584	1600	1550	1476
1300	1456	1781	1716	1937	1781	1716	1730	1680	1600
1400	1568	1918	1848	2086	1918	1848	1860	1810	1722
1500	1680	2055	1980	2235	2055	1980	2000	1935	1845
1600	1792	2192	2112	2384	2192	2112	2130	2064	1968
1700	1904	2329	2244	2533	2329	2244	2260	2195	2091
1800	2016	2466	2376	2682	2466	2376	2400	2322	2214
1900	2128	2603	2508	2830	2603	2508	2530	2450	2337
2000	2240	2740	2640	2980	2740	2640	2660	2580	2460
2100	2352	2877	2772	3130	2877	2772	2800	2710	2583
2200	2464	3014	2904	3278	3014	2904	2930	2840	2706
2300	2576	3151	3036	3427	3151	3036	3060	2970	2829
2400	2688	3288	3168	3576	3288	3168	3200	3100	2952
2500	2800	3425	3300	3725	3425	3300	3325	3225	3075
2600	2912	3562	3432	3874	3562	3432	3460	3354	3198
2700	3024	3700	3564	4023	3700	3564	3590	3483	3321
2800	3136	3836	3696	4172	3836	3696	3724	3610	3440
2900	3248	3973	3828	4321	3973	3828	3860	3740	3567
3000	3360	4110	3960	4470	4110	3960	3990	3870	3690

Note: For types of roof decking not shown, use Roof Surface Area column.
Add the appropriate milling and cutting waste.

TABLE 10–21

Roof Decking for Gable and Hip Roofs, Including Intersecting Roofs, with
Pitch of 1/3 (milling and cutting waste is included)

Ground Floor Area	Roof Surface Area	Shiplap 1" by		Tongue & Groove Boards 1" by			Square Edge Boards 1" by		Plywood. Insulation & Particleboard
		6" & 8"	10"	4"	6" & 8"	10"	4"	6" & 8"	
Sq Ft	Sq Ft	Board Ft	Board Ft	Board Ft	Board Ft	Board Ft	Board Ft	Board Ft	Square Feet
200	240	292	284	320	292	284	286	276	264
300	360	438	426	480	438	426	429	414	396
400	480	584	569	640	584	568	572	552	528
500	600	730	710	800	730	710	715	690	660
600	720	876	852	960	876	852	858	828	792
700	840	1022	994	1120	1022	994	1000	966	924
800	960	1170	1136	1280	1170	1136	1144	1106	1056
900	1080	1314	1278	1440	1314	1278	1287	1242	1188
1000	1200	1460	1420	1600	1460	1420	1430	1380	1320
1100	1320	1610	1562	1760	1610	1562	1573	1518	1452
1200	1440	1752	1704	1920	1752	1704	1716	1656	1584
1300	1560	1900	1846	2080	1900	1846	1859	1794	1716
1400	1680	2044	1990	2240	2044	1990	2000	1932	1848
1500	1800	2190	2130	2400	2190	2130	2145	2070	1980
1600	1920	2336	2272	2560	2336	2272	2288	2208	2112
1700	2040	2480	2414	2720	2480	2414	2430	2346	2244
1800	2160	2630	2556	2880	2630	2556	2574	2484	2376
1900	2280	2774	2700	3040	2774	2700	2717	2622	2508
2000	2400	2920	2840	3200	2920	2840	2860	2760	2640
2100	2520	3066	2982	3360	3066	2982	3000	2898	2772
2200	2740	3212	3124	3520	3212	3124	3146	3036	2904
2300	2760	3358	3266	3680	3358	3266	3290	3174	3036
2400	2880	3504	3410	3840	3504	3410	3430	3312	3168
2500	3000	3650	3550	4000	3650	3550	3575	3450	3300
2600	3120	3800	3692	4160	3800	3692	3720	3588	3432
2700	3240	3942	3834	4320	3942	3834	3860	3726	3564
2800	3360	4090	3976	4480	4090	3976	4000	3864	3696
2900	3480	4234	4118	4640	4234	4118	4150	4000	3828
3000	3600	4380	4260	4800	4380	4260	4290	4140	3960

Note: For types of roof decking not shown, use Roof Surface Area column.
Add the appropriate milling and cutting waste.

TABLE 10–22

Roof Decking for Gable and Hip Roofs, Including Intersecting Roofs, with
Pitch of 1/2 (milling and cutting waste is included)

Ground Floor Area Sq Ft	Roof Surface Area Sq Ft	Shiplap 1" by 6" & 8" Board Ft	Shiplap 1" by 10" Board Ft	Tongue & Groove Boards 1" by 4" Board Ft	Tongue & Groove Boards 1" by 6" & 8" Board Ft	Tongue & Groove Boards 1" by 10" Board Ft	Square Edge Boards 1" by 4" Board Ft	Square Edge Boards 1" by 6" & 8" Board Ft	Plywood, Insulation & Particleboard Square Feet
200	284	346	336	378	346	336	338	326	312
300	426	520	504	567	520	504	507	489	468
400	568	690	672	756	690	672	676	652	624
500	710	865	840	945	865	840	845	815	780
600	852	1040	1008	1134	1040	1008	1014	978	936
700	994	1210	1176	1323	1210	1176	1183	1141	1092
800	1136	1385	1344	1512	1385	1344	1352	1304	1248
900	1278	1560	1512	1700	1560	1512	1521	1467	1404
1000	1420	1730	1680	1890	1730	1680	1690	1630	1560
1100	1562	1900	1848	2080	1900	1848	1859	1793	1716
1200	1704	2076	2016	2268	2076	2016	2028	1956	1872
1300	1846	2250	2184	2457	2250	2184	2200	2119	2028
1400	1990	2422	2352	2646	2422	2352	2370	2282	2184
1500	2130	2595	2520	2835	2595	2520	2535	2445	2340
1600	2272	2770	2688	3024	2770	2688	2705	2608	2500
1700	2414	2940	2856	3213	2940	2856	2875	2771	2652
1800	2556	3115	3024	3400	3115	3024	3040	2934	2810
1900	2700	3290	3192	3590	3290	3192	3210	3097	2964
2000	2840	3460	3360	3780	3460	3360	3380	3260	3120
2100	2982	3633	3528	3970	3633	3528	3550	3423	3276
2200	3124	3810	3696	4158	3810	3696	3720	3586	3432
2300	3266	3980	3864	4350	3980	3864	3890	3749	3590
2400	3410	4150	4032	4536	4150	4032	4056	3912	3744
2500	3550	4325	4200	4725	4325	4200	4225	4075	3900
2600	3692	4500	4368	4914	4500	4368	4395	4238	4056
2700	3834	4670	4536	5100	4670	4536	4565	4400	4212
2800	3976	4845	4704	5292	4845	4704	4730	4564	4368
2900	4118	5020	4872	5480	5020	4872	4900	4727	4524
3000	4260	5190	5040	5670	5190	5040	5070	4890	4680

Note: For types of roof decking not shown, use Roof Surface Area column.
Add the appropriate milling and cutting waste.

TABLE 10-23

Rough and Finish Floor Materials, Including Milling and Cutting Waste. Quantities May Be Read Directly for All Room and Floor Areas 200 to 3,000 Square Feet. Interpolate for Areas in Between.

Floor Area Sq Ft	Rough Flooring, Incl. Milling and Cutting Waste				Fin. Floor Incl. Milling & Cutting Waste	
	Shiplap & Tongue & Groove Bds.	Square Edge Boards		Plywood, Particle-Board, etc.		
	1"×6", 1"×8"	1"×4"	1"×6"		1"×3"	1"×4"
	Board Ft	Board Ft	Board Ft	Square Ft	Bd Ft	Bd Ft
200	242	238	228	220	276	258
300	363	357	342	330	414	387
400	484	476	456	440	552	516
500	605	595	570	550	690	645
600	726	714	684	660	828	774
700	847	833	800	770	966	900
800	968	952	912	880	1104	1030
900	1089	1071	1026	990	1242	1160
1000	1210	1190	1140	1100	1380	1290
1100	1331	1309	1254	1210	1518	1420
1200	1452	1428	1368	1320	1656	1550
1300	1573	1547	1482	1430	1794	1680
1400	1694	1666	1600	1540	1932	1800
1500	1815	1785	1710	1650	2070	1935
1600	1936	1904	1824	1760	2208	2064
1700	2057	2023	1938	1870	2346	2195
1800	2178	2142	2052	1980	2484	2322
1900	2300	2261	2166	2090	2622	2450
2000	2420	2380	2280	2200	2760	2580
2100	2541	2500	2400	2310	2898	2710
2200	2662	2618	2500	2420	3036	2840
2300	2783	2737	2622	2530	3174	2970
2400	2904	2856	2736	2640	3312	3100
2500	3025	2975	2850	2750	3450	3225
2600	3146	3094	2964	2860	3580	3354
2700	3267	3213	3078	2970	3726	3483
2800	3388	3332	3200	3080	3864	3610
2900	3509	3450	3300	3190	4000	3740
3000	3630	3570	3420	3300	4140	3870

ESTIMATING LABOR—SHEATHING,
SUBFLOORING, ROOF DECKING

The cost to apply sheathing, subflooring, and roof boards is estimated by the board foot or square foot for plywood, particle board and composition boards. The labor required varies with the kind of work, the physical conditions under which the carpenters are working and the amount of cutting and fitting. When carpenters are working from ladders or scaffolds, or on a steeply pitched roof, the number of hours is increased. Patching small areas is time consuming and, in most cases, the labor is best determined on a job basis instead of a rate of hours per 1,000 FBM, or 1,000 square feet.

Except for small patch jobs (when the carpenters simply cut and nail boards over clear areas, however small), they generally can apply materials at an average rate. Table 10–15 shows the average number of hours of carpentry labor required for different kinds of work under normal working conditions.

Portable and Mechanical Nailers

Figures 10–11 and 10–12 show workmen using the portable and the mechanical nailers. Both of these tools save considerable time in the laying of subflooring and roof decking and in applying sheathing.

Figure 10–11
Pneumatic nailing tool being used to apply plywood sub-flooring. Courtesy Senco Products, Inc., Cincinnati OH 45244

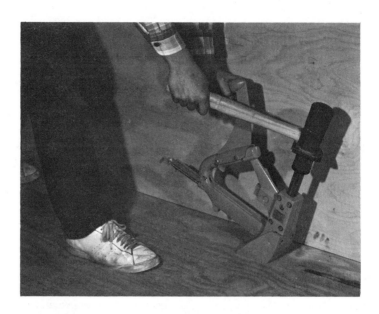

Figure 10-12(A)

Porta-Face Nailer Model 501 used to nail down the first two and last six courses of flooring, through the face of the board. Courtesy Porta-Tools, Inc., Wilmington, NC 28401

Figure 10-12(B)

Porta-Nailer model 401, with universal nailing shoe attached, being used to nail down subflooring. Can also be used for roofing and sheathing. Courtesy, Porta-Tools, Inc., Wilmington, NC 28401

Unit Costs for Sheathing, Subflooring
and Roof Boards

A unit cost for these materials is developed from a unit of 1,000 FBM. It includes the material and labor to install or apply it. The prices for material and the rate of wages should be obtained locally. The number of hours for the particular kind of work may be taken from Table 10–15.

The following basic formula is used to develop a unit cost for 1 inch sheathing, subflooring, and roof boards:

Basic Formula

1,000 FBM lumber	@	**$**	= **$**	
Lbs of nails	@	**$**	=	
Hours carpenter labor	@	**$**	=	
Unit cost per 1,000 FBM			= **$** _____	= **$**
Unit cost per board foot			**1,000**	

Illustrative Examples
(Prices and Wages Assumed)

Sheathing, applied horizontally

1,000 FBM 1″ × 6″ matched boards	@	$ 350.00 =	$ 350.00
30 lbs nails	@	.60 =	18.00
16 hours carpenter labor	@	12.00 =	192.00
Unit cost per 1,000 FBM		=	$ 560.00
Unit cost per board foot		=	$ .56

Subflooring laid diagonally

1,000 FBM 1″ × 6″ matched boards	@	$ 350.00 =	$ 350.00
30 lbs nails	@	.60 =	18.00
17 hours carpenter labor	@	12.00 =	204.00
Unit cost per 1,000 FBM		=	$ 572.00
Unit cost per board foot		=	$ .57

Roof boards (hip roof)

1,000 FBM 1″ × 6″ matched boards	@	$ 350.00 =	$ 350.00
30 lbs nails	@	.60 =	18.00
22 hours carpenter labor	@	12.00 =	264.00
Unit cost per 1,000 FBM		=	$ 632.00
Unit cost per board foot		=	$ .63

Roof boards (plain gable)

1,000 FBM 1″ × 6″ matched boards	@	$ 350.00 =	$ 350.00
30 lbs nails	@	.60 =	18.00
20 hours carpenter labor	@	12.00 =	240.00
Unit cost per 1,000 FBM		=	$ 608.00
Unit cost per board foot		=	$ .61

Roof boards (flat roof)
1,000 FBM 1" × 8" matched
boards   @ $ 350.00 = $ 350.00
30 lbs nails   @   .60 =   18.00
14 hours carpenter labor   @   12.00 =   168.00
    Unit cost per 1,000 FBM   = $ 536.00
    Unit cost per board foot   = $ .54

Unit Costs for Plywood and Composition Materials

The following basic formula may be used to develop unit costs for plywood and composition materials used for sheathing, subflooring and roof decking. It is based on 1,000 square feet of material plus the labor to install or apply that quantity. Nails are omitted because their effect on the unit cost is minor.

Basic Formula

1,000 sq ft material	@	**$**	**= $** _____
Hours carpenter labor	@	**$**	**= $** _____
Unit cost per 1,000 sq ft			**= $**
Unit cost per sq ft $ _____			**= $** _____
1,000			

Illustrative Examples
(Prices and Wages Assumed)

5/16" Plywood sheathing
1,000 sq ft 5/16" plywood sheathing   @ $ 250.00 = $ 250.00
14 hours carpenter labor   @   12.00 =   168.00
    Unit cost per 1,000 sq ft   = $ 418.00
    Unit cost per sq ft   = $ .42

½" Plywood roof decking on cut-up roof
1,000 sq ft ½" plywood decking   @ $ 300.00 = $ 300.00
18 hours carpenter labor   @   12.00 =   216.00
    Unit cost per 1,000 sq ft   = $ 516.00
    Unit cost per sq ft   = $ .52

5/8" Plywood subflooring
1,000 sq ft 5/8" subflooring   @ $ 450.00 = $ 450.00
12 hours carpenter labor   @   12.00 =   144.00
    Unit cost per 1,000 sq ft   = $ 594.00
    Unit cost per sq ft   = $ .59

25/32" Insulation board sheathing
1,000 sq ft insulation board   @ $ 150.00 = $ 150.00
14 hours carpenter labor   @   12.00 =   168.00
    Unit cost per 1,000 sq ft   = $ 318.00
    Unit cost per sq ft   = $ .32

The unit costs for other materials may be computed by using the same formula and inserting the cost of material, the number of pounds of nails from Table 10–14, and the hours of labor from Table 10–15. The hours of labor should be carefully select-ed from the Tables depending on the size and type of the material used.

Application of Unit Costs

In the following illustrative example, the small one story dwelling in Figure 10–13 requires rough flooring, sheathing and roof boards.

Figure 10–13

Quantities are obtained as shown, and the unit costs per square foot used are the same as those developed for illustrative purposes. For example, subflooring laid diagonally $.57, sheathing applied horizontally $.56, roof boards on plain gable type $.61.

Roofboards
(1″ × 6″ matched boards)

One side 15′ × 31′ =	465 sq ft
	× 2
Two sides	930 sq ft
Add 20% waste	186
	1,116 FBM @ $.61 = $680.76

Exterior wall sheathing
(1″ × 6″ matched boards)

Sides	2(10′ × 31′) =	620 sq ft
Ends	2(10′ × 24′) =	480 sq ft
2-Gables	9′ × 24′ =	216 sq ft
		1,316 sq ft

Deduct openings
(Same for sides not shown)

10—3′ × 5′ wds =	150 sq ft
2—3′ × 7′ drs. =	42 sq ft
	192

	192 sq ft	
	1,124 sq ft	
Add 20% waste	225	
	1,349 FBM @ $.56	= $755.44

Rough flooring
(1″ × 6″ matched boards diagonal)
   24′ × 31′   = 744 sq ft
   Add 25% waste   = 186
                  930 FBM @ $.57   = $ 530.10
                                         $1,966.30

ESTIMATING WOODEN FENCES

The cost to erect fences generally is developed on a lineal foot basis. The cost of the labor and material for a section between two posts is first obtained; then the spacing in feet between posts is divided into this cost to determine the unit cost per lineal foot. This unit cost is then multiplied by the total number of lineal feet of fencing required. When determining the cost of one section, the second post is omitted.

Illustrative Example

To find the cost per linear foot of a solid board fence using 4″ × 4″ posts, 2″ × 4″ rails and 1″ × 6″ matched boards. The posts are 8 feet on center and the fence is 5 feet above ground with posts 3 feet in the ground. Material prices and wage rates are assumed.

Figure 10–14

1- 4″ × 4″ × 8′ post   = 11 FBM
2- 2″ × 4″ × 8′ rails   = 11 FBM
                             22 FBM @ $400.00 per M = $8.80
5′ × 8′ - 1″ × 6″ matched boards   40
        add 20% waste   8
                             48 FBM @ $350.00 per M = $16.80
2 lbs nails (estimated)   .60 ″ lb   = 1.20
                             Total material cost $26.80

Labor for one Section

Allow 1/2 hour to dig hole
plumb and set one post   @ $9.00 per hr = $4.50
Allow 1/3 hour to cut and
fit rails   @ $12.00 " " =   4.00
Allow labor for boards at same rate as
sheathing, 16 hrs per MBF. 12 ×
$12.00 = $192.00
40 FBM × $192 per MBF   =   7.68

Total labor cost $ 16.18

Total labor and material for 8' section $42.98

Cost per lineal foot $42.98 = $ 5.37

8

Note: Waste allowance should be increased depending on height of fence. Painting is to be added as required.

If a yard is to be enclosed with this type of fence, and the measured distance is 120 lineal feet, the cost would be 120' × $5.37 = $644.40.

This principle applies to picket fences, wire fences, and straight post rail fences. The material costs will be governed by the kind of materials and the sizes specified. The labor to dig post holes will vary with the type of soil as to whether it is sandy, loamy, rocky or hard clay. When many post holes are to be dug by machine, the cost can be greatly reduced.

ESTIMATING SCAFFOLDING

Most contractors and builders keep scaffold plank on hand as part of their equipment. It is moved from one job to another as it is needed. When a large quantity is required for a particular job, it may be purchased and part of the original cost charged against the specific job. Sometimes the materials find their way into other construction jobs in the form of floor joists or other structural members.

Interior scaffolds such as those used by masons for plastering ceilings, or by brick layers to lay up brick walls, are make by placing 2″ plank on wooden horses. The time for setting up and shifting such scaffold is included in the labor rate of the mason or bricklayer's helper.

Sometimes an exterior scaffold one or two stories high is needed for workmen to install cornices, lay up veneer brick or stone, or apply exterior siding. These are erected by carpenters and consist of posts, ledgers, and bracing. Scaffold plank is laid loose, or is nailed to horizontal ledgers. The posts are usually two 2″ × 4″ spiked together. The ledgers are 1-inch or 2-inch stock and the braces are 1-inch boards.

The cost of this type of scaffold may be estimated by the lineal foot, or computed for the entire scaffold if a moderate amount is required.

Illustrative Example

The cost of erecting the scaffold shown in Figure 10–15 would be estimated as follows. An average price per MBF of material is assumed to be $400.00 and carpenter's wages are arbitrarily taken at $12.00 per hour.

Uprights	3—4″ × 4″ × 18′	72 FBM	
Ledgers	6—2″ × 6″ × 8′	48 FBM	(for two levels)
Braces	4—1″ × 6″ × 12′	24 FBM	
		144 FBM	
Planking	8—2″ × 10″ × 12′	160 FBM	
		304 FBM	@ $400.00 = $121.60

Labor—2 carpenters
2 hours   =   4 hours @ $12.00 =   48.00
                                                              $169.60

The total cost of erecting this 16-foot scaffold at these prices for material and labor is $169.60 or $10.60 per lineal foot.

Figure 10–15

The scaffold will have to be taken down when it has served its purpose and a labor charge should be make for this work. Some of the materials may be used on the job later or, if none of the materials are used, they will go back to the contractor's yard for future use as construction material or scaffolding. Whether or not salvage of scaffold materials should be allowed will depend on the circumstances in each situation.

When a competitive position permits, the contractor will charge all or as much of the cost of the material as is possible to the job including *all* of the labor for transporting, erecting, and dismantling the scaffold.

Scaffolding made of wood for more than two stories is seldom used. Steel tubular scaffold can be bought or rented and is more economical. The labor for erecting and dismantling steel tubular scaffolding averages between 2 and 3 hours per 100 square feet of wall surface. To this should be added the cost of rental and the trucking charges to and from the job. The cost of large quantities of tubular scaffold to be erected should be obtained from subcontractors who rent, erect and dismantle scaffolding on a contract basis.

Figure 10–16 shows workmen on a movable aluminum "Ladder Scaffold."

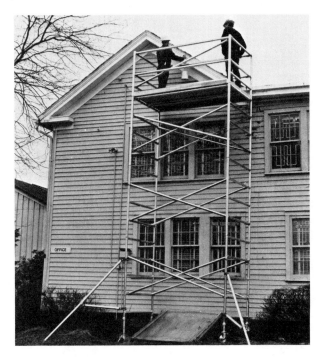

Figure 10–16

Men working from movable aluminum ladder scaffold. Courtesy Patent
Scaffold, Co., Fort Lee, NJ 07024

CHAPTER 11

Finish Carpentry

While there is general agreement among estimators on what constitutes "rough" and what constitutes "finish" carpentry, there are some types of work that could be placed in either one or the other classification. For purposes of convenience in discussion, and for consistency, all carpentry which requires finished materials and perhaps greater skill in installation than rough carpentry, will be classified here as finish carpentry. Such items include interior and exterior trim, doors and windows including frames, screens, wood siding, stair building, cabinet work and finish floors.

A further division is made here by separating interior and exterior finish carpentry. All doors and windows are treated under the classification of interior trim.

FINISH WOOD FLOORS

The top layer of flooring, which is the wearing surface, is referred to as the finish flooring. In dwelling construction the finish flooring may be pine, oak, maple or beech depending on the appearance and quality desired. There are different grades of each kind of wood flooring, so it is important when estimating costs to be certain of the kind, size, and the quality involved.

Finish wood floors can be laid over concrete slabs satisfactorily. The system meets FHA requirements and has been thoroughly tested by the National Oak Manufacturers' Association. A double layer of 1 × 2 inch wood sleepers are nailed together with a moisture barrier of 4 mil polyethylene film between them. The bottom layer of sleepers is secured to the slab by mastic and concrete nails. The strip hardwood flooring is nailed at right angles to the sleepers, with one nail at each bearing point. The slab when poured should also have a 4 mil or 6 mil polyethylene film over the base or fill.

In buildings of joist construction, the finish wood floors are laid over subflooring, (see Rough Carpentry on page 161.)

Nails

Table 11-1 shows a Nail Schedule for finished flooring. The type of nail follows the recommendations of the National Oak Flooring Manufacturers' Association.

TABLE 11–1

Nail Schedule for Flooring

Finished Size in Inches	Type and Size of Nails	Lbs of Nails Needed Per 1,000 FBM Spaced 10 to 12 Inches
25/32 × 3¼	7d or 8d screw type	35
" × 2¼	or	50
" × 1¼	cut nails*	70
½ × 2 or ½ × 1½	5d screw type, cut or casing nail	20
3/8 × 2 or 3/8 × 1½	4d bright casing (8-inch spacing)	

*Machine driven barb fasteners are equally acceptable; follow manufacturer's recommendation.

Strip flooring is the most popular style of wood flooring used today. It comes in various widths but the 25/32" × 2 1/4" (nominally 1' × 3") is the most popular. It is tongue and grooved and is available in red or white oak and in yellow pine. There is very little difference in red and white oak as to quality and utility. There are several grades of oak flooring and the price varies with these grades. The same is true of yellow pine flooring.

Maple flooring is seldom used in homes except in areas such as New England where there is a fair supply left. Clear maple flooring is the top grade. Select No. 1 is next.

Prefinished strip oak flooring, a newer development in flooring, is prefinished, buffed and polished at the factory. While the cost of prefinished flooring is a third or more higher than unfinished for the same grade, there is less waste and considerable time is saved in completing the job.

Prefinished block oak flooring is the most expensive type floor and is usually laid in adhesive over wood subflooring or concrete slabs. Standard size is 9" × 9" and 5/16" thick. Other sizes are available and also other thicknesses suitable for nailing. This type of flooring is delivered in exact quantities needed and there is no percentage to be added to the area to be covered.

Estimating Quantities of Flooring

Table 11–2 shows the percentage that must be added to the *measured* floor area to obtain the quantity that is needed to cover the area. These figures include both milling and cutting and fitting waste. See also Table 10–23

TABLE 11–2

Estimating Flooring Quantities

Nominal Size in Inches	Acutal Size in Inches	Milling + 5% Cutting Waste	Multiply Area by	Floor Needed Per 1,000 FBM
1 × 2	25/32 × 1½	55%	1.55	1,550
2 × 2½	" × 2	42	1.42	1,420
1 × 3	" × 2¼	38	1.38	1,380
1 × 4	" × 3¼	29	1.29	1.290
1 × 2	3/8 × 1½	38	1.38	1,380
1 × 2½	" × 2	30	1.30	1,300
	½ × 1½	38	1.38	1,380
	" × 2	30	1.30	1,300

The labor rates in Table 11–3 are for quality work done by carpenters—no hammer or nail marks. When subcontractors with experienced floor layers do the work the time can be substantially reduced.

TABLE 11–3

Average Hours to Lay Flooring

Nominal Size and Type Flooring Layed	Per 100 FBM	Per 1000 FBM	FBM Per Hour
1″ × 2″ Hardwood strip	4.5	45	22
1″ × 3″ Hardwood strip	3.0	30	34
1″ × 3″ Softwood strip	2.5	25	40
1″ × 4″ Softwood strip	2.0	20	50
3/8″ × 2″ Hardwood strip	4.5	45	22
3/8″ × 3″ Hardwoof strip	4.0	40	25
3/8″ × 2″ Prefinished strip	4.5	45	22
9″ × 9″ Prefinished block	3.0	30	34

Use of automatic hammers also reduces the time to lay floors and do a better job. Figure 11–1 shows the use of a nailing machine on strip flooring.

Unit Cost for Finish Wood Flooring

The unit of material for developing a unit cost for finish flooring is 1,000 FBM of the flooring at the local delivered price. Table 11–1 gives the pounds of nails and Table 11–3 gives the number of hours to lay 1,000 board feet. The following basic formula is used to develop a unit cost.

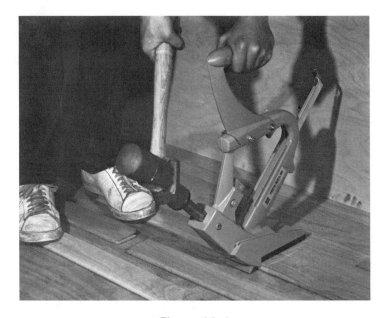

Figure 11-1

Worker using a Porta-Nailer to nail strip Hardwood finish flooring. Courtesy Porta-Tools, Wilmington, NC 28401

Basic Formula

1,000 FBM flooring	@ $	= $
Lbs 8d cut nails	@	=
Hours labor	@	= _____
	Unit cost per 1,000 FMB	$

$$\text{Unit cost per BF} = \frac{\$}{1,000} = \$$$

Assuming prices and wages as indicated, the following are illustrative examples of computing unit costs (7d or 8d screw type nails may be used):

1" × 3" Oak flooring

1,000 FBM oak flooring	@ $ 1,200.00 =	$ 1,200.00
50 lbs 8d nails	@ .60 =	30.00
30 hours labor	@ 12.00 =	360.00
	Unit cost per 1,000 FBM =	$ 1,590.00
	Unit cost per BF =	$ 1.59

1" × 3" Pine flooring

1,000 FBM pine flooring	@ $ 1,000.00 =	$ 1,000.00
50 lbs 8d nails	@ .60 =	30.00
25 hours labor	@ 12.00 =	300.00
	Unit cost per 1,000 FBM =	$ 1,330.00
	Unit cost per BF =	$ 1.33

1" × 4" porch flooring

1,000 FBM fir flooring	@ $ 1,000.00 =	$ 1,000.00
35 lbs 8d nails	@ .60 =	21.00
20 hours labor	@ 12.00 =	240.00
	Unit cost per 1,000 FBM =	$ 1,261.00
	Unit cost per BF = $	1.26

Application of Unit Costs

The area over which finish flooring is to be laid is measured and the number of square feet is obtained. Openings such as stairwells are deducted. Areas to be patched should be measured liberally to allow for cutting back so that all joints will not be along one line or joist.

Illustrative Example

Area of A = 10' × 24' =	240	sq ft
Area of B = 9' × 10' =	90	sq ft
	330	sq ft
Add 38% waste, Table 11–2	125	
	455	FBM
1" × 3" Oak flooring		
(as determined) Unit cost	$× 1.59	
Total cost	$723.45	

SANDING FINISH FLOORING

After finish floors are laid, they are usually sanded by machine before varnish, shellac or other finishes are applied. This work is done either by the floor layer or by a floor sanding subcontractor. In most sections of the country quotations can be obtained for the cost per square foot. The cost varies with the kind of flooring, the size of the rooms and the area involved.

While it is recommended that a unit cost per square foot be obtained from local sources, there are occasions when it is not possible and the unit cost then must be estimated. The basis for determining an approximate cost would be the number of hours of labor required plus the cost of the abrasive paper and rental of a sanding machine. An operator can average 100 square feet per hour on new floors including the finishing along the base. On old floors and ridged floors, the rate may drop down to as little as 50 square feet per hour. Good judgment should be used in any case.

Assuming the following wage rate, machine rental cost, and abrasive paper cost, the unit cost per square foot may be developed.

Cost of floor sanding per hour

Operator's wages	$12.00 per hour
Machine rental	3.00 per hour
Abrasive paper	1.00 per hour
	$16.00 per hour

If an operator sands at a rate of 100 square feet per hour, the unit cost would be $16.00 ÷ 100 = $.16 per square foot.

ESTIMATING WINDOWS, STORM SASH
AND SCREENS

The principal parts of a window in a building are the *frame, sash, interior trim* and the *hardware*. These components may be purchased individually, or a window can be delivered to the job as a completely assembled unit ready to install.

Of the several different styles of windows in use, the following ones are most common:

> *Double Hung*—The sash slide up and down in the frame.
> *Casement*—The sash are hinged to the side of the frame.
> *Jalousie*—The sash pivot and swing up and down like louvres.
> *Stationary*—The sash is fixed permanently in the frames.
> *Gliding*—The sash slide horizontally.

Each type is available in either wood or aluminum. Casement, gliding, and stationary windows are also available in steel. Before estimating the cost of windows, or the particular parts that need to be replaced, it is necessary to determine the style, quality, dimensions and kind of material.

Figure 11–2 shows several typical window styles available for residential construction. (Courtesy of Andersen Corporation, Bayport, Minnesota 55003) see page 224.

Sash should be described according to size, number of lights of glass, thickness and so forth. When wooden sash is broken or burned, new sash can be installed without replacing the window frame if it is undamaged. Expensive non-stock wooden sash can often be repaired by replacing the dividing bars, or a side of the sash-frame as may be required, at less cost than ordering new sash. Such matters need to be investigated carefully before concluding that a damaged window is a total loss.

Windows, whether wood, primed, vinyl covered, or aluminum; whether awning type, casement, double hung or sliding, are generally sold as complete units and frequently with necessary hardware and screens. Prices vary by type, size, quality and locality. Therefore, it is essential that the estimator obtain an exact description of the unit involved to establish both price and cost of installation.

Table 11–4 shows the average number of hours of labor to install window parts and complete units, also screens and storm sash. Where difficult or unusual working conditions exist, the labor hours should be adjusted (see Estimating Labor, page 20–23). Examples of unit costs for windows, screens and storm sash are shown below. Prices and wage rates are assumed for illustrative purposes.

Example 1.

Double hung window, wood frame and sash, single glazed including hardware and trim, 2' - 6" × 4' - 0" Facotry assembled	$ 50.00
2 Hours of labor to install completely	24.00
Total cost	$ 74.00

Example 2.

Double hung aluminum window, double glazed 2' - 4" × 4'	$ 80.00
2 Hours labor to install complete with hardware and screen	24.00
Total cost	$104.00

Example 3.

Wood storm window and fiberglass screen 1' - 4″ × 4' - 0″	$ 28.00
1 Hour labor to install	12.00
Total cost	$ 40.00

Example 4.

Aluminum combination screen and storm sash 2' - 4″ × 4'	$ 38.00
2 Hours of labor to install	24.00
Total cost	$ 62.00

Figure 11–2

Typical window styles readily available in residential construction. 1. Gliding door. 2. Double-hung window. 3. Double-hung angle bay window. 4. Awning window with fixed sash above. 5. Casement angle bay window. 6. Casement window. 7. Gliding window. 8. Bow window. Windows and gliding door shown are available in low-maintenance Perma-Shield line from Andersen Corporation, Bayport, MN 55003.

TABLE 11–4

Average Hours Required to Install Windows, Storm Sash and Screens

Type of Work Done	Hours Required
Assemble wooden frames on the job	1
Install single wooden frame in frame building	.5
Install single wooden frames in masonry building	1
Additional time for double frames with mullion	.5
Fit and hang wood sash complete (pair)	1.5
Fit and hang casement sash complete (each)	1.5
Interior window trim installed—softwood	1
Interior window trim installed—hardwood	1.5
Complete window, wood frame and sash—factory assembled	2
Complete window, wood frame and sash—knockdown	4
Complete basement window	.5
Wood window screens	1
Wood storm windows	1
Wood shutters—stationary—per pair	2
Wood shutters—hinged—per pair	3
Complete D.H. Aluminum window	2
Complete Aluminum window with sliding sash	2
Picture window, wood frame, trim and sash	4
(Thermopane same labor plus cost of material)	
Aluminum combination screens and storm sash	2
Installing factory steel sash (each)	3
Glazing factory steel sash (see Chapter 11)	5
Weather stripping D.H. windows	2
Placing back band per window	.75

ESTIMATING INTERIOR AND EXTERIOR DOORS, SCREENS, ETC.

The component parts of a door are the jambs (called the bucks in exterior doors), the door, door stop, casings, hardware and, for some exterior doors, a saddle. The parts are generally purchased separately, although suppliers furnish doors as a complete unit with all parts including hardware.

Doors should be described as to kind of wood, size and style of door, thickness, and also style of casings. Hardware for doors varies in cost depending on type, material and manufacturer. Sometimes only part of a door assembly may be damaged such as a casing, the door itself, the hardware; in other situations the entire assembly may be involved. Table 11–5 shows the average hours of labor required to install the parts of a door, or the entire unit. It also shows the hours required to fit and hang garage doors, screen doors, storm doors, and to weather-strip exterior doors. (See page 226.)

TABLE 11–5

Hours Required to Install Interior and Exterior Doors, and Their Parts

Type of Work Done	Hours Required
Set interior or exterior jambs (frames)	.75
Install mortise lock set	2.0
Cylinder lock on front door	.75
Fit and hang interior door—softwood	1.5
Fit and hang interior door—hardwood	2
Put on stops and casings (two sides)—softwood	1
Put on stops and casings (two sides)—hardwood	1.25
Install frame, fit and hang door trim complete—interior	4
Install frame, fit and hang door trim complete—exterior	5
Install one pre-assembled interior door complete	1.5
Installing several units of pre-assembled doors	1.
Make and hang 3' × 7' batten door	3
Garage swinging doors 8' × 8' fit and hang per pair	4
Garage overhead door 8' × 8' × 13/8"—hang complete	6
Garage overhead door 15' × 7' × 1¾—hang complete	8
Storm door—fit and hang complete, or combination storm and screen	2
Screen door—fit and hang complete	2
Weather-stripping exterior door	2
Trimmed opening, set jams and trim 7' × 8'	2
Placing back-band per one side door opening	.75
Placing back-band per one side trimmed opening	1

Unit Costs for Doors Including Storm and Screen Doors

Unit costs for interior and exterior doors, storm doors and screen doors, are estimated by obtaining the price of the units delivered to the job plus the labor to install them. Because of the variety of styles, construction, material and quality, it is important to have a complete description of each unit being considered. Examples of unit costs are shown below. The prices and wage rates are assumed for illustrative purposes. Table 11–5 shows the average hours of labor for the various operations.

Example 1.

Interior 5-panel pine door, 2' - 8" × 6' - 8" × 1 3/8"	$	30.00
Door frame, 2' - 8" × 6' - 8" × 5¼"		16.00
Pine casings, 2-sides, 5/8" × 2½"		9.60
1 Pr. 3½" × 3½" hinges		3.40
Passage lock set		15.00
Material cost	$	74.00
Labor, 4 hours @ $12		48.00
	$	122.00

Example 2.

Pre-Assembled flush birch door complete with hardware		
Interior, 2' - 4" × 6' - 8" × 1 3/8"	$	60.00
Labor to install, 1.5 hours @ $12		18.00
	$	78.00

Example 3.
Aluminum storm and screen door with hardware, 3' × 6'-8" $ 36.00
Labor to hang, 2 hours @ $12   24.00
                                                                 $ 60.00

Example 4.
Garage door, 4-section, 4-panel, wood 8' × 7' complete $ 160.00
Labor to install 6 hours @ $12   72.00
                                                                 $ 232.00

The unit cost for other types of doors or their component parts may be computed by obtaining local material costs and adding the labor cost based on the average hours shown in Table 11–5 multiplied by the local wage rate.

ESTIMATING MISCELLANEOUS INTERIOR TRIM

There are several items of interior trim, the materials for which are bought by the lineal foot, and the unit cost is figured in the same unit. These items consist mainly of the various base moldings, baseboards, floor moldings, picture moldings, chair and plate rails.

Baseboard may consist of one of the standard single-member molded stock styles, ranging from 2 to 4 inches high, or it may be a so-called three-member base found in older or more expensive homes. This type base is made up of a 4-, 6- and sometimes 8-inch board, 7/8-inch thick. A quarter-round floor molding, and a wall molding on top of the baseboard complete the three members. Occasionally a two-member base is found with the wall molding omitted.

Plate rails are usually found in older homes that were built in a period when it was a custom to display attractive plates and platters on a plate rail 5 or 6 feet from the floor and extending around the dining room wall. The rail consists usually of two members, and occasionally three members.

Ceiling Molding, usually a cove type; and *picture moldings,* are one-member units.

Chair Rails are either one-member or three-member units applied to dining room walls at chair-back height to prevent the chairs from marring the walls.

Unit Costs for Miscellaneous Interior Trim

The basic formula (with assumed figures below) for developing the unit costs for interior trim items, is expressed in terms of cost per lineal foot.

Basic Formula

100 lin ft of material	@	**$** = **$**
Number of hours to install 100 lin ft	@	=
Unit cost per 100 lin ft		= **$**

This cost divided by 100 equals the cost per lineal foot. Nails are relatively a minor item of cost when installing 100 lineal feet of trim and for that reason are omitted in figuring the unit cost.

Table 11–6 shows the average number of hours of labor that are required to install 100 lineal feet of different types of interior trim. By substituting these hours in the formula, and multiplying by the local wage rate, the labor cost per 100 lineal feet can be obtained.

TABLE 11–6
Hours Required to Install 100 Lineal Feet of Miscellaneous Items of Interior Trim

Type of Work Done		Kind of Wood	Hours Required Per 100 Lineal Feet	Lineal Feet Per Hour
Baseboard	1 member	Softwood	6	17
Baseboard	2 member	Softwood	7	14
Baseboard	3 member	Softwood	8	13
Baseboard	3 member	Hardwood	10	10
Chair rail	1 member	Softwood	5	20
Chair rail	1 member	Hardwood	6	17
Plate rail	2 member	Hardwood	12	8
Plate rail	3 member	Hardwood	16	6
Picture molding		Softwood	5	20
Ceiling molding		Softwood	6	17
Wood paneling strips		7/8″ Hardwood	8	13

Note: On cheap work, hours may be reduced.

Illustrative Examples

1-member baseboard-softwood
100 lin ft 1″ × 3″ pine base	@ $ .35 =	$ 35.00
6 hours labor	@ 12.00 =	72.00
	Unit cost per 100 lin ft =	$ 107.00
	Unit cost per lin ft =	$ 1.07

Chair rail-hardwood
100 lin ft 1″ × 3″ oak chair rail	@ $ .50 =	$ 50.00
5 hours labor	@ 12.00 =	60.00
	Unit cost per 100 lin ft =	$ 110.00
	Unit cost per lin ft =	$ 1.10

Picture molding-softwood
100 lin ft 1½″ molding	@ $ .30 =	$ 30.00
5 hours labor	@ 12.00 =	60.00
	Unit cost per 100 lin ft =	$ 90.00
	Unit cost per lin ft =	$ .90

Application of Unit Costs for Interior Trim

When estimating the number of lineal feet of baseboard, plate rail, picture molding, and so forth, to be installed in a room, it is customary to ignore window and door openings less than 8 or 10 feet wide as their inclusion compensates for waste in cutting and fitting. Generally such items extend around the perimeter of a room and the lineal feet of perimeter is the quantity needed.

Illustrative Example

In a room 14' × 22' a baseboard, a chair rail, and a picture molding are to be installed. Using the unit costs for each just developed and a perimeter of 2 (14' + 22') = 72 lin ft:

72 lin ft baseboard	@	$	1.07	=	$ 77.04
72 lin ft chair rail	@	$	1.10	=	79.20
72 lin ft picture molding	@		.90	=	64.80
Total cost				=	$ 221.04

ESTIMATING INTERIOR STAIRS

Stair building is a specialty in carpentry, and today most of the residential staircases are factory built. They are delivered knockdown, or with the stringers, risers and treads assembled.

The main parts of a staircase are the *treads*, the vertical *risers* and the *stringers*, which are the side pieces to which treads and risers are attached. When one or both sides of the stairs are open, a balustrade, or handrail and balusters is installed.

An open, or plain, stringer is one where the stringers are notched out and the treads and risers are nailed to them. This type of stairs is found in less expensive construction. The rough, open-string basement stairs are typical examples.

Box stairs are those which have the stringers routed out, or housed out. The treads and risers are slid into place from the back side, wedged and glued with hardwood wedges. This type of stair construction is sturdy and squeak-proof. It is more expensive to build than open-string stairs.

Because of the many different architectural designs in staircases, and the different kinds of wood used in their construction, it is recommended that local stair builders be consulted for accurate estimates. Where an approximate cost is required, the materials should be listed and priced. Table 11–7 shows the average hours of labor required to cut and assemble and install the various parts of staircases.

TABLE 11–7

Hours Required to Cut and Erect Staircases

Type of Work Done	Hours Required Per Tread
Cutting out open stringer	.3
Housing out box stringer	.8
Erecting box or open stringers	.8
Placing treads or risers on open stringer	.1
Placing treads or risers on housed stringer	.4
Placing rail and balusters	1.0
*Complete erection of open staircase	1.5
*Complete erection of boxed stairs	2.5
Erection of plank cellar stairs without risers	1.0

*Add handrail and balusters as required.

Many estimators compute the cost of the labor and material required in a typical staircase and from that obtain the cost per tread. This unit cost per tread then may be increased or decreased depending upon the quality and design of a particular staircase under consideration.

Illustrative Example

To estimate the cost of labor and material to cut out and erect an oak staircase in a residence with 8' 6" ceiling assuming material prices and labor wages as shown.

Material

11-10 1/2 " × 1-1/8 " × 3'-0" oak treads	@ $	9.00 =	$ 99.00
12-7 1/2 " × 3/4 " × 3'-0" oak risers	@	5.00 =	60.00
2 end Newels	@	15.00 =	30.00
22 balusters	@	2.00 =	44.00
14 lin ft handrail	@	2.50 =	35.00
48 lin ft scotia molding	@	.20 =	9.60
1-5/4" × 12" × 14' wall stringer	@	1.75 =	24.50
1-1" × 12" × 14' outside stringer	@	1.75 =	24.50
2-2" × 12" × 14' rough stringer	@	.80 =	22.40
	Total material cost =		$349.00

Labor (from Table 11-7)

Complete cutting out and erecting			
11 treads @ 2 1/2 hrs each = 27 1/2 hrs	@ $12.00 =		$330.00
Placing rail and balusters,			
11 treads @ 1 hr each = 11 hrs	@	12.00 =	132.00
	Total cost of stairs	=	$811.00

$$\text{Cost per tread} = \frac{\$811.00}{11} = \$ 73.73$$

EXTERIOR WOOD TRIM

Water table	Wooden gutters
Drip cap	Cornice trim
Corner boards	Verge or barge trim

Most items of exterior trim are both ornamental, and functional. The woods used are usually fir, pine, cedar, cypress and redwood. For less expensive construction local woods like spruce and a lower grade yellow pine may be used. Quantities of trim are customarily estimated by the lineal foot per member. Unit costs are developed from quantities of 100 lineal feet. Because the cost of nails used to apply exterior trim is trifling when compared to the cost of the labor and lumber, it is unnecessary to include them in unit costs.

Water Table is the baseboard at which exterior siding begins. It is at the base of the building where the foundation ends and the frame wall begins. Actually, a water table is a projection which permits rain water that runs down the outside surface of the building to drip away from the foundation. Frequently, it consists of two parts; a 1" × 6" or 1" × 8" board and a beveled cap on top called a *drip cap.*

Corner Boards, as their name implies, attach vertically to the outside and inside corners of a house, forming the finish up against which the exterior horizontal siding

butts. The stock used is generally 1 1/8 by 3 or 4 inches in width. In some cases, a
4-inch board laps a 3-inch board at the corner, forming a square edge. In other cases
two equally wide boards meet at their inside edges, and a quarter round molding is
nailed where they meet to form a rounded corner.

Wooden Gutters are used in many sections of the country in place of galvanized
or copper gutters. They are less expensive and, for some types of construction, are better suited architecturally. They are purchased in lengths at the lumber yard. Wood gutters are attached with galvanized screws or nails.

Cornices can be generally classified into three types:

1. Closed or Flush cornices.
2. Open cornices.
3. Box cornices.

A cornice which has no overhang, the rafters being cut off flush with the exterior
sheathing, is called a *closed or flush cornice*. The members consist of a simple frieze
board and a crown molding.

A cornice which has the open or exposed rafters projecting is known as an *open
cornice*.

There are two common kinds of *box cornice*. Both are projecting cornices that are
boxed in with a frieze board on the face of the house. There is a soffit underneath the
cornice, and a facia board on the face of the cornice. The soffit and the facia are nailed
to the rafter ends. When the cornice has a wide overhang, it is necessary to nail cleats
from the end of the rafter to the wall of the house for the soffit boards. These cleats
are called lookouts. Examples of typical cornices are shown in Figures 11–3, 11–4, 11–5.

Figure 11–3

Construction details of a "narrow box cornice" (without lookouts).

Figure 11–4

Construction details of a "wide box cornice" (with horizontal lookouts).

Figure 11–5

Construction details of a "closed cornice."

When estimating the cost of cornice work, the lineal feet of the members are listed individually. A unit cost is developed for each member using 100 lineal feet as a base. The number of hours of labor for installing the members are shown in Table 11–8. In all cases, the kind of wood, and the exact size of each member should be measured with a rule to determine its thickness and width.

Verge or Barge Board is the member used to finish off the gable end of a building. It may be planted on top of the siding.

Estimating Quantities of Exterior Trim

Each member of trim should be measured for thickness and width. The thickness varies with the type of trim and quality of construction. The number of lineal feet required should be measured, making appropriate allowance for cutting waste.

Unit Costs for Exterior Trim

A unit cost for exterior trim is developed from a quantity of 100 lineal feet of material for each member. The labor to install the material is shown in Table 11–8.

TABLE 11–8

Approximate Hours of Labor to Install 100 Lineal
Feet of Various Kinds of Exterior Work

Trim	Hours Labor Per 100 Lin Ft	Lineal Feet Per Hour
Water table and drip cap	4	25
Corner boards (2 members)	5	20
Verge board	4	25
2-member closed cornice	8	13
3-member boxed cornice	12	8
Wood gutters	10	10

Basic Formula
100 lin ft trim = $
Hours carpenter labor @ $ = _____
Unit cost per 100 lin ft $

Illustrative Example

Assume carpenter wages at $12.00 and the price of materials as used in the examples

Unit Cost for 1" × 8" Water Table

100 lin ft 1" × 6" pine	@ $	.60 =	$ 60.00
100 lin ft drip cap	@	.20 =	20.00
4 hours labor	@	12.00 =	48.00
	Unit cost per 100 lin ft =		$128.00
	Unit cost per lin ft =		$ 1.28

Unit for 2-Member Corner Board

100 lin ft 1 1/8″ × 4″	@ $ .50 =	$ 50.00
100 lin ft 1 1/8″ × 3″	@ .40 =	40.00
5 hours labor	@ 12.00 =	60.00
	Unit cost per 100 lin ft =	$150.00
	Unit cost per lin ft =	$ 1.50

Unit Cost for Closed Cornice
Frieze Molding-(2 member)

100 lin ft 1 1/8″ × 8″	@ $ .90 =	$ 90.00
100 lin ft 3″ crown molding	@ .60 =	60.00
8 hours labor	@ 12.00 =	96.00
	Unit cost for 100 lin ft =	$246.00
	Unit cost per lin ft =	$ 2.46

Unit Cost for Boxed Cornice
Frieze, Soffit and Facia-(3 member)

100 lin ft 1 1/8″ × 10″ frieze	@ $ 1.15 =	$115.00
100 lin ft 1″ × 10″ soffit	@ .90 =	90.00
100 lin ft 1″ × 6″ facia	@ .60 =	60.00
12 hours labor	@ 12.00 =	144.00
	Unit cost for 100 lin ft =	$409.00
	Unit cost per lin ft =	$ 4.09

DAMAGEABILITY OF ROUGH AND
FINISHED CARPENTRY

Damage by Fire*

When wood is exposed to heat, it loses moisture which causes it to shrink, warp, or check (split). These conditions may occur even though the heat is insufficient to actually burn or char the wood. The damaging effect of fire or heat on wood will be one of either appearance, structural, or both. The scorching or slight charring of millwork and trim, will usually require replacement, because the spoiled appearance is more important than lack of any structural injury.

Damage by Water

While wood will shrink as it loses moisture, it also swells when it absorbs moisture. Air-dried lumber normally contains 12 to 15 percent moisture by weight. Kiln-dried lumber, intended for interior millwork, has approximately a 7 percent moisture content. Unprotected wood that is exposed to water or heavily moisture-laden air in a building, expands across the grain. The manner in which the wood is attached or held in place will determine to some extent the amount of deformation. Window and door casings, baseboard, shelving, stair treads and risers may show no signs of warping or twisting when wet. Much will depend on the amount of water and the duration of ex-

---

* See also Chapter 1, pages 2, 3.

posure or submersion. Finish floors, particularly edge-grain, having little or no room to expand sideways, will frequently *ridge* along the edges of each piece producing a washboard effect. In some instances the expanding can result in large areas of the finish floor being thrust upward in great mounds or waves. A floor that is highly finished with paint, varnish, or oil, has less tendency to deform because the water cannot penetrate to the raw wood. If the water comes in contact with the raw wood flooring from the underside, however, there is little to prevent it from soaking into the cells. Because of the susceptibility of finish floors to water damage, prompt measures should always be taken immediately to remove excess surface water by mopping or spreading sawdust. The circulation of air by opening windows or using fans, and also furnishing a moderate amount of heat, generally speeds up the drying out process in order to reduce damage.

The deformation of wood by swelling begins to show up within a matter of hours after exposure, and the ultimate damage may take a few days to disclose itself. For this reason it is well not to attempt to write the specifications for repair until a sufficient period of time has elapsed. If no damage is visible within a week after the water has been removed, it is unlikely that any will occur after that.

Plywood made for interiors does not have waterproof glue between the plies and the veneer and frequently separates if it is permitted to remain directly in water very long, or if it is wet down continuously. Plywoods designed for exterior application are not so affected.

Interior plywood doors are especially susceptible to damage at the bottoms or tops where the water penetrates the unfinished edges.

Exterior woodwork and trim are not usually damaged by water unless submerged, as might be the case during flooding. Damage is then mainly the result of water coming in contact with the unfinished back of the wood.

Wood that is constantly submerged or is continuously kept away from water or moisture will not be attacked by decay or rot. However, intermittent wetting and drying as often occurs to wood near the ground, or to framing members of a roof or to flooring exposed to periodic leaks, will eventually cause rot.

Water in contact with wood may also cause staining which affects the appearance of paneling, trim, and other millwork that has a light or natural finish. The staining may be caused by clean water, or water which carries with it colors or dirt from contents or other parts of the building.

Damage by Breakage or Marring

Breakage of carpentry items includes damage caused by numerous perils other than fire or flood. This includes lightning, explosion, collapse, settlement, slides, earthquake, vehicles, sonic boom, falling trees, windstorm, hail, animals, or occupants. The extent of damage may range from that done by a firefighter who chops through a door while gaining entrance to a burning building, to the complete destruction of a frame building by an explosion or a tornado.

METHODS OF REPAIR

Carpentry items that have been damaged will require repairing or replacing, and the specifications will fall into one of the following categories:

1. Replacing completely.
2. Replacing part or parts.
3. Reinforcing.
4. Concealing.
5. Resurfacing.

When taking off the specifications to restore items that are damaged, it is very important to recognize situations where it will be necessary to include the removal and replacement of undamaged adjacent or contiguous structural items. For example, when charred floor joists are to be removed, the subflooring and finished flooring, though slightly or not at all damaged, will also have to be replaced. When rafters are charred, and must be replaced, the roof boards and the roofing will also have to be replaced. Frequently, damaged flooring which runs under the baseboard cannot be properly relaid unless the baseboard is removed and replaced. When charred studding must be replaced, the exterior sheathing and siding, or stucco, though undamaged, will usually have to be replaced also. Each situation should be examined carefully to avoid omitting items.

Replacing Completely

The complete replacement of damaged carpentry needs no special explanation, except to point out the need to include all materials and labor necessary to effect proper replacement. For example, when a wall plate or sill has been charred, requiring replacement and necessary shoring, the removal and putting back of siding or cornice, should be included in the estimate. Unless damaged structural members such as joists, girders, studding and rafters can be satisfactorily concealed or reinforced, it will be necessary to replace them. Millwork and trim items or finished floors that cannot be properly repaired to restore them to their original condition should be replaced.

Replacing Part or Parts

Whenever a door, window, cabinet, or some other type of millwork can be satisfactorily repaired by replacing a stile, panel, or part at less cost than replacing the entire unit, the estimate should be made accordingly. A door, for example, with a broken panel or stile frequently can be repaired for less than the cost of a new door, especially if it is an expensive one.

When the casings or sash bars in a window have been damaged, the cost of replacing those parts should be compared to the cost of replacing the entire unit.

Finished floors, paneling, or other surface materials that are damaged over a part of the area can usually be repaired by carefully cutting out the damaged section and replacing with new materials.

Reinforcing

The reinforcement of damaged carpentry items applies to structural items. Split or moderately charred timbers frequently can be reinforced by splicing new members alongside those that are damaged. This method of repair is generally considered wherever the cost of replacing damaged timbers greatly exceeds the cost of reinforcing and where such repairs are not objectionable.

Concealing

Many times exposed structural members in a building are heavily smoked, slightly charred or otherwise injured in appearance but not in structural strength. A method of repair which is frequently satisfactory is to conceal the members by scraping and painting, or by covering with a suitable wallboard. Sometimes, when exposed timbers are reinforced, the damaged area is also concealed in this manner.

Damaged rafters and joists lend themselves to this treatment. When the timbers are smoked or charred, the surface is scraped and whitewashed or painted before applying wallboard in order to eliminate any possible odor that might seep through later.

Heavy girders, posts, and truss cords may be scraped and painted when slightly charred, and then boxed-in with 7/8 inch boards to conceal any unsightly appearance.

Resurfacing

Resurfacing by sanding is generally a satisfactory method of repairing floors that have been cupped or ridged by water. The sanding operation is preceded by any replacement of flooring or patching which is necessary.

Trim and other millwork that has had the surface damaged by heat, smoke, or water may also be sanded down prior to refinishing.

Specific Methods of Repair

The following check-list will serve as a guide when contemplating the repairs to specific carpentry items:

Framing
1. Remove and install new framing members.
2. Reinforce by installing new members alongside.
   (a) Scrape and paint or whitewash charred areas.
   (b) Cover area with wallboard or similar material.
3. Cut out section damaged (as with corner post, sill, plate, etc.) and splice in new member.

Wood Trusses
1. Remove and replace entirely.
2. Remove member (cord, rafter, strut, or brace) and replace with new member.
3. Scrape char (if no structural damage), paint, or box with 7/8 inch boards.
4. Paint or whitewash when only smoked.

Sheathing, Rough Flooring and Roof Boards and Siding
1. Remove and replace entirely.
2. Cut out damaged area and replace with new wood.
3. Paint or whitewash when only smoked.

Windows

1. Remove and replace entirely.
2. Replace sash and hardware only.
3. Replace trim only.
4. Glaze sash only.
5. Refit and hang sash.

Doors

1. Remove and replace frame, door, trim and hardware completely.
2. Replace door and hardware only.
3. Replace door stile or panel.
4. Replace trim only.
5. Glaze door only.
6. Refit door.

Finish Floors

1. Remove and replace entirely.
2. Cut out and patch damaged areas.
3. Sand entire area.
4. Paint, shellac, or varnish.

Baseboard, Picture Molding, Chair Rail

1. Remove and replace.

Built-in Cabinets

1. Remove and replace completely.
2. Replace damaged doors, and hardware as required.

Paneling and Wainscot

1. Remove and replace entirely.
2. Remove and replace damaged stiles, rails, and panels as required.

Stairways

1. Remove and replace entirely.
2. Replace damaged treads or risers.
3. Replace hand rail and balusters as required.
4. Sand treads.

Miscellaneous Exterior Trim

1. Remove and replace units damaged.

CHAPTER 12

Interior Wall Finish

INTERIOR WALL BOARDS

This chapter deals with the various types of interior wall finish commonly encountered, excluding wet plaster finishes which are discussed in Chapter 16 under Lathing and Plastering. These wall finishes include gypsum board; the different fibre boards made of wood fibre, cane fibre and wood pulp; plastic coated masonite; the several kinds of plywood finishes; and solid wood paneling.

Most wallboards are available in sheets 4 feet wide by 8 to 10 feet long. Plywoods generally run 8 feet long, and gypsum board runs 6 to 14 feet in length.

Application to studs or furring strips is by nailing, adhesives, or both.

Estimating Materials

An estimate of the materials required to apply composition wallboards should include the furring and blocking that is needed. In many cases where repairs are contemplated, existing furring may be undamaged or simply may require renailing.

The area to be covered is computed after deducting door and window openings to obtain the actual number of square feet of wall and ceiling area. An allowance is then made for fitting and cutting waste; generally 10 percent is adequate for average conditions. Much will depend on the shape, number of openings, size and layout of the room. In actual practice a sketch of each wall is made, from which the number and size of each piece of wallboard may be determined.

Estimating Labor

Consideration must be given to several factors in estimating the labor necessary to apply wallboard. For example, a worker can put on wallboard at a faster rate in a large room than in a small one; cutting and fitting around doors and windows will slow down the rate; and the worker will put on large sheets at a faster rate per square foot than small sheets, or tile-size pieces. Wallboards with high-grade finishes require more care and consequently more time than the less expensive types where joints are to be covered with wood strips that conceal poor joining or hammer marks. Placing wallboards on ceilings takes more time per square foot than on sidewalls. Light insulating type wallboards are easier to handle than heavier kinds, some of which take two people to carry and put in place.

DRYWALL CONSTRUCTION

Interior walls and ceilings that are finished with gypsum wallboard in place of lath and wet plaster is called "drywall" construction in the building trade. More than seventy-five percent of all new residential buildings have drywall construction.

Sheetrock SW Wallboard

Sheetrock, or gypsum board, is essentially the same type of material described under gypsum lath in Chapter 16 except the paper on the finished side is less porous and is harder and smoother for wearing quality and appearance. It is made in four thicknesses:

> 5/8", recommended for the finest single layer drywall construction. The greater thickness provides increased resistance to fire exposure and transmission of sound.
> 1/2", for single layer application in new residential construction.
> 3/8", lightweight, applied principally in the double wall system and in repair and remodel work. Not recommended over metal framing.
> Width: 4'; length; 8', 9', 12' or 14'; edges: rounded, tapered; finish: ivory manila paper, suitable for paint or other decoration.

Tapered Edge SHEETROCK has long edges tapered on the face side in order to form a shallow channel for the joint reinforcement which provides smooth, continuous wall and ceiling surfaces. Made in the same thicknesses as SHEETROCK SW Wallboard and in one other thickness:

> 1/4", a lightweight, low cost, utility gypsum wallboard, for use over old wall and ceiling surfaces.
> Width: 4'; length: 8', 10', edges: tapered; finish: ivory manila paper, suitable for paint, wallpaper or other decoration.

SHEETROCK W/R Gypsum Wallboard is a water-resistant gypsum wallboard that provides an excellent base for the adhesive application of ceramic, metal, and plastic tile. It is water-resistant all the way through: (1) multi-layered and back paper is chemically treated to combat penetration of moisture; (2) the gypsum core is made water-resistant with a special asphalt composition. It was developed for application in bathrooms, powder rooms, kitchens, utility rooms, and other high moisture areas. SHEETROCK W/R Gypsum Wallboard is easily recognized because of its distinctive green face.

In addition to its use as a superior tile base in new construction, SHEETROCK W/R Wallboard is a cost-saver in modernization work. It permits new tilework to be installed over existing surfaces without tearing out old walls.

This back-up for wall tile is available in plain core, 1/2" and 5/8" thickness.

Wallboard Trim Accessories

In order to provide true and straight lines for smooth finishing at outside wallboard corners, metal trim accessories are available as corner reinforcements, casing beads, metal trims, control joints and decorative moldings.

Predecorated Panels

Gypsum panels are available, predecorated with a vinyl faced lamination in a variety of patterns making on the job decorating unnecessary.

Wallboard Application

Gypsum board may be applied to sidewalls vertically or horizontally; the latter is recommended in rooms where full-length sheets can be used, as it reduces the number of vertical joints. Nails for 3/8 and 1/2-inch sheets are five-penny (1 5/8 inches). Spacing of nails is 6 to 8 inches. Where the adhesive nail-on application is used, nail spacing is considerably more. Table 12–1 shows nails needed.

TABLE 12–1

Nail Schedule for Gypsum Board

Thickness	Type Nail	Lbs per 1000 Sq Ft
¼ "	1 ¼ " coated	3
3/8" and ½ "	1 5/8" coated	5
5/8"	1 7/8" coated	5

Joint Treatment

Joints are concealed by perforated tape and joint compound. On large jobs the work is done with a taping machine whereas on small jobs it is done by hand. Approximately 50 lbs. of joint compound and 370 feet of PERF-A-TAPE are required to finish 1,000 sq. ft. of SHEETROCK Gypsum Wallboard (See Table 12–2 for quantities needed).

TABLE 12–2

Tape and Compound Schedule

Square Feet Wallboard	Joint Compound	Rolls of Tape
100–200	1 gallon	2-60 ft rolls
300–400	2 gallon	3-60 ft rolls
500–600	3 gallon	1-250 ft roll
700–800	4 gallon	1-250 ft roll and 1-60 ft roll
900–1000	5 gallon	1-500 ft roll

Note: An alternative of 50 lbs of powder joint compound per 1,000 sq. ft. may be used.

Labor Applying Gypsum Wallboard

An experienced person should place 3/8" or 1/2" gypsum wallboard at the rate of 100 sq. ft. per hour on side walls, and at 100 sq. ft. per 1 1/2 to 2 hours on ceilings in average size rooms. In small rooms or where much cutting and fitting is necessary, the rate of application will range from 50–75 sq. ft. per hour on side walls and ceilings. A rule of thumb for average work is 2 hours per 100 sq. ft. including taping on walls and ceilings.

Most prefinished gypsum wallboards are estimated at the same labor rates as unfinished depending upon the amount of cutting and matching necessary.

Two-Ply Gypsum Wallboard

Two-ply gypsum wallboard consists of two layers of 3/8" board with the second layer applied with an adhesive. Labor costs can be estimated at about twice that of single-ply.

Unit Costs for Drywall

A unit cost is developed for drywall by adding to the cost of 100 square feet of material, the cost of labor based on local hourly wages. Adjust the rate of installation for unusual job conditions.

Illustrative Examples

(Labor and Material prices assumed)

½-inch Gypsum wallboard on studding

100 sq ft wallboard	@ $10.00 =	$10.00
1 hour labor	@ 12.00 =	12.00
Unit cost per 100 sq ft	=	$22.00
Unit cost per sq ft	=	$ .22

Taping and Finishing Joints

Material per 100 sq ft	@ $ 1.50 =	$ 1.50
1½ hours labor per 100 sq ft	@ 12.00 =	18.00
Unit cost per 100 sq ft	=	$19.50
Unit cost per sq ft	=	$ .20

Average For ½-inch Gypsum Wallboard on Walls and Ceilings Including Taping Joints

100 sq ft wallboard	@ $10.00 =	$10.00
2½ hours labor	@ 12.00 =	30.00
Unit cost per 100 sq ft	=	$40.00
Unit cost per sq ft	=	$ .40

When the adhesive nail-on method is used, the adhesive is applied to joists or studs before positioning the wallboard. Approximately 8 quart-size tubes per 1,000 sq. ft. are needed.

Drywall Ceiling Finish

A number of products are available for finishing drywall ceilings economically and artistically. These are either sprayed on, or applied with a roller and stippled, swirled or raked to form pleasing patterns. They come in powdered form and are mixed with water.

Perfect Spray contains shredded polystyrene aggregate and is packaged in 50 lb bags. Mixed with 8 gallons of water this will cover 8 to 10 sq. ft. per lb or 400 to 500 sq. ft. per bag.

Quick Spray—Bestex C is sprayed on and comes in medium-fine, medium and coarse powder form.

Texture contains no aggregate and can be sprayed or applied with a roller.

Stilts And Taping Machines Save Time

Adjustable stilts for workers which replace scaffolding are shown in use in Figure 12–1. The workers quickly become accustomed to these and they considerably reduce the time to put on sheetrock. A taping machine is also shown which greatly reduces the hours to apply the tape and compound (Figure 12–2 on page 244).

Figure 12–1
Applying tape to drywall, working on adjustable stilts. Courtesy Goldblatt Trowel Trade Tools, Kansas City, KA 66110.

Figure 12–2

Taping machine used to apply tape to drywall. Courtesy Goldblatt Trowel Trade Tools, Kansas City, KA 66110.

FIBERBOARD

The various types and finishes on fiber base wallboards are too numerous to mention. The sheets are prefinished and are usually 4 ft. wide and 8 ft. long with 6, 10, and 12 ft lengths available on order. Thicknesses range from 1/8″ to 3/8″ with insulating type sheets 1/2″ thick. Some of the well known trade names are Celotex, Fir-Tex, Nuwood, Presdwood, Masonite, Upson-Board, Tile-board, Asbestos-Board, and Hardboard.

Application varies and may be by nails colored to match the finish, by adhesive, or by both.

Matching inside and outside corner moldings, base molding, and joint strips are available in plastic or wood; metal is also used.

Estimating Costs

Rough estimates of the quantity of fiber wallboard needed can be made by computing the area of wall space and adding a judgment percentage for waste in cutting and fitting. A more accurate determination can be made by laying out each room on paper and determining the exact number of panels and moldings required.

Labor costs will vary with the type and quality of fiber board, the size of room, and the amount of cutting and fitting. On average work a carpenter should apply 100 sq ft in 1 1/2 to 2 hours on side walls. On smaller rooms and where much cutting and fitting is required the time per 100 sq ft should be increased to 2 1/2 to 3 hours. Application of moldings over joints and corners will require an average of 2 hours per 100 lin ft.

WALL AND CEILING INSULATION TILE

Fiber composition tiles are used for insulating as well as decorative purposes on walls and ceilings. They come in sizes ranging from 12″ × 12″ to 16″ × 32″, by 1/2″ thick. They are applied by use of adhesives or by concealed nails or staples. Numerous designs and finishes are available.

Tiles may be applied directly over old walls with adhesives or over furring nailed to studs, rafters, or over old walls and ceilings.

Furring Strips on Ceilings

Furring strips are usually 1″ × 2″ nailed across joists 12 inches on center. A carpenter should average 100 lin. ft. of furring every 3 hours. (Table 12–3 on page 246). A unit cost, using the following assumed prices may be developed as follows:

Illustrative Example

1″ × 2″ Wood furring applied over ceiling joists

100 lin ft furring	@ $ .06 =	$ 6.00
3 hours labor	@ 12.00 =	36.00
Unit cost per 100 lin ft	=	$42.00
Unit cost per lin ft	=	$ .42

Furring Strips on Studding

Furring on side walls over studs or over old walls can be applied at an average rate of 100 lin ft every 2 hours. If much leveling and shimming is necessary to obtain a smooth surface the time should be increased accordingly (Table 12–3).

Unit Costs for Ceiling Tile on Furring

A unit cost may be developed for ceiling tile by adding to the cost of 100 square feet of material, the cost of labor based on the local hourly wages times the number of hours per 100 square feet shown in Table 12–3. The rate of installation may be adjusted for unusual job conditions other than average.

Installing 12″ × 12″ insulating stapled tile on ceiling furring

100 sq ft insulating tile	@ $ .30 =	$30.00
5 hours labor	@ 12.00 =	60.00
Unit cost per 100 sq ft	=	$90.00
Unit cost per sq ft	=	$ .90

PLYWOOD WALLBOARD

Prefinished plywood, a popular interior finish comes in sheets 4 feet wide and 6 to 12 feet long. Thicknesses range from 3/16 to 1-inch but the 1/4 inch is most commonly used.

Plywood may be placed over bare studding, studding with 3/8″ or 1/2″ gypsum board backing—the joints of which may or may not be taped—or it may be placed over furring on studs or over old walls. Nails, colored to match the finish, are used, or adhesive, or both nails and adhesive.

Prices per panel vary considerably and it is very important when estimating to be certain of the thickness, and quality of plywood under consideration.

Joints may be flush or V-grooved, but are very often concealed with panel strips at joints and corners.

TABLE 12–3
Hours Required To Install Composition Wallboard, Plywood, and Ceiling Tile.

Type of Work Done	Hours Per 100 Sq. Ft.
Furring on studding 12" center	2
Furring on clg. joists 12" center	3
3/8" or ½" Gypsum wallboard on studs	1
3/8" or ½" Gypsum wallboard on clg. joists	1½–2
2-Ply, 3/8" Gypsum wallboard on studs	2
Taping and finishing joints	1½
Wood strips over joints	1
Fiber board (Celotex etc)	1½–2
Ceiling tile (12" × 12") on furring, stapled	5
Ceiling tile (12" × 12") on flat surface-adhesive	2
(for large size tile reduce time accordingly)	
Plywood, prefinished, ¼" on studs	
including stripped joints and corners	2–3
Plywood, prefinished, ½" flush joint on studs	6–8
Solid wood paneling, 1" × 8"	3–4

Unit Costs

The following are illustrations of unit costs for prefinished plywood panels using assumed prices for materials and wages (Table 12–3).

Inexpensive ¼-inch plywood paneling on studding-
stripped joints

100 sq ft ¼" plywood	@ $	.50 =	$50.00
2 hours labor	@	12.00 =	24.00
Unit cost per 100 sq ft		=	$74.00
Unit cost per sq ft		=	$ .74

High grade ½-inch plywood paneling on studding-
flushed joints

100 sq ft ½" plywood	@ $	1.80 =	$180.00
8 hours labor	@	12.00 =	96.00
Unit cost per 100 sq ft		=	$276.00
Unit cost per sq ft		=	$ 2.76

SOLID WOOD PANELING

Solid wood paneling of knotty pine, cherry, cypress and so forth is usually nailed to furring applied over studding or masonry. There may or may not be an underlayer of composition wallboard for sealing and insulating purposes.

Most paneling consists of 3/4 or 7/8-inch boards 6, 8, or 10 inches wide. They may all be in one width or of random widths. The boards have tongue and grooved or shiplapped edges for joining and concealed nailing. Where it is necessary to facenail, 6 or 8 penny finish nails are used.

The material is estimated by obtaining the net wall area, deducting windows and door openings, and adding the appropriate amount of waste for the width of boards used.

The labor to apply solid wood paneling varies with the size and layout of the area involved, and with the kind of wood used. For average conditions a carpenter should be able to install 1,000 board feet in approximately 40 hours or at the rate of 4 hours per 100 board feet. (Table 12–3)

Illustrative Example

Installing 1″ × 8″ Knotty Pine Paneling

1,000 FBM knotty pine	@ $800.00 =	$ 800.00
40 hours labor	@ 12.00 =	480.00
Unit cost per 1,000 FBM	=	$1,200.00
Unit cost per BF	=	$ 1.20

CHAPTER 13

Exterior Wall Finish

Exterior wall finishes are generally of wood, metal or composition materials. In colder climates the materials are put on over an insulating underlayment of wood or composition sheathing. In temperate climates, on summer bungalows, and secondary buildings, it is common to apply the exterior finish directly over the studs, without any sheathing.

WOOD SIDING

There are several types of wood siding, the most common being *bevel* siding, *drop* siding, *vertical matched* siding, *batten or rough sawed* siding, and plywood of various finishes.

Bevel siding comes in nominal widths from 4 to 10 inches. The wood may be pine, redwood, cedar or fir. As its name implies, bevel siding is bevel-cut, tapering from about 3/16 inch at the top edge, to 9/16 inch or 11/16 inch at the butt edge; the latter is for boards over 6 inches wide.

It is applied horizontally, one board overlapping the one below by 1/2 inch for narrow siding, to 3/4 inch for wider boards. This leaves an exposure which varies from 3 to about 8 inches. The exposure is the distance from the butt edge of one board to the butt edge of the one which overlaps it. Bevel siding is usually applied with 6d or 8d common nails depending on the thickness of the siding. For high-grade work, the nails are set with a nail set, and the nail holes are later puttied before painting. Aluminum nails need no setting or puttying.

At the corners of a building, bevel siding may be joined either by mitering, overlapping ends, or by the use of a patented metal clip that conceals the square end-cuts of both boards. The use of corner boards is not as common in applying bevel siding as when drop siding or rustic siding is used. It takes more time to miter corners, and, therefore, that method is more expensive than either using metal clips or installing corner boards against which the bevel siding butts.

The quantity of bevel siding needed to cover an area is determined by computing the number of square feet in the area, deducting the openings, and adding sufficient waste to take care of the loss in milling, lapping, cutting, and fitting on the job. The percentage of milling and lapping waste to be added *increases* as the width of the siding *decreases*. Bevel siding 6 inches wide has a waste for milling, lapping and cutting of about 35 percent, while bevel siding that is 4 inches wide has a waste of about 50 percent. Table No. 13–1 shows the quantity of nails required, and the approximate waste to be added for various sizes and types of siding.

Drop siding is made in a number of patterns, and it is known by various names including novelty siding and rustic siding. Log cabin siding is a modification of drop siding which is designed to give a log cabin appearance. Drop siding comes in nominal widths ranging from 4 to 9 1/2 inches, with thicknesses from 9/16 to 3/4 inches. The exposure is fixed by reason of the shiplapped or tongue and groove edge where the boards are joined.

Corner boards are used with the various types of drop siding rather than mitering or using metal clips as in the case of bevel siding. The nails used are generally 8d common and they should be set and puttied before painting.

As with bevel siding, the quantity of drop siding required is obtained by deducting the window and door openings from the area and adding a percentage for waste. Table No. 13–1 shows the normal percentage to be added.

Vertical matched siding is the application of matched boards vertically. This is usually nailed over horizontal furring strips to provide good nailing. The waste in milling, cutting, and fitting for vertical siding depends on the width of the boards. Table 13–1 shows the approximate percentage to be added. Each board is toe-nailed with 6d

TABLE 13–1

Percentage of Milling and Cutting Waste, and Nails Required Per 1,000 FBM of Various Types of Wood Siding

Type of Siding	Nominal Size in Inches	Lap in Inches 1" lap	Pounds Nails Per 1,000 FBM	Percentage of Waste
Bevel Siding	1 × 4	1	25-6d common	63
	1 × 6	1	25-6d common	45
	1 × 8	1¼	20-8d common	35
	1 × 10	1½	20-8d common	30
Rustic and drop siding	1 × 4	Matched	40-8d common	33
	1 × 6	Matched	30-8d common	25
	1 × 8	Matched	25-8d common	20
Vertical siding	1 × 6	Matched	25-8d finish	20
	1 × 8	Matched	20-8d finish	18
	1 × 10	Matched	20-8d finish	15
*Batten siding	1 × 8	Rough	25	5
	1 × 10	Rough	20	5
	1 × 12	Rough	20	5
	1 × 8	Dressed	25	13
	1 × 10	Dressed	20	11
	1 × 12	Dressed	20	10
Plywood siding	¼	Sheets		5–10
	3/8	Sheets	15 per MSF	5–10
	5/8	Sheets		5–10

*For 1″ × 10″ boards allow 1,334 lineal feet 1″ × 2″ joint strips for each 1,000 FBM of batten siding. Add 12 pounds 8d common nails.

or 8d finishing nails. Wide boards which are inclined to cup, are surface nailed and the nails are set before finishing.

Batten siding is a type that is very common on farm buildings. Wide 1-inch, square-edge boards are placed vertically. The joints are covered with 1" × 2" strips that are nailed to only one of the vertical boards to permit them to expand and contact without splitting. The amount of waste to be added for milling will depend upon the kind of boards used; i.e., whether they are rough sawed, or square edge dressed. Waste for cutting and fitting will run about 5 to 10 percent. Rough sawed lumber has no milling waste. For milling and cutting waste for dressed square edge boards see Table 13–1.

Plywood siding is becoming more popular as an exterior finish on houses. The sheets of plywood should be a minimum of 3/8 inch thick where the walls are not sheathed. Where studding is spaced 20 or 24 inches on center thickness of 1/2 to 5/8 inch are recommended. Nailing is done with 6d or 8d nails depending on the thickness of the material. The only waste to be computed in figuring the quantity of plywood siding is what is lost in cutting and fitting. Approximately 5 to 10 percent cutting waste is adequate in most cases. The quantity of siding required is computed by measuring the over-all area and deducting the window and door openings.

UNIT COSTS FOR WOOD SIDING

The quantity of solid siding used to develop a unit cost is 1,000 FBM. The cost of the nails is added, and also the cost of labor, determined from Table 13–2 which shows the number of hours required to install 1,000 FBM.

TABLE 13–2

Hours of Labor Required to Install 1,000 FBM
of Various Types of Wood Siding

Type of Siding	Nominal Size in Inches	Hours Labor Per 1,000 FBM
Bevel siding (add 10% for mitered corners)	1 × 4	40
	1 × 6	38
	1 × 8	36
	1 × 10	34
Rustic and drop siding	1 × 4	32
	1 × 6	30
	1 × 8	28
Vertical matched siding	1 × 4	40
	1 × 6	36
	1 × 8	32
*Batten siding	1 × 8	22
	1 × 10	20
	1 × 12	20
Plywood siding	¼" thick	12 per 1,000 sq ft
	3/8" thick	14 per 1,000 sq ft
	5/8" thick	14 per 1,000 sq ft

*Allow 15 hours labor per 1,000 FBM of siding to apply joint strips when 1" × 10" boards are used.

Illustrative Examples

(Prices and Wages Assumed)
½ " × 8-Inch bevel siding

1,000 FBM 1" × 8" cedar siding	@ $	800.00 =	$ 800.00
20 lbs nails	@	.60 =	12.00
36 hours labor	@	12.00 =	432.00
	Unit cost per 1,000 FBM =		$ 1,244.00
	Unit cost per BF =		$ 1.24

6-Inch drop siding

1,000 FBM 1" × 6" fir drop siding	@ $	650.00 =	$ 650.00
25 lbs nails	@	.60 =	15.00
30 hours labor	@	12.00 =	360.00
	Unit cost per 1,000 FBM =		$ 1,025.00
	Unit cost per BF =		$ 1.03

1/2" × 8-Inch redwood siding

1,000 FBM 1/2" × 8" redwood siding	@ $ 1,000.00 =		$ 1,000.00
20 lbs nails	@	.60 =	12.00
36 hours labor	@	12.00 =	432.00
	Unit cost per 1,000 FBM =		$ 1,444.00
	Unit cost per BF =		$ 1.44

5/8-Inch plywood siding

1,000 Sq ft plywood siding	@ $	750.00 =	$ 750.00
20 lbs nails	@	.60 =	12.00
20 hours labor	@	12.00 =	240.00
	Unit per 1,000 sq ft =		$ 1,002.00
	Unit cost per sq ft =		$ 1.00

10-Inch batten siding

1,000 FBM 1" × 10" sq edge batten siding	@ $	500.00 =	$ 500.00
25 lbs nails	@	.60 =	15.00
1,334 lin ft 1" × 2" joint strips	@	.10 =	133.40
12 lbs nails	@	.60 =	7.20
20 hours labor for boards	@	12.00 =	240.00
15 hours labor for strips	@	12.00 =	180.00
	Unit cost per 1,000 FBM =		$ 1,075.60
	Unit cost per BF =		$ 1.08

Application of Unit Costs for Wood Siding

Assuming that the over-all dimensions of the windows in Figure 13–1 are 3′ × 6′, and that the door is 4′ × 7′; the net area of the two sides shown would be:

Front: 12′ × 34′	= 408 sq ft	
Less openings:		
2(3′ × 6′) = 36 sq ft		
4′ × 7′ = 28 sq ft	64 sq ft	
		344 sq ft
Gable end: 12′ × 22′	= 264 sq ft	
Plus gable 11′ × 8′	= 88 sq ft	
	352 sq ft	
Less Openings:		
3(3′ × 6′)	= 54 sq ft	

Total net area, end and one side = $\dfrac{298}{642}$ sq ft

Figure 13–1

Using the unit costs previously developed, the cost of installing wood siding on the two sides of this house would be:

	8-Inch bevel siding
Area	= 642 sq ft
Add waste (Table 13–1) 35%	= 225 sq ft
Amount of siding required	867 FBM
Unit cost @	$ 1.24
Total cost	= $1,075.08

	6-Inch drop siding
Area	= 642 sq ft
Add waste (Table 13–1) 25%	= 161 sq ft
Amount of siding required	= 803 FBM
Unit cost @	$ 1.03
Total cost	= $827.09

	5/8-Inch Plywood siding
Area	= 642 sq ft
Add waste 10%	= 64 sq ft
Amount of siding required	= 706 sq ft
Unit cost @	$ 1.00
Total cost	= $706.00

	½ × 8-Inch Redwood siding
Area	= 642 sq ft
Add waste (Table 13–1) 35%	= 225 sq ft
Amount of siding required	= 867 FBM
Unit cost @	$ 1.44
Total cost	= $1,248.48

	10-Inch Batten siding
Area	= 642 sq ft
Add waste (Table 13–1) 20%	= 128 sq ft
Amount of siding required	= 770 FBM
Unit cost @	$ 1.08
Total cost	= $831.60

Wood Shingle Siding*

Wood shingles are a very popular type of finish for both exterior side walls on frame buildings, and as a roofing surface. In congested areas of many of the larger cities, wood shingle roofing is prohibited by ordinance because of its combustible nature and the possibility of the spreading of a fire by embers, from one residence to another. However, in most suburban and rural areas they are widely used because of their long life (25 to 35 years) and architectural attractiveness.

Wood shingles are manufactured from several kinds of wood, the best of which are western red cedar, redwood and cypress. All of these have a high decay resistance, low shrinkage, and a tendency to lie flat on the surface. The better grades of shingles are of clear verticle (edge) grain wood that consists entirely of heartwood with no defects. The intermediate and the economical grades are clear on the lower 6 inches but may contain defects such as knots, feather tips, flat grains, or some sapwood. These are recommended more for farm buildings, sidewalls of garages, summer homes, and secondary buildings. They are also used for undercourses where double coursing of side walls is contemplated. Grading rules of wood shingles, as with all lumber products, have been formulated by the Bureau of Standards of the U.S. Department of Commerce in cooperation with the industry.

The three standard lengths of wood shingles are 16-inch, 18-inch, and 24-inch. The standard exposure for 16-inch shingles is 5 inches; for 18-inch it is 5 1/2 inches; and for 14-inch, 7 1/2 inches. There are four bundles of shingles to a square. The Red Cedar Shingle Bureau recommends for roofs with a pitch less than 5/24ths, that the shingle exposure be reduced to 3 3/4 inches, 4 1/4 inches and 5 3/4 inches respectively for 16-, 18-, and 24-inch shingles. On all roofs, there should be a minimum of *three* layers of wood at every point to prevent leakage in wind-driven rain storms. On side walls, two layers of wood shingle at every point are considered adequate.

The standard thickness of four butts of 16- and 24-inch shingles is 2 inches. The standard thickness of four butts of 18-inch shingles is 2 1/4 inches.

Table 13–3 shows the covering capacities, for Certigrade Red Cedar Shingles. Since the manufacturer specifies 5, 5 1/2 and 7 1/2 inches respectively for the exposure to apply one square of shingles, any variation of the exposure so specified will result in a square of shingles covering more, or less, than 100 square feet of surface. For example, a 16-inch shingle is designed for a 5-inch exposure and will cover 100 square feet (page 254).

---

*Handsplit shakes are discussed in Chapter 14, Roofing.

TABLE 13-3

Percentage of a Square of Shingles Required to Cover
100 Square Feet for Exposures Shown

(1) Shingle Length	(2) Exposure In Inches	(3) Percentage of 1 Square of Shingles For Each 100 Square Feet of Area
16 Inch	4	125
	*5	100
	6	84
	7	72
	**7½	67
18 Inch	4	123
	*5½	100
	6	92
	6½	85
	7	79
	8	70
	**8½	65
24 Inch	6	125
	*7½	100
	8	94
	9	84
	10	75
	11	69
	**11½	66

*Maximum exposure recommended for roofs by the Red Cedar Shingle & Handsplit Shake Bureau.
**Maximum exposure recommended for side walls by the Red Cedar Shingle & Handsplit Shake Bureau.

When exposure for a 16-inch shingle is increased to 7 1/2 inches one square will cover 150 square feet. Putting it another way, if an exposure of 7 1/2 inches is used, the quantity of shingles needed to cover 100 square feet will be 100/150=67% of a square. It becomes very important, therefore, when estimating the wood shingles required, to make certain of the exposure to be used for the particular length shingle. *Both the quantity of material and the required labor to cover 100 square feet decrease as the exposure is increased and vice versa.*

The length and type nail used to apply wood shingles is important, not more than two nails being used to a shingle. Nails must be rust-resistant, and may be zinc-coated or aluminum. For new work the 3d nail is used on 16- and 18-inch shingles, while the 4d nail is recommended for the 24-inch shingle. When new shingles are being applied over old shingles, the 5d or 6d nails are used. For double coursing on side walls the 5d nail is considered best. If shingles are applied on side walls directly over composition sheathing without nailing strips, a special self-clinching nail should be used in accordance with the specifications of the manufacturer.

Shingles are available in either natural wood or stained for roofs, and painted a prime coat for side walls.

APPLYING WOOD SHINGLES TO SIDE WALLS

Wood shingles can be applied on side walls over open sheathing, solid sheathing, or composition types of sheathing.

Open, or spaced sheathing, on centers equal to the exposure used, is common in the warm climates and on seasonal bungalows or farm buildings. The strips may be 1" × 2", 1" × 3", or 1" × 4" nailed to the studs. Saturated felt building paper is first applied between the nailing strips and the studs.

When side wall sheathing is solid boarding, or plywood, a single layer of saturated felt paper is applied before the shingles are put in place.

Composition sheathing requires either stripping, as with open sheathing, or the use of clincher nails if the sheathing is designed for their use. Where composition sheathing has a waterproof surface, it is unnecessary to apply saturated felt paper under the shingles.

There are two general methods of side wall application. The *Single Coursing* and the *Double Coursing*. Because the sides of a building are less exposed to the direct action of rain, snow, and sun, the shingle exposure can be considerably greater (Table 13–3) than on roofs.

In Single Coursing the first or lowest course is doubled, and succeeding courses are applied above, using a tacked lath as a guide for the butts or by striking a chalk-line.

Double Coursing is principally used where a deep shadow line is desired for architectural appeal. Considerably greater exposures are possible with this form of application and, for the under course, a lower grade shingle is used. The starting course should be tripled. The outer course shingles are applied about 1/2 inch lower than the under course. In other words the butt of the outer shingle is about 1/2 inch below the butt of the under shingle.

Where shingles join on an outside corner, they may be mitered or they may be laced by one overlapping the other. Inside corners are finished by jointing the shingles up against a vertical strip which is installed prior to shingling. Mitering of the outside corners takes a little more time than lacing, making the labor cost slightly higher.

Side wall shingles can be applied over old shingles, and also over old stucco or brick, provided horizontal furring or nailing strips are securely attached on centers equal to the shingle exposure.

Estimating Material

The quantity of shingles needed for the side walls of a building is determined by measuring the actual area to be covered, and deducting all window and door openings to obtain the net area in square feet. By consulting Table 13–3 the portion of a square of shingles needed to cover each 100 square feet can be determined for the corresponding exposure. The quantity should be carried out to the next 1/4 of a square (bundle), as shingles are sold four bundles to the square.

TABLE 13-4

Covering Capacities and Approximate Nail Requirements of Certigrade Red Cedar Shingles

Shingle Exposure in Inches	No. 1 GRADE SIXTEEN INCH SHINGLES NAIL SIZE: 3d, 1¼-inch long				No. 1 GRADE EIGHTEEN INCH SHINGLES NAIL SIZE: 3d, 1¼-inch long				No. 1 GRADE TWENTY-FOUR INCH SHINGLES NAIL SIZE: 4d, 1½-inch long			
	Four-Bundle Square		One Bundle		Four-Bundle Square		One Bundle		Four-Bundle Square		One Bundle	
	Coverage in Sq. Ft.	Pounds Nails	Coverage in Sq. Ft.	Pounds Nails	Coverage in Sq. Ft.	Pounds Nails	Coverage in Sq. Ft.	Pounds Nails	Coverage in Sq. Ft.	Pounds Nails	Coverage in Sq. Ft.	Pounds Nails
3½	70	27/8	17½	¾								
4	80	2½	20	5/8	72½	2½	18	5/8				
4½	90	2¼	22½	5/8	81½	2¼	20	5/8				
5	100*	2	25	½	90½	2	22½	½				
5½	110	1¾	27½	½	100*	1¾	25	½				
6	120	1½	30	3/8	109	1½	27	3/8	80	2½	20	5/8
6½	130	1½	32½	3/8	118	1½	29½	3/8	86½	21/8	21½	½
7	140	12/5	35	⅓	127	12/5	31½	3/8	93	2	23	½
7½	‡150	1⅓	37½	⅓	136	1⅓	34	⅓	100*	17/8	25	½
8	160		40		145½	1¼	36	⅓	106½	1¾	26½	½
8½	170		42½		‡154½	1¼	38½*	⅓	113	1¾	28	½
9	180		45		163½		40½		120	1½	30	3/8
9½	190		47½		172½		43		126½	1½	31½	3/8
10	200		50		181½		45		133	1½	33	3/8
10½	210		52½		191		47½		140	1½	35	⅓
11	220		55		200		50		146½	1¼	36½	⅓
11½	230		57½		209		52		‡153	1¼	38	⅓
12	‡240		60		218		54½		160		40	
12½					227		56½		166½		41½	
13					236		59		173		43	
13½					245½		61		180		45	
14					‡254½		63½		186½		46½	
14½									193		48	
15									200		50	
15½									206½		51½	
16									‡213		53	

* Maximum exposure recommended for roofs.
† Maximum exposure recommended for single-coursing on side walls.
‡ Maximum exposure recommended for double-coursing on side walls. Figures in italics are inserted for convenience in estimating quantity of shingles needed for wide exposures in double-coursing, with butt-nailing. In double-coursing, with any exposure chosen, the figures indicate the amount of shingles for the outer courses. Order an equivalent number of shingles for concealed courses. Approximately 1½ lbs. 5d small-headed nails required per square (100 sq. ft. wall area) to apply outer course of 16-inch shingles at 12-inch weather exposure. Plus ½ lb. 3d nails for under course shingles. Figure slightly fewer nails for 18-inch shingles at 14-inch exposure.

From *Certigrade Handbook of Red Cedar Shingles*, published by Red Cedar Shingle Bureau (Red Cedar Shingle and Handsplit Shake Bureau), Seattle, Washington, 1957.

The normal waste allowance for starting courses and for cutting and fitting around windows and doors is 10 percent.

Illustrative Example

To obtain the number of squares of 18 inch shingles required to cover a measured area, using an 8½ inch exposure.

Measured Area: 20' × 30'		600 sq ft
Deduct 2 wds. (3' × 5') 30 sq ft		
Deduct 1 dr. 3' × 7' 21 sq ft		51 sq ft
		549 sq ft
Add 10% waste		55
		604 sq ft

Table 13–3 shows that it takes 65% of a square of 18-inch shingles laid with an 8½ inch exposure to cover 100 square feet.

.65 × 604 = 4.0 squares required

The number of squares of shingles is obtained by multiplying the net area to be covered by the percentage shown in column No. 3, Table 13–3.

Labor for Wood Shingles on Sidewalls

Under average conditions a *carpenter* can apply wood shingles to sidewalls at the rate of about 5 hours per square for 16- and 18-inch shingles, and 4 hours per square for 24-inch shingles. At this rate, and using the exposures recommended by the manufacturer, he will cover 100 square feet of wall area. When conditions are encountered that affect his rate of application, the hours per square may be adjusted accordingly. For example, specifications may call for staggering the butts, or there may be an unusual number of openings, jogs or setbacks to cut and fit around. These factors should be given consideration in the labor rate.

Experienced shinglers can apply shingles as much as 25 percent faster than carpenters who are not accustomed to that type of work. When experienced shinglers do the work, the labor rate may be reduced.

Unit Costs for Wood Shingles on Sidewalls

Based on the labor rates given, unit costs for shingles may be developed as shown in the following illustrations. Prices and wages are assumed.

*Unit Cost for 16- and 18-Inch Shingles on
Sidewalls with 5 and 5½ Inch Exposures*

1 square shingles	@ $60.00 =	$ 60.00
2 lbs nails	@ .60 =	1.20
5 hours carpenter labor	@ 12.00 =	60.00
	Unit cost per square =	$121.20
	Unit cost per sq ft =	$ 1.21

Unit Cost for 24-Inch Shingles on
Sidewalls with 7½ Inch Exposure

1 square shingles	@ $70.00 =	$ 70.00
2 lbs nails	@ .60 =	1.20
4 hours carpenter labor	@ 12.00 =	48.00
	Unit cost per square =	$119.00
	Unit cost per sq ft =	$ 1.19

The illustrated unit cost may be applied to the number of square feet of wall area to be covered only when the manufacturer's specified exposure is being used; otherwise the following rules should be noted.

1. When the developed *unit cost per square* is used, the proper number of squares required must be determined from Table 13–3 as illustrated under "Estimating Materials."

2. When the developed *unit cost per square foot* is used, the *unit cost* is multiplied by the factor in Table 13–3 corresponding to the exposure. This adjusted unit cost per square foot may then be applied to the square feet of wall area including the waste to be added.

Illustrative Example

The wall area to be covered, including waste, is 2,000 square feet. The shingles are 18-inch with an 8-inch exposure. Table 13–3 column 3 shows that 70 per cent of a square of 18-inch shingles laid with an 8-inch exposure will cover 100 square feet.

Therefore (.70 × 2,000 sq ft) 1,400 sq ft or 14 squares are needed. The unit cost per square developed is $100.00 which, multiplied by 14 square, is $1400.00.

If it is desired to express the unit cost in terms of cents per square foot, the developed unit cost, for standard exposure, of $100.00 per square is $1.00 per square foot. This is corrected from Table 13–3 by multiplying the unit cost per square foot by 70 per cent. This adjusted unit cost per square foot may now be applied to the 2,000 square foot area.

$$.70 \times 1.00 = .70 \times 2,000 = 1,400.00$$

COMPOSITION SIDING

There are various kinds of composition siding. The most commonly used are asphalt shingles, asbestos-cement shingle siding, and rolls of asphalt and felt similar to roll-roofing, manufactured in imitation brick and other patterns. The rigid asbestos-cement siding is used on many new homes. The asphalt shingles and rolls are found mostly on older buildings where it is desired to cover worn siding that has served its life.

Asbestos-cement Siding

(See also Roofing—Chapter 14)

Asbestos siding, shingles or board, is made by combining asbestos fibers and portland cement under pressure. The material is fire resistive, rigid, and comes in several colors. It may also be painted with a special type of paint.

Asbestos-cement shingles are generally made in a rectangular shape, 12 inches wide and 24 inches long. They are put on like other shingles, lapping 1 1/2 inches. This type of siding is sold by the square. Each square contains three bundles of 57 pieces each. The weight of a square is approximately 180 pounds.

The shingles are applied over solid wood or plywood sheathing and saturated felt paper, or over composition sheathing. Galvanized nails, which are furnished by the manufacturer, come in different lengths. Barbed nails are used to provide better holding in old wood. About 2 pounds of nails will take care of a square of siding. When applied over composition sheathing, patented fasteners are supplied by the manufacturer. A special tool is supplied for cutting, and for punching holes in asbestos siding.

The quantity of asbestos-cement shingles required is computed by measuring the area, deducting window and door openings, and adding 10 percent waste for cutting and fitting around windows, doors, and corners.

The hours of labor required to apply a square of asbestos-cement siding varies with conditions, but for average straight work where there is a moderate amount of cutting and fitting, two men should be able to apply a square in about 2 hours or a total of 4 hours for a square. When working on the second floor of a two-story building from a scaffold, the time required should be increased to about 6 hours a square.

Illustrative Example

Assuming asbestos-cement shingles cost $65.00 per square, nails $.60 per pound, and wages are $12.00 per hour.

1 square shingles	@ $65.00 =	$ 65.00
2 lbs nails	@ .60 =	1.20
4 hours labor	@ 12.00 =	48.00
	Unit cost per square =	$114.20
	Unit cost per sq ft =	$ 1.14

Asphalt Shingle Siding

Asphalt shingles are frequently applied to the exterior of frame buildings, generally over old siding. A more detailed discussion of the characteristics of asphalt shingles, and also asphalt roll siding will be found in Chapter 14 on "Roofing."

Asphalt shingles are applied over saturated felt paper on either sheathing or old wood siding. It requires about 2 pounds of asphalt shingle nails per square. It takes more shingles per 100 square feet than when they are applied as roofing. The reason is that, on side walls, the exposure used is generally less than specified by the manufacturer. It varies between 3 and 4 inches. If the shingles are manufactured for an exposure of 5 inches, then one square will cover exactly 100 square feet at that exposure, because there are 80 shingles to the square and each shingle is 36 inches long.

$$\frac{5'' \times 36''}{144} = \frac{180}{144} \times 80 = 100 \text{ sq ft}$$

Asphalt shingles manufactured for a 4-inch exposure have 100 shingles to the square and cover exactly 100 square feet.

$$\frac{4'' \times 36''}{144} = \frac{144}{144} \times 100 = 100 \text{ sq ft}$$

By reducing the specified exposure when applying a 4-inch or 5-inch shingle, the area that a square will cover is proportionately reduced. The percentage to be added to the area to be covered is shown in Table 13–5.

TABLE 13–5

Percentage to Be Added to Area to Obtain the Quantity
Required for 36-Inch Strip Shingles

Exposure When Applied	4-Inch Exposure Specified by Manufacturer	5-Inch Exposure Specified by Manufacturer
3	.34	.67
3½	.15	.43
4	—	.25
4½	—	.11
5	—	—

Illustrative Example

If the net area to be covered is 3,000 square feet and the shingle used has a specified exposure of 5 inches, when applied with an exposure of 3½ inches, the quantity needed to cover an area 3,000 square feet is:

Area	3,000 sq ft
Add 43%	1,290 sq ft
	4,290 sq ft

It is customary to add a percentage to the quantity of shingles computed, to allow for waste in cutting and fitting around windows and doors, and also for the bottom starting course which has two layers of shingles. The percentage varies with each job and depends on the number of openings, corners and so forth. Customarily 10 percent is considered adequate. As with other types of siding, the areas of windows and doors are deducted from the gross area.

The hours of labor to apply asphalt shingles to side walls vary, but a good average rate is 3 hours per square.

Illustrative Example

(Wages and Prices Assumed)

1 square asphalt shingles	@ $20.00 =	$20.00
2 lbs nails	@ .60 =	1.20
3 hours labor	@ 12.00 =	36.00
	Unit cost per square =	$57.20
	Unit cost per sq ft =	$ .57

Whatever unit cost is developed, is applied to the quantity required as adjusted from Table 13.5.

Asphalt Roll Brick Siding

Rolls of asphalt siding come in different weights, and patterns. The most common is the brick pattern which weighs about 105 pounds to the roll. A roll is 31 inches wide and 43 feet long, and contains 111 square feet. It is intended to cover one square with approximately 11 percent waste for lapping. The nails and asphalt adhesive for joining edges are supplied inside each roll.

The amount of material needed to cover an area is determined by computing the number of square feet in the overall area, deducting the window and door openings, and adding a percentage for waste in cutting and fitting. While 10 percent is considered adequate for ordinary straight work, each individual job should be examined to judge best the proper amount of waste to be added.

The hours of labor necessary to apply a roll of brick siding varies but under average working conditions two hours per square is considered adequate.

Illustrative Example

(Wages and Prices Assumed)

1 roll siding	@ $13.00 =	$13.00
2 hours labor	@ 12.00 =	24.00
	Unit cost per square =	$37.00
	Unit cost per sq ft =	$ .37

ALUMINUM SIDING

Aluminum siding is sold in panels or strips 8″ wide, or double 4″-widths for a narrow clapboard effect. The length of the panels is generally 12 1/2 feet. They may be backed with an insulating material, or unbacked. Panels are factory-finished in acrylic enamel modified with silicones to resist fading. Color selection varies with the manufacturer. Color-matching nails, inside and outside corners, window trim, and moldings are available. Metal thicknesses vary, but generally .024 is considered standard. A slightly curved panel is designed for vertical application.

Aluminum shingles or shakes, though primarily designed for roofing, are sometimes used as exterior siding. (See Roofing, Chapter 14.)

Labor to apply aluminum panel siding horizontally, once the surface has been prepared, will range between 25 and 30 hours per 1,000 square feet. When it is applied vertically the labor will range between 30 and 35 hours per 1,000 square feet. A unit cost, excluding any corner and trim accessories, may be developed as follows, using the assumed price and wages:

1,000 sq ft insulated aluminum siding	@ $800.00 =	$ 800.00
25 hours labor to apply	@ 12.00 =	300.00
	Cost per 1,000 sq ft =	$1,100.00
	Cost per sq ft =	$ 1.10

VINYL SIDING

Solid vinyl siding is made of extruded polyvinyl Chloride (PVC). It is produced in several colors that are *solid* all through the material, and is available with a full line of color-matching accessories. The surface finishes are usually natural wood grain texture.

Widths are 8" to simulate the wider clapboards; double 4" to simulate the narrow clapboards; double 6" to provide more coverage per panel; and triple 4" vertical solid vinyl siding. Panels are 12' 6" in length, except triple 4" which is approximately 10 feet long. There are 8 pieces (100 linear feet) per square; 16 pieces to the carton of 2 squares. The approximate weight per square is 50-lbs or 1/2 lb per square foot.

Vinyl siding and accessories are similar in many respects to aluminum siding in both shapes and in application procedures. A unit cost, excluding necessary corners and trim accessories, may be developed as follows, using the assumed price and wages:

1,000 sq ft solid vinyl siding	@ $650.00 =	$650.00
25 hours labor to apply horizontally	@ 12.00 =	300.00
	Cost per 1,000 sq ft =	$950.00
	Cost per sq ft =	$ .95

HARDBOARD LAP SIDING

This is a composition of great density for use on the exterior. It is 7/16" × 12" × 16' long and is delivered to the job prime-coated.

Application is the same as for wood bevel siding using metal corner clips. Lapping should be at least 1 1/2 inches which, on a width of 12 inches would produce an area waste of 12 1/2 percent. The overall area plus cutting and fitting waste will average 20 percent. In other words to cover 100 square feet, 120 square feet of hardboard lap siding is required.

A unit cost would be developed as follows:

Illustrative Example

(Prices and Wages Assumed)

1,000 sq ft hardboard siding	@ $400.00 =	$400.00
20 lbs nails	@ .60 =	12.00
25 hours labor	@ 12.00 =	300.00
	Unit cost per 1,000 sq ft =	$712.00
	Unit cost per sq ft =	$ .71
	(Add cost of aluminum metal corner clips)	

CHAPTER 14

Roofing

The material that is applied to the surface of a roof to make it waterproof and tight against the weather is called *roofing*. There is a variety of material used for this purpose, but well over 75 percent consists of asphalt roofing products. These are in the form of built-up roofing, roll roofing, and shingles. Other materials that are commonly used for shingles are wood, slate, tile, and aluminum. Flat sheet or corrugated aluminum, and galvanized iron roofing are frequently applied to the commercial, industrial, and rural structures. Corrugated glass-fiber panel roofing has become popular for patios, garages, and structures where the admission of light is desired. Asbestos-cement shingles are also used.

QUALITY OF ROOFING

In estimating, the designation of a particular roofing material by type is not sufficient to determine its cost. There are different grades of each type and the identification of the quality, thickness, or size is equally as important as the kind of material. Asphalt shingles, roll roofing, and felts are made in several weights. Wood and slate shingles are sold in different sizes and grades. It makes considerable difference whether a built-up roof is 3-ply or 5-ply, and whether it has a smooth surface, slag, or crushed marble. Spanish tile may be terra cotta, or it may be made of colored cement to simulate terra cotta. Metal roofs, either corrugated or sheet metal, are made in several thicknesses or gauges, and the cost of the material varies accordingly.

In many instances the quality of a particular kind of roofing material is not easily determined, and it may be necessary to remove a sample to have it identified by a supplier. A properly prepared estimate of the cost of roofing should clearly indicate the type and quality of the particular roofing material.

FACTORS AFFECTING LABOR

While all of the factors which affect labor in general, as discussed in Chapter 2, should be given consideration in estimating the cost of roofing, the following are of special importance:

1. Are the workers experienced in applying the particular type of roofing?
2. Is the roof surface of irregular shape, or is it cut up with dormers and intersecting roofs?

3. Is the slope of the roof steep enough to require foot scaffolds, or is it low pitched?
4. Is it necessary to erect scaffolding?
5. Will it be necessary to hoist materials by hand or mechanically? Will materials have to be carried up to the roof from the inside of the building by stairs, ladders, or elevators?
6. Is it a patch job which requires matching in the new roofing at the perimeter of the damaged area?
7. Do the old shingles have to be removed, and does the roof decking have to be repaired, or prepared, for the new roofing?

Experience of Workers

Some of the experienced roofers, who apply asphalt strip shingles every day, can put them on at the rate of about one square an hour. The average roofer can apply them at the rate of 1 1/2 to 2 hours per square on an ordinary roof without too many irregularities. But a carpenter unaccustomed to such work may take as long as 3 hours per square. Roofers whose trade is applying asbestos-cement shingles work much more rapidly than a novice at the business who requires time to learn the art of cutting, punching nail-holes, and fitting the material into place. The experience of those who will apply the roofing should be carefully considered.

Shape of Roof

The presence of dormers, intersecting gables, and other architectural features that create a cut-up roof surface will slow down a roofer's rate of applying roofing; no matter how experienced, it takes additional time to cut and fit around the valleys, hips, dormers and eaves.

Applying shingles to a Gothic roof, or to the steep sides of a Gambrel or Mansard roof, takes more hours per square than putting them on a building with a low-pitched roof.

Slope of Roof

On roofs that have a steep incline, the roofer generally sets up foot scaffolds to keep from slipping. The rate of work is measurably reduced due to lack of freedom of movement, and care must be exercised to keep materials at hand and maintain position. The steeper the slope, the slower the work.

Scaffolding

There is usually no need to erect scaffolding on one and two-story buildings, particularly if the roof is not too steep. Sometimes scaffold brackets are used on extension ladders. After the first several rows of shingles are laid, a foot or roof scaffold is used. It consists of a 2″ × 4″ laid on the roof and supported at each end by stirrups, or by wire loops taken from the shingle bundles. These are nailed to the roof at a point that will be covered by the shingles. As the roofer moves up toward the ridge, the position of the foot scaffold is shifted by slipping the stirrup, or wire loop, off the nail. The nail which held the scaffold support is then driven the rest of the way in and covered by the shingle.

When conditions require the building of a light scaffold from the ground to the eaves, allowance must be made for its cost, less any anticipated salvage. (See "Scaffolding," Chapter 10.) The cost of the scaffolding should not be included in the *unit cost* of the roofing as it distorts the true cost of application.

Hoisting

Hoisting is an item to be considered, especially in applying built-up roofs when rolls of roofing, asphalt or gravel must be raised from the ground to the area that is being worked on. Circumstances of each job will indicate what additional allowances are to be made for the cost of hoisting materials. In the case of built-up roofs the hoisting charges are included in the unit cost inasmuch as it is a continuous operation during the application of the material.

Patch-roofing

The patching of roofs, either over a small area or spotting-in shingles here and there, cannot be estimated on a unit cost basis. The most reliable method is to estimate the overall time required, and determine the amount of material it will take to cover the area.

When estimating the labor, ample time should be allowed for lacing-in and matching-in the roofing, particularly shingles, where the new and old join. On shingle roofs of all types such work is time-consuming.

Where the patch area is in excess of a couple of squares, depending on the conditions, unit costs may be used provided the extra labor is added for joining the new work with the old.

Removing Old Roofing

The cost of removing the old roofing, prior to applying the new, should be estimated as a separate item since it consists only of labor and trucking away from the premises. Some builders and roofers compute the cost to remove the old roofing, and then reduce it to a cost per square or square foot. This cost is then added to make a total unit cost for removing and replacing the roofing.

LABOR RATES TO APPLY ROOFING

Table 14–5 (page 292) shows average hours of labor to remove and install various types of roofing. The rates are based on the average roof surface of flat to medium pitch. The figures in the table contemplate that the roof may have a dormer or two but is not considered cut-up. The rates are for the average worker who is experienced although not necessarily specializing as a roofer. Adjustments in these labor rates are in order where substantial variations from the preceding conditions are encountered. In the majority of cases, however, the rates shown will be found adequate.

MEASUREMENT OF ROOF AREAS

The unit of measurement for roofing is the *square*, which is an area $10' \times 10'$, or 100 square feet. The quantity of roofing material needed is obtained from the measured area of the roof surface.

The method of obtaining measurements of any roof depends on the shape, and whether or not conditions permit access to it so that the measurements can be taken in the field, as in the case of a flat or low-pitched one-story roof. When it is not possible to get up on the roof, there are several methods used to obtain the necessary dimensions for computing area. Horizontal measurements may be taken from the ground, adding the cornice overhang at the gable or eaves. The distance from the ridge to the eaves, sometimes called the *rafter length*, is frequently obtained by counting the number of shingles from the lower edge of the roof to ridge, and multiplying by the exposure in inches. The product is divided by 12 to obtain the lineal feet.

Illustrative Example

Rows of shingles	36
Exposure in inches	$\times 4\frac{3}{4}$
	171 inches

$$\text{Rafter length} = \frac{171}{12} = 14'3''$$

It is not always easy to estimate the vertical height of roofs and sometimes it is impossible. The following suggestions may be helpful:

1. If the gable end is finished in shingles, brick, horizontal siding, concrete block, or other materials of uniform courses, the number of courses may be counted and multiplied by the number of inches in one course. The product is divided by 12 to obtain the height in feet.

Illustrative Examples

Common brick courses	36
Inches per course	$2\frac{3}{4}$
	99 inches

$$\text{Vertical height} = \frac{99}{12} = 8'3''$$

2. When the attic is accessible, the vertical height may be measured directly from the interior after establishing any difference in level between the attic floor and the eave line.
3. In some instances the vertical height can be obtained by leaning out of a window in a gable end and measuring down to the eave line, and up to the ridge.
4. Actual measurements frequently can be obtained from inside a garage, barn, shed, or other structure, where rafters and the interior of the roof structure are exposed to view and can be reached.

5. When a structure has been badly damaged by fire or windstorm, and enough of the roof, trusses etc., are available in the debris, measurements of vertical heights may be obtained directly.

6. In buildings of masonry, the wood roof may be completely destroyed, but enough of the outline may remain on the walls to determine its shape and measurements.

7. When a building is completely destroyed, the data for computing the roof area is obtained from existing plans, pictures, and a full description from the occupant or owner as to the number of floors, average ceiling heights, existence of an attic, and shape of the roof.

SPAN, RISE AND PITCH

Roof Pitch—The pitch of a roof is the slope expressed as a ratio of the rise to the span. The rise is the vertical distance between the ridge and the supporting plates. The span is the distance between the sidewalls that support the roof.

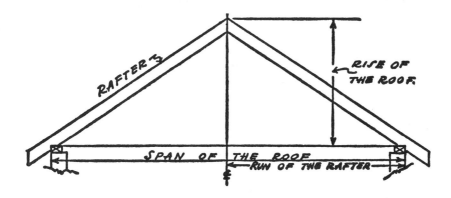

Illustrative Example

Where the rise of a roof is 7 feet and the span is 21 feet.

$$\text{Pitch} = \frac{\text{Rise}}{\text{Span}}$$

$$\text{Pitch} = \frac{7}{21} = \frac{1}{3}$$

When the pitch and the span are known, the rise can be obtained by the following formula:

$$\text{Rise} = \text{Pitch} \times \text{Span}$$

When the pitch and the rise are known, the span is obtained by the following formula:

$$\text{Span} = \frac{\text{Rise}}{\text{Pitch}}$$

Measurement of Pitch From Ground

The pitch of a roof can be fairly well determined from the ground using a carpenter's 6-foot folding rule. The method is shown below, courtesy of the Asphalt Roofing Manufacturers Association, and reproduced from their 1974 Manual.

Pitch - The span, rise and run of a simple gable roof are shown in Fig. 87. The pitch or slope of the roof is most often stated as the relation between the rise and the span. If the span is 24'0" and the rise is 8'0" the pitch will be 8/24 or 1/3. If the rise were 6'0" then the pitch would be 6/24 or 1/4. The 1/3 pitch roof rises 8" per foot of horizontal run, and the 1/4 pitch roof rises 6" per foot of run.

Fig. 87 - Pitch relations

It is possible to determine the pitch of any roof without leaving the ground by using a carpenter's folding rule in the following manner.

Form a triangle with the rule. Stand across the street or road from the building, and holding the rule at arms length, align the roof slope with the sides of the rule, being sure that the base of the triangle is held horizontal. It will appear within the triangle as in Fig. 88. Take a reading on the base section of the rule (note the "reading point", Fig. 88). Then locate on Fig. 89 in the top line headed "Rule Reading"

Fig. 88 - Use of Carpenters' Rules to Find Roof Pitch.

the point nearest your reading. Below this point will be found the pitch and the rise per foot of run. In the case illustrated the reading is 22. Under the figure 22 in Fig. 89, the pitch is designated as 1/3, or a rise of 8" per foot of horizontal run.

RULE READING	$20\frac{1}{2}$	$20\frac{7}{8}$	$21\frac{1}{4}$	$21\frac{5}{8}$	22	$22\frac{3}{8}$	$22\frac{3}{4}$	$23\frac{1}{16}$	$23\frac{3}{8}$	$23\frac{5}{8}$	$23\frac{13}{16}$	$23\frac{15}{16}$
PITCH FRACTIONS	$\frac{1}{2}$	$\frac{11}{24}$	$\frac{5}{12}$	$\frac{3}{8}$	$\frac{1}{3}$	$\frac{7}{24}$	$\frac{1}{4}$	$\frac{5}{24}$	$\frac{1}{6}$	$\frac{1}{8}$	$\frac{1}{12}$	$\frac{1}{24}$
RISE-INCHES PER FT.	12	11	10	9	8	7	6	5	4	3	2	1

Fig. 89 - Reading Point converted to pitch.

Measurement of roof pitch from the ground using a carpenter's 6-foot rule. Courtesy Asphalt Roofing Manufacturers Association, New York, N.Y.

GABLE ROOF AREAS

The area of a gable roof may be obtained by determining the *slope* length from the ridge down to the lower edge or eave line including the overhang. This measurement is multiplied by the horizontal length of the roof. When the rafter length cannot be physically measured, two methods for computing the length are suggested.

Method 1. The rafter length is equal to the square root of the sum of the square of the run and the square of the rise.

Illustrative Example

In Figure 14–1, the run is 14 ft and the rise is 8 ft.

$$\text{Rafter length} = \sqrt{14^2 + 8^2}$$
$$= \sqrt{196 + 64}$$
$$= \sqrt{260}$$
$$= 16.12 \text{ ft}$$

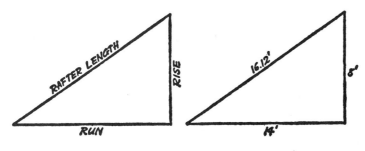

Figure 14–1

Method 2: When the rise and run are known, the rise-per-foot-of-run is obtained by dividing the rise in inches by the run in feet. Table 14–1 shows the factor for rise-per-foot-of-run by which the run is multiplied to obtain the rafter length.

Illustrative Example

In Figure 14–1, the rise is 96 inches (8′ × 12″), and the run is 14 ft.

$$\text{Rise-per-foot-of-run} = \frac{96''}{14'} = 6.85 \text{ inches}$$

The closest factor shown in Table 14–1 for a rise-per-foot-of-run of 6.85 inches is 1.160. (page 270)

$$\text{Rafter length} = 1.160 \times 14 = 16.24 \text{ ft}$$

When the rafter length has been obtained by methods 1 or 2, it should be carried to the next half or full foot. The area of the roof is then computed by multiplying the rafter length, plus any overhang, by the horizontal length of the roof. Gabled dormers with the same pitch as the main roof, except for their overhang, may be ignored for all practical purposes as the *dormer roof area is the same as that taken up by the dormer in the main roof.*

TABLE 14–1

Factor by Which Run Is Multiplied to Obtain Rafter Length When
Rise-per-Foot-of-Run is Known

No. Inches Rise-per-foot-of-run	*Pitch	Rafter length in inches per foot of run	To get rafter length multiply run in ft. by
3.0	1/8	12.37 ÷ 12 =	1.030
3.5			1.045
4.0	1/6	12.65	1.060
4.5			1.075
5.0	5/24	13.00	1.090
5.5			1.105
6.0	1/4	13.42	1.120
6.5			1.140
7.0	7/24	13.89	1.160
7.5			1.185
8.0	1/3	14.42	1.201
8.5			1.230
9.0	3/8	15.00	1.250
9.5			1.280
10.0	5/12	15.62	1.301
10.5			1.335
11.0	11/24	16.28	1.360
11.5			1.390
12.0	1/2	16.97	1.420

* Pitch is determined by dividing the inches rise-per-foot-of-run by 24.
(Most roofs of dwellings are constructed with a pitch of 1/8, 1/6, 1/4,
1/3 or 1/2.)

RAFTER LENGTHS FROM TABLES

Table 14–2 gives the length of rafters for gable roofs with the rise (height) and span (width) shown. Any overhang must be added.

Interpolating is necessary where the width under consideration falls between those given in the table. Heights to the nearest 6 inches should be used in using the table.

Illustrative Example

To find the rafter length where the height or rise is 8'0" and the width of the building is 31'0".

Rafter length for 8'0" and 30'0" width = 17'0"
Rafter length for 8'0" and 32'0" width = 17'10"

Use the 30'0" width and add 1/2 the difference between the 30' and 32' widths = 17'5". Add any overhang and use next nearest even foot.

DETERMINING ROOF AREA FROM GROUND AREA

The area of a gable, hip or intersecting gable roof can be obtained sufficiently close, for practical purposes, by first measuring the horizontal or ground area. The inches rise-per-foot-of-run is computed, and from Table 14–1 the nearest corresponding

TABLE 14–2
Rafter Lengths

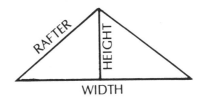

Width of Building in Feet

Height (Feet & Inches)	18	20	22	24	26	28	30	32	34	36
	Length of Rafters in Feet and Inches									
4'6"	10 2	10 11	11 9	12 10	13 9	14 9	15 8	16 8	17 7	18 7
5'0"	10 4	11 3	12 1	13 0	13 11	14 11	15 10	16 9	17 9	18 8
5'6"	10 6	11 5	12 4	13 3	14 2	15 1	16 0	16 11	17 11	18 10
6'0"	10 9	11 8	12 6	13 5	14 4	15 3	16 2	17 1	18 1	19 0
6'6"	11 0	11 11	12 9	13 8	14 6	15 5	16 4	17 3	18 3	19 2
7'0"	11 4	12 3	13 0	13 11	14 8	15 8	16 7	17 5	18 5	19 4
7'6"	11 8	12 6	13 4	14 2	15 0	15 11	16 9	17 8	18 7	19 6
8'0"	12 0	12 10	13 8	14 5	15 3	16 2	17 0	17 10	18 9	19 8
8'6"	12 4	13 2	13 11	14 8	15 6	16 5	17 3	18 2	18 11	19 11
9'0"	12 9	13 5	14 3	15 0	15 10	16 8	17 6	18 5	19 3	20 2
9'6"	13 2	13 10	14 6	15 4	16 1	16 11	17 9	18 8	19 6	20 5
10'0"	13 6	14 2	14 10	15 8	16 5	17 3	18 1	18 11	19 9	20 7
10'6"	13 10	14 6	15 3	15 11	16 9	17 6	18 4	19 3	20 0	20 10
11'0"	14 3	14 10	15 7	16 4	17 0	17 10	18 8	19 6	20 3	21 1
11'6"	14 7	15 3	15 11	16 8	17 4	18 2	18 11	19 9	20 6	21 4
12'0"	15 0	15 8	16 4	17 0	17 9	18 5	19 3	20 0	20 10	21 8
12'6"	15 5	16 0	16 8	17 4	18 0	18 9	19 7	20 4	21 1	21 11
13'0"	15 10	16 5	17 0	17 8	18 5	19 2	19 10	20 8	21 5	22 2
13'6"	16 3	16 10	17 5	18 0	18 9	19 6	20 3	20 11	21 9	22 6
14'0"	16 8	17 3	17 10	18 5	19 2	19 10	20 7	21 3	22 0	22 10
14'6"	17 1	17 8	18 3	18 10	19 6	20 2	20 10	21 7	22 4	23 2
15'0"	17 6	18 0	18 8	19 3	19 11	20 7	21 3	21 11	22 8	23 6
15'6"	17 11	18 5	19 0	19 8	20 3	20 11	21 7	22 4	23 0	23 10
16'0"	18 4	18 10	19 5	20 0	20 8	21 4	21 11	22 8	23 4	24 1
16'6"	18 10	19 4	19 10	20 5	21 0	21 8	22 4	23 0	23 8	24 5
17'0"	19 3	19 9	20 3	20 11	21 5	22 1	22 8	23 4	24 1	24 9
17'6"	19 9	20 2	20 8	21 3	21 10	22 5	23 0	23 8	24 5	25 1
18'0"	20 1	20 7	21 0	21 8	22 3	22 10	23 5	24 1	24 9	25 6
18'6"	20 7	21 0	21 6	22 0	22 8	23 3	23 10	24 5	25 2	25 10
19'0"	21 0	21 5	22 0	22 6	23 1	23 8	24 3	24 10	25 6	26 2
19'6"	21 5	21 11	22 5	22 11	23 5	24 0	24 8	25 3	25 10	26 7
20'0"	21 11	22 4	22 10	23 4	23 10	24 5	25 0	25 8	26 3	26 11
20'6"	22 5	22 10	23 3	23 9	24 4	24 10	25 5	26 0	26 8	27 4

factor is selected. The roof area is then obtained by multiplying this factor by the ground area.

Figure 14-2

Figure 14-3

Illustrative Example

In Figure 14-2, the rise-per-foot-of-run is:

$$\frac{96"}{14'} = 6.85 \text{ inches}$$

The nearest factor for 6.85 inches in Table 14-1 is 1.160
The ground area = 28' × 36' = 1,008 sq ft
The roof area = 1.160 × 1,008 = 1,169 sq ft

Where there is an overhang at the eaves, the measurement for the run is taken from the center of the gable to the eave as in Figure 14–3. *The length measurement of the roof should include the overhang at the gable end.*

INTERSECTING GABLE ROOF AREAS

Where one gable roof joins another, and each has the same pitch or slope, the area of the main roof is computed without deducting the area taken up by the intersecting roof. The area of the intersecting roof is computed by taking its length as the distance from its gable end to the point where it joins the main roof at the eaves. (See Figure 14–4.)

Illustrative Example

Roof Area Obtained from Ground Area (Figure 14–4)
Main roof ground area	28' × 60'	= 1,680 sq ft
Intersecting roof ground area	20' × 28'	= 560 sq ft
Total ground area		= 2,240 sq ft

Inches rise-per-foot-of-run $\frac{96"}{14'}$ = 6.85

From Table 14–1, nearest factor	= 1.160 in
Total area of roof (2,240 sq ft × 1.160)	= 2,598 sq ft

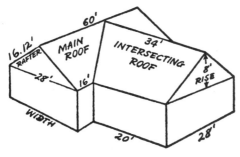

Figure 14–4

Actual measurement of the roof in Figure 14–4 would produce an area of 2,598 sq ft as follows:

One side of main roof:
16.24′ × 60′   = 974.4 sq ft
Intersected side of roof:
$$\frac{2(16' + 30') \times 16.24'}{2}$$ = 747.0 sq ft
Intersecting roof:
$$\frac{2(20' + 34') \times 16.24'}{2}$$ = 876.9 sq ft
Total area   = 2,598.3 sq ft

DORMER ROOF AREAS

The area of a *flat dormer*, excluding overhang, is slightly less than the area which the dormer itself occupies in the main roof. Some estimators determine the gross area of the entire roof without considering the dormer area separately except to make allowance for any overhang. Where exact measurement of area is desired, the area of each segment of the roof is determined.

When the pitch of either a gable or hip dormer is the same as that of the main roof, the area of the dormer roof is equal to that which the dormer itself occupies in the main roof, plus the area of any overhang. In other words, where dormers are encountered, the computation of their roof areas may be ignored except for their overhang. The main roof area is determined as though the dormers did not exist.

HIP ROOF AREAS

The pitch of all sides of a hip roof are usually the same, as a matter of good construction. *The total area of a hip roof is identical to that of a gable roof of the same pitch and covering the same horizontal or ground area.*

Figure 14–5

Area of hip roof (Figure 14–5)

2 end areas $= \dfrac{2(28' \times 16.24')}{2}$ $=$ 454.72 sq ft

2 sides areas $= \dfrac{2(12' + 40') \times 16.24'}{2}$ $=$ 844.48 sq ft

Total area $=$ 1,299.20 sq ft

Area of gable roof (Figure 14–5)

$2(40' \times 16.24')$ $=$ 1,299.20 sq ft

A quick method of obtaining the area of a hip roof is to treat it as though it were a gable roof. The area is computed by multiplying the longest side by the rafter length, or by determining the ground area (including overhang) and multiplying by the factor in Table 14–1 for the rise-per-foot-of-run.

Illustrative Example

In Figure 14–5 the ground area is:
$$28' \times 40' = 1,120 \text{ sq ft}$$
The rise-per-foot-of-run is:
$$\frac{96''}{14'} = 1.160$$
The roof area $= 1.160 \times 1,120 = 1,299.20$ sq ft

GAMBREL ROOF AREAS

Gambrel roofs are frequently found on barns, and also on older dwellings. They are sometimes called "Dutch Colonial" roofs. In a gambrel roof, the slope or pitch is broken between the plate and the ridge to form a double-pitched roof as in Figure 14–6.

While the proportions of this type of roof vary, the outer steeply sloped sides usually have a pitch of 1. This means that the rise is twice the run, which is frequently one fifth of the width of the building. The upper roof has a pitch of 1/4, which means that the rise is one half of the run.

The roof area can best be determined by calculating the length of the rafters A and B (Figure 14–6), by measuring the rise and the run of both slopes.

Figure 14–6

Illustrative Example

Roof 40' long (Figure 14–6)
Lower slope length	$= \sqrt{4^2 + 8^2}$	$= \sqrt{80}$	$=$ 8.94 lin ft
Upper slope length	$= \sqrt{3^2 + 6^2}$	$= \sqrt{45}$	$=$ 6.71 lin ft
Area of sides	$= 2(40' \times 8.94')$		$=$ 715 sq ft
Area of top	$= 2(40' \times 6.7')$		$=$ 536 sq ft
			1,251 sq ft

MANSARD ROOF AREAS

A mansard, sometimes referred to as a French roof, is one that is approximately flat on top, and has four steeply sloped sides. The top deck is usually surfaced with a type of roofing suitable for flat roofs such as roll roofing, built-up roofing or tin. The sides are often finished in some type of shingle.

The sides of a mansard roof appear as trapezoids in profile. The height is the length of the slope from the lower edge to the deck. The lower base of the trapezoid is the distance along the eaves. The upper base is the distance along the deck edge (Figure 14–7).

The area of the sides is obtained by determining the area of each of the trapezoidal sides, and adding them together.

Figure 14–7

Illustrative Example

Area side one	$= \dfrac{(20' + 26') \times 5'}{2}$	$=$ 115 sq ft
Area side two	$=$ (opposite side the same)	$=$ 115 sq ft
Area side three	$= \dfrac{(36' + 30') \times 5'}{2}$	$=$ 165 sq ft
Area side four	$=$ (opposite side the same)	$=$ 165 sq ft
		Total area 560 sq ft

(or)

Add the perimeter of the lower base to the perimeter of the upper base; divide by 2 and multiply by the slope length.

Lower base-2(26' + 36') = 124 lin ft
Upper base-2(20' + 30') = 100 lin ft
                          224 ÷ 2 = 112 × 5 = 560 sq ft

The area of the top deck is obtained by multiplying the length by the width (Figure 14–7).

$$20' \times 30' = 600 \text{ sq ft}$$

Because of the additional work involved in cutting and fitting around dormers or windows in a mansard, they are usually not deducted when computing the area in order to determine the quantity of shingles.

FLAT AND SHED ROOF AREAS

The areas of flat and shed type roofs are the least difficult to obtain, as in most cases the person making the estimate can get up on the roof to take measurements. The roof area is also relatively easy to determine from the ground measurements of the building.

A shed roof is a single-pitch roof, and the length along the slope can be obtained from the rise and the span (Figure 14–8).

Figure 14–8

Illustrative Example

Span = 12 ft
Rise = 3 ft
Inches-rise-per-foot-of-run = $\dfrac{3 \times 12"}{12'}$ = 3 inches

From Table 14–1 the factor by which the run is multiplied to determine the length of the rafter is 1.03.

Rafter length = $12' \times 1.03 = 12.36'$
Area of roof = $12.36 \times 20' = 247$ sq ft
(or)
Rafter length = $\sqrt{3^2 + 12^2} = \sqrt{153} = 12.36$ lin ft
Area of roof = $12.36' \times 20' = 247$ sq ft
(or)

Using the ground area multiplied by the factor 1.03 (Table 14–1).

Ground area = $20' \times 12' \times 1.03$
= $240' \times 1.03$
Area of roof = 247 sq ft

GOTHIC SHAPE ROOF AREAS

A gothic roof (Figure 14–9) is formed by two circles passing through the wall plates and intersecting at the ridge. This type of roof is found principally on barns and is very common in some of the midwestern dairy states. One of the most frequently used rules to lay out the proportions of a gothic roof is to scribe two circles, each one with a radius equal to two-thirds of the width of the building, with the centers of the circles at A and A1 as shown.

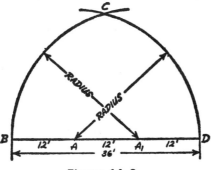

Figure 14–9

To find the roof area, the length of the building including any overhang at the gables is multiplied by the distance BC + CD.

Distance BC is equal to the circumference of the circle times .2087. The circumference of the circle is the radius (2/3 of the width of the building) times 6.2832. Since BC and CD are equal, the formula for obtaining the roof area would be:

$$\text{Area} = 2(\text{Radius} \times 6.2832) \times .2087 \times \text{length}$$

In terms of the width and length of the building, the area for a roof of these proportions may be expressed for practical application as follows:

W = width
L = length
Area = 1.75 W × L

Figure 14–10

Illustrative Example

Width of building = 42'
Radius of circles = 42' × ²/₃ = 28 lin ft
Using the formula:
Roof area = 2(Radius × 6.2832) × .2087 × Length
          = 2(28' × 6.2832) × .2087 × 60' = 4,406 sq ft
Using the formula:
Roof area = 1.75 × W × L
          = 1.75 × 42' × 60'
          = 4,410 sq ft

CIRCULAR ROOF AREAS

The area of a circular roof is obtained by determining 1/2 of the circumference of a circle where the width of the building is the diameter, and multiplying by the length of the roof.

Figure 14–11

Illustrative Example

Where W = Width of building
      L = Length of roof

$$\text{Area of roof} = \frac{3.1416 \times W}{2} \times L$$

(Figure 14–11) $\text{Area of roof} = \dfrac{3.1416 \times 30'}{2} \times 80' = 3{,}770$ sq ft

HEMISPHERICAL ROOF AREAS

The area of a hemispherical roof (Figure 14–12), commonly found on silos, would be equal to one half the area of a complete sphere. It is also equal to twice the area of its base.

$$\text{Area} = \frac{(3.1416)}{2} \times d^2 \ (\text{Where } d = \text{diameter})$$

$$(\text{or})$$

$$= 2(3.1416) \times r^2 \ (\text{Where } r = \text{radius})$$

Illustrative Example

Area of roof $= \frac{1}{2}(3.1416) \times 20^2$
$= 1.5708 \times 400$
$= 628.32$ sq ft
(or)
$= 2(3.1416) \times 10^2$
$= 628.32$ sq ft

Figure 14–12

CONICAL ROOF AREAS

Roofs of silos, water tanks, and so forth, are often conical in shape. The area of a conical roof is 1/2 the slope length "s" (Figure 14–13) × "d" the diameter × 3.1416.

Illustrative Example

Area $= \frac{1}{2}$ s $\times$ (3.1416 × d)
Example
Area $= \frac{1}{2}(10) \times (3.1416 \times 20')$
$= 314$ sq ft

Figure 14–13

SHINGLE ROOFING IN GENERAL

While at one time the word *shingle* referred to a tapered piece of wood, one end being thinner than the other, used to cover roofs, it now includes other materials which are similarly applied. The more commonly used shingles, other than wood, are asphalt, asbestos-cement, slate, tile, or its cement imitation, and aluminum. All shingles, for reasons which are apparent, are suitable only for inclined roof surfaces with sufficient pitch to prevent water from getting under the lapped ends or butts.

In application of the various kinds of shingles, there are several important factors to be given consideration:

Type of Shingle.	Starter Shingles.
Type of Roof Decking.	Ridge and Hip Caps.
Felt Underlayment.	Waste.
Exposure.	

Type of Shingle

Each type of roofing shingle is available in different grades, which is reflected in the cost. The type and grade are governed by the size and shape of the shingle, by the material of which it is made, the color, and the weight. Wood shingles, for example, are made in three sizes: 16-, 18-, and 24-inch, and each size can be obtained in three or more grades. Asphalt shingles are made in several sizes, shapes, and colors; and the weights per square vary from 135 to 350 pounds. Slate shingles are sold in different sizes, thicknesses, colors and grades. To identify a shingle roof solely by the material of which it is made is, therefore, no real indication of its cost.

When a roof has been damaged by hail or windstorm, or partly damaged by a fire, a fairly positive verification of the grade of shingle is possible. When the roof is completely destroyed by fire, the problem is more difficult. Much will depend on the investigative resourcefulness of the individual making the estimate.

Type of Roof Decking

All shingles, with the exception of wood, should be laid over solid roof boards or roof decking. When nominal 1-inch boards are used, they should be matched to provide rigidity and firmness between rafters. Plywood roof decking should be a minimum of 5/16-inch thick, and the joints of the sheets must be reinforced by blocking. The surface should be clean, and all raised nails in old decking should be either replaced or driven in so that they cannot puncture the new shingles.

It is generally recommended that wood shingles be laid over open sheathing. This may be nominal 1-inch square-edge boards spaced an inch or so apart to provide air for the back of the shingles to aid the drying-out process when they have been wet.

Felt Underlay

A layer of roofing paper of asphalt-saturated felt is applied over roof decking before laying all but wood shingles. This seals the roof surface against moisture. Under asphalt and asbestos-cement shingles a 15-pound felt is adequate. Under heavier shingles like tile or slate, a 30-pound felt is recommended.

Exposure

The term *exposure* as applied to roof shingles refers to the number of inches that a shingle is exposed to the weather. It is the distance from the butt edge of one shingle to the butt edge of the shingle above or below it.

Shingles are manufactured and sold by the *square*. When they are laid in accordance with the exposure specified by the manufacturer, one square will cover 100 square feet. If the shingles are laid with an exposure less than specified, a square of them will cover less than 100 square feet. Conversely, as the exposure is increased, the number of square feet one square will cover increases.

The percentage by which the coverage of one square is to be increased or decreased may be obtained by the following formula:

$$\frac{\textbf{Difference in Specified and Laid Exposure}}{\textbf{Actual Exposure Laid}} = \%$$

If the shingle is laid with an exposure *greater* than specified, the coverage of one shingle should be *increased* by the percentage obtained. When the actual exposure of the laid shingle is less than specified, the coverage should be *decreased* by the percentage obtained.

Illustrative Example

Manufacturer's specified exposure   = 5 inches
Actual exposure shingle is laid   = 4 inches
                                   5 inches—4 inches   = 1 inch
                                        $\dfrac{1}{4}$ = 25%

The shingle is laid with less exposure and, therefore, 1 square will cover 100 sq ft — 25 sq ft = 75 sq ft.

(or)

Manufacturer's specified exposure   = 5 inches
Actual exposure shingle is laid   = 5½ inches
                             5½ inches — 5 inches   = ½ inch
                                        $\dfrac{.5}{5.5}$ = 9%

The shingle is laid with greater exposure and, therefore, 1 square will cover 100 sq ft + 9 sq ft = 109 sq ft.

The amount of the exposure is important in estimating the *quantity* of shingles required. The labor factor is automatically taken care of by the quantity. A roofer lays shingles at a particular rate. If the rate of laying shingles is based on the number of hours per square, the overall labor cost increases or decreases with the number of squares to be laid.

Starter Shingles

Where shingling begins at the eave line of a main roof or dormer, *starting* shingles, or a *starting strip*, are laid first. The usual practice is to reverse the shingle so that the butt end is toward the ridge. After laying the first course in this manner, the shingling is begun with the first exposed course laid directly over the starters. This method is necessary because wood, asphalt, and asbestos-cement shingles are laid with a spacing between them and the starters seal the roof under the spacing of the first course.

In the case of asphalt shingles a starter roll of similar material and color is available to be used in place of shingles. For asbestos-cement shingle roofs, special starting shingles are made.

Hip and Ridge Caps

To seal the ridge and hips of a roof, and also to give them a finished appearance, they are capped with the same shingles used on the roof, or with special caps provided by the roofing manufacturer.

Asphalt shingle roofs are generally capped with pieces cut out of the roof shingles. Asbestos-cement, and slate and tile shingle roofs are usually capped with pieces of special shape and size made available by the manufacturer.

The quantity of caps to be ordered is determined from the lineal feet of the ridges and hips. When caps are to be made from the roof shingles, the quantity is included in the overall percentage added for waste.

WASTE

The term *waste* in estimating roofing includes the material lost by cutting and fitting around gables, ridges, hips and valleys, plus starters and caps, when the shingles are used for those purposes. While some roofers estimate waste by adding a square of roofing, more or less depending on circumstances, the common practice is to add a percentage to the net roof area to be covered.

The percentage to be added is a matter of an educated guess based on the shape and size of the roof. More waste is added for a cut-up surface than a plain one. More waste is also required for a hip roof than for a gable roof.

Table 14–3 shows the approximate waste to be added to the net area of a roof. The figures include cutting waste, starter waste and an allowance for caps for ridges and hips. Where ridge and hip caps are ordered separately, the waste percentage may be reduced up to one-half of the figure shown.

TABLE 14–3
Approximate Waste to Be Added to the Net Area
of Variously Shaped Roofs

Roof Shape	Plain	Cut-up
Gable	10	15
Hip	15	20
Gambrel	10	20
Gothic	10	15
Mansard (sides)	10	15
Porches	10	—

WOOD SHINGLES

Wood shingles are discussed fully in Chapter 13, pages 253 to 258, including Table 13–3 which gives the percentage of a square of shingles required to cover 100 square feet. Table 13–4 shows the covering capacities and nail requirements of Centigrade Red Cedar Shingles for various exposures.

Open or solid sheathing under wood shingles on roofs is mainly a matter of choice. Open sheathing is more common as it not only permits the shingles to "breathe" in drying out, but also is less expensive because a third to one-half of the material is needed; there is a saving in labor as well as material. Open sheathing is applied on centers equal to the exposure of the shingles. Solid sheathing may be of matched or unmatched boards. Plywood may also be used with 5/8 inch thickness on rafters that are 16 inches on center, or 3/4 inch for rafters on 24-inch centers. Wood shingles are not recommended directly on 5/16 inch plywood or on composition roof sheathing.

Rosin-sized building paper is sometimes placed under wood shingles on roofs for insulation, but saturated-felt paper is undesirable as it retards the drying of the shingles when they are wet.

The recommended maximum exposure for wood shingles on *roofs* is 5 inches, 5 1/2 inches, and 7 1/2 inches for 16-, 18-, and 24-inch shingles respectively. These exposures are specified by the manufacturer, and a square of 16-, 18-, or 24-inch shingles laid with these exposures will cover 100 square feet of surface. If the exposure is reduced, it will require more shingles to cover 100 square feet, and any increase in the exposure will decrease the quantity of shingles required for 100 square feet (Table 13–3). Shingles on roofs with a steep pitch can have a greater exposure than on roofs with a low pitch.

In applying wood shingles, the first course or row at the eaves is doubled to prevent water or snow from leaking between the shingles; also to induce a slight tension in succeeding courses. Roof shingles should have a space of approximately 1/4 inch between them to allow for swelling when they take up moisture.

To prevent dripping of water from the gables, or the formation of icicles, a roofer frequently nails a strip of 6-inch bevel siding along the edge with butt side out. The shingles are then laid in the usual manner. The result is to tilt the shingles in toward the center of the roof and direct the water away from the edge of the roof.

Ridges and hips are made watertight by applying "caps." These are available in pre-assembled form, or can be made on the job from shingles of uniform width. This capping gives a finished effect to the roof.

Wood shingles may be laid directly over an old wood shingle roof without removing the old shingles. First, all decayed shingles should be removed and new shingles inserted to form a solid bearing and nailing surface; longer nails are required for penetration, 5d or 6d being recommended. Applying new wood shingles over old is frequently preferred because of the added strength and insulating qualities. Laying wood shingles over composition shingles is not recommended as condensation may accumulate beneath the wood shingles.

Estimating Quantities of Wood Shingles on Roofs

The quantity of shingles needed for a roof is determined by computing the square foot area to be covered as discussed under "Measuring Roof Areas." From column 3 in Table 13–3 the percentage of a square of shingles needed to cover each 100 square feet may be obtained for the corresponding exposure and size of shingle. (See example on page 257.) The waste to be added to the quantity of shingles for cutting, the starting course, and for ridge and hip caps is normally about 10 percent, unless the roof is considerably cut up with dormers and intersecting gables. Some roofers add 5 percent for cutting waste, and one bundle of shingles (1/4 of a square) for each 25 lineal feet of starters, hip and ridge caps.

Estimating Costs of Wood Shingles on Roofs

The method for determining the quantity and cost of applying wood shingles to roofs is, in general, the same as outlined for sidewalls. Carpenters can apply 16- and

18-inch shingles on roofs under average conditions at a rate of about 4 hours per square. The rate for 24-inch shingles is about 3.2 hours per square. Where cut-up, irregular roof surfaces are encountered, the time should be adjusted upward accordingly. Where experienced shinglers are employed, the rate of application may be reduced somewhat.

Unit Costs for Wood Shingles on Roofs

Using assumed prices and wages, unit costs may be developed as follows:

Unit Cost for 24-Inch Shingles on Roofs With 7½ Inch Exposure

1 square shingles	@ $ 60.00	=	$ 60.00
2 lbs nails	@ .60	=	1.20
3.2 hours carpenter labor	@ 12.00	=	38.40
	Unit cost per square	=	$ 99.60
	Unit cost per sq ft	=	$ 1.00

Where the roof shingle exposures are greater or less than those given in the preceding illustration of unit costs, it is important to make an adjustment in either the unit cost or in the area to be covered. This is done by multiplying by the percentage shown in Column 3 of Table 13–3. The method is explained under sidewall shingling, pages 255 to 257.

HANDSPLIT SHAKES

There are few materials used on the roofs or side walls of residences to compare with *shakes* for architectural attractiveness. There are also few that enjoy the length of life of shakes when properly applied.

Handsplit shakes are available in three forms.

*Handsplit and Resawn Shakes have split faces and sawn backs, and are produced by running cedar blanks or boards of proper thickness diagonally through a bandsaw to produce two tapered shakes from each blank.

Tapersplit Shakes are produced mainly by hand, using a sharp-bladed steel froe and a wooden mallet. A natural shingle-like taper, from butt to tip, is achieved by reversing the block, end-for-end, with each split.

Straight-Split Shakes are split in the same manner as tapersplit shakes, except that the splitting is done from one end of the block only, producing shakes which are the same thickness throughout.

How Handsplit Shakes Are Made

Selected cedar logs are cut to specified lengths, the resulting blocks or "bolts" being trimmed to remove bark and sapwood. In the case of tapersplit or straight-split types, individual shakes are then rived from the blocks with a heavy steel blade called a "froe." To obtain a tapered thickness for tapersplit shakes, blocks are turned end-for-end after each split.

---

*Data for discussion of shakes by courtesy of Red Cedar Shingle and Handsplit Shake Bureau.

To produce handsplit and resawn shakes, blocks are split into boards of the desired thickness, then passed through a thin bandsaw to form two shakes, each with a handsplit face and a sawn back. Expert sawyers guide the boards diagonally through the saw to produce thin tips and thick butts.

Roof Application

Roof sheathing may be solid or spaced, $1'' \times 4''$ boards on centers equal to the weather exposure the shakes are to be laid, but not greater than 10 inches. At the eave line a 36-inch strip of 30-lb roofing felt is laid. As in regular shingle application, starter courses should be doubled. After applying each course, an 18-inch wide strip of 30-lb felt is laid over the top portion of the shakes, extending onto the sheathing. This roofing felt interlay is not necessary when straight-split or tapersplit shakes are applied in snow-free areas at weather exposures less than one-third the total shake length (a 3-ply roof).

Individual shakes should be spaced about 1/4 to 1/2 inch apart to allow for expansion.

Roof Pitch and Exposure

Proper weather exposure is important. As a general rule, a 7 1/2 inch maximum exposure is recommended for 18-inch shakes, and 10-inch maximum exposure for 24-inch shakes. Minimum pitch for handsplit shakes is 1/6 (4-in-12).

Table 14–4 (page 286) gives a Summary of Sizes, Packing Regulations and Coverage for Handsplit Shakes, courtesy of Red Cedar Shingle and Handsplit Shake Bureau.

Hips and Ridges

Pre-manufactured hip and ridge units are recommended. One bundle covers 16 2/3 lineal feet.

Sidewall Application

Maximum recommended weather exposure on sidewalls with single-course construction is 8 1/2 inches for 18-inch shakes and 11 1/2 inches for 24-inch shakes. Double-course application requires an underlay of shakes or regular cedar shingles. Weather exposures up to 14 inches are permissible with 18-inch handsplit/resawn or tapersplit shakes, and 20 inches with 24-inch shakes. If straight-split shakes are used, the double course exposure may be 16 inches for 18-inch shakes and 22 inches for 24-inch shakes.

Nailing

Two 6d rust-resistant nails should be driven at least one inch from each edge, and one or two inches above the butt line of the course to follow. Longer nails should be used if thickness of shake indicates them, to achieve proper penetration. It takes 2 lbs of nails per square at standard exposures.

TABLE 14-4

Shake Type, Length and Thickness	No. of Courses per Bundle	No. of Bundles per Square	Approximate coverage (in sq. ft.) of one square when shakes are applied with ½" spacing, at following weather exposures (in inches):								
			5½	6½	7	7½	8½	10	11½	14	16
18" x ½" to 3/4" Resawn	9/9 (a)	5 (b)	55(c)	65	70	75(d)	85(e)	100(f)			
18" x 3/4" to 1¼" Resawn	9/9 (a)	5 (b)	55(c)	65	70	75(d)	85(e)	100(f)			
24" x 3/8" Handsplit	9/9 (a)	5		65	70	75(g)	85	100(h)	115(i)		
24" x ½" to 3/4" Resawn	9/9 (a)	5		65	70	75(c)	85	100 (j)	115(i)		
24" x 3/4" to 1¼" Resawn	9/9 (a)	5		65	70	75(c)	85	100 (j)	115(i)		
24" x ½" to 5/8" Tapersplit	9/9 (a)	5		65	70	75(c)	85	100 (j)	115(i)		
18" x 3/8" True-Edge Straight-Split	14 (k) Straight	4								100	112(l)
18" x 3/8" Straight-Split	19 (k) Straight	5	65(c)	75	80	90 (j)	100(i)				
24" x 3/8" Straight-Split	16 (k) Straight	5		65	70	75(c)	85	100 (j)	115(i)		

(a) - Packed in 18"-wide frames.

(b) - 5 bundles will cover 100 sq. ft. roof area when used as starter-finish course at 10" weather exposure; 6 bundles will cover 100 sq. ft. wall area when used at 8½" weather exposure; 7 bundles will cover 100 sq. ft. roof area when used at 7½" weather exposure; see footnote (m).

(c) - Maximum recommended weather exposure for three-ply roof construction.

(d) - Maximum recommended weather exposure for two-ply roof construction; 7 bundles will cover 100 sq. ft. roof area when applied at 7½" weather exposure; see footnote (m).

(e) - Maximum recommended weather exposure for sidewall construction; 6 bundles will cover 100 sq. ft. when applied at 8½" weather exposure; see footnote (m).

(f) - Maximum recommended weather exposure for starter-finish course application; 5 bundles will cover 100 sq. ft. when applied at 10" weather exposure; see footnote (m).

(g) - Maximum recommended weather exposure for application on roof pitches between 4-in-12 and 8-in-12.

(h) - Maximum recommended weather exposure for application on roof pitches of 8-in-12 and steeper.

(i) - Maximum recommended weather exposure for single-coursed wall construction.

(j) - Maximum recommended weather exposure for two-ply roof construction.

(k) - Packed in 20"-wide frames.

(l) - Maximum recommended weather exposure for double-covered wall construction.

(m) - All coverage based on ½" spacing between shakes.

Figure 14-14

Workmen applying handsplit shakes, using pneumatic nailing tools. Courtesy Senco Products, Inc., Cincinnati, OH 45244

Estimating Quantities

Obtain the area of the roof as for other types of roof covering and divide by 100 to compute the number of squares needed at *standard exposures*. Allow for double coursing at eaves. One square will provide about 240 lineal feet of starter course. Allow one extra square for every 100 lineal feet of valleys. Check table for coverage when shakes are applied at exposures less than standard exposures.

Estimating Labor

As with other roof shingle application, the labor per square varies with pitch of the roof, how badly cut up the surface is with dormers and intersecting roofs, and the experience of the roofers.

A fairly experienced shingler can apply shakes at the rate of about 1 square an hour. A highly experienced one can apply 1 1/2 squares an hour on uncomplicated roofs using automatic nailing tools as shown above.

A unit cost per square would be developed from the following formula.

1 square handsplit shakes   **$**
1 hour labor @ $ per hour
2 lbs of 6d aluminum nails   _____
    Total cost per square   **$**

Illustrative Example

1 Square 24″ handsplit shakes	@ $ 80.00	=	$ 80.00
Labor to apply, 1 hour	@ 12.00	=	12.00
Nails 2 lbs	@ 1.00	=	2.00
	Cost per square	=	$ 94.00
	Cost per sq ft	=	$ .94

ASPHALT SHINGLES

The three major types of asphalt roofing products are the saturated felt used for built-up roofs, shingles, and roll or ready roofing.

Asphalt is a bituminous material found either in native form, or as a byproduct of petroleum. The latter source is the one from which the majority of asphalt for roofing is obtained. All three types of asphalt roofing are made by impregnating, with asphalt, a felt paper made from such organic fibres as rag, wood, and jute; asbestos fibres are also used.

To help resist weathering, the asphalt is combined with stabilizers of finely divided minerals. Silica, talc, slate dust, mica, and trap rock are the materials most commonly used. Coarser minerals of the same kind are applied to the surface of shingles and roll roofing to protect the asphalt from the rays of the sun, to improve the fire-resistive qualities of the roofing, and to provide attractive colors for architectural purposes. It has also been found that the light colors reflect the heat of the sun to a greater extent than darker colors, thereby reducing the temperature of the interior of the building during warm seasons.

Figure 14–15

Workman applying asphalt shingles with pneumatic stapler. Courtesy Senco Products, Inc., Cincinnati, OH 45244.

1	2	3			4		5	6
		Per Square			Size			
PRODUCT	Configuration	Approximate Shipping Weight	Shingles	Bundles	Width	Length	Exposure	Underwriters' Listing
Wood Appearance Strip Shingle More Than One Thickness Per Strip Laminated or Job Applied	Various Edge, Surface Texture & Application Treatments	285# to 390#	67 to 90	4 or 5	11-1/2" to 15"	36" or 40"	4" to 6"	A or C - Many Wind Resistant
Wood Appearance Strip Shingle Single Thickness Per Strip	Various Edge, Surface Texture & Application Treatments	Various 250# to 350#	78 to 90	3 or 4	12" or 12.25"	36" or 40"	4" to 5-1/8"	A or C - Many Wind Resistant
Self-Sealing Strip Shingle	Conventional 3 Tab	205#- 240#	78 or 80	3	12" or 12.25"	36"	5" or 5-1/8"	A or C - All Wind Resistant
	2 or 4 Tab	Various 215# to 325#	78 or 80	3 or 4	12" or 12.25"	36"	5" or 5-1/8"	
Self-Sealing Strip Shingle No Cut Out	Various Edge and Texture Treatments	Various 215# to 290#	78 to 81	3 or 4	12" or 12.25"	36" or 36-1/4"	-5"	A or C - All Wind Resistant
Individual Lock Down Basic Design	Several Design Variations	180# to 250#	72 to 120	3 or 4	18" to 22-1/4"	20" to 22-1/2"	-	C - Many Wind Resistant

Other types available from some manufacturers in certain areas of the Country.
Consult your Regional Asphalt Roofing Manufacturers Association manufacturer.

Figure 14–16

Typical asphalt shingles and typical asphalt rolls. Courtesy Asphalt Roofing Manufacturers Association, New York, NY 10017.

1	2		3	4		5		6	7
PRODUCT	Approximate Shipping Weight		Sqs. Per Package	Length	Width	Side or End Lap	Top Lap	Exposure	Underwriters' Listing
	Per Roll	Per Sq.							
Mineral Surface Roll	75# to 90#	75# to 90#	One	36' 38'	36" 36"	6"	2" 4"	34" 32"	C
	Available in some areas in 9/10 or 3/4 Square rolls.								
Mineral Surface Roll Double Coverage	55# to 70#	55# to 70#	One Half	36'	36"	-	19"	17"	C
Coated Roll	50# to 65#	50# to 65#	One	36'	36"	6"	2"	34"	None
Saturated Felt	60# 60# 60#	15# 20# 30#	4 3 2	144' 108' 72'	36" 36" 36"	4" to 6"	2"	34"	None

Figure 14–17

Typical asphalt shingles and typical asphalt rolls. Courtesy Asphalt Roofing Manufacturers Association, New York, NY 10017.

Figures 14–16 and 14–17 show the weight and quantities per square, the sizes and exposures for Typical Asphalt Shingles and Asphalt Rolls as furnished by the Asphalt Roofing Manufacturers Association.

When asphalt shingles are to be laid directly over an old roof, the galvanized nails should be 1 3/4 inches long. Some property owners consider that putting new asphalt shingles on top of old roofing will provide better insulation. However, more than two thicknesses of roofing is not recommended because of the weight and also the unevenness of the surface that results. If asphalt shingles are laid over old wood shingles, a tapered strip like bevel siding, called a "horse feather" is nailed between the butts of the wood shingle to form a flat surface. The cost of the labor and material to do this work is estimated separately from the cost of shingling.

Estimating Material

The number of squares of asphalt shingles required is obtained by computing the area to be covered, adding the appropriate percentage of waste as shown in Table 14–3, and dividing by 100. The quantity should be carried out to the next half or third of a square, depending on whether the shingles come two or three bundles to the square.

Large-head, rust-resistant nails 1 1/4 inches long are used on new work. The quantity of nails needed per square of shingles varies slightly, but for estimating, 2 pounds is adequate.

When 15-lb saturated felt is used, 100 square feet (approximately 1/4 of a standard roll) will be required for each square of shingles.

Estimating Labor

Table 14–5 shows the approximate number of hours of labor required to lay a square of various types of roofing. Under average working conditions a moderately experienced shingler will lay a square of 3-tab shingles in 2 hours. If the roof is badly cut-up it may take slightly longer. A thoroughly experienced roofer will lay 3-tab shingles on a plain roof at a rate close to 1 1/2 hours per square. (See page 292).

Individual-type shingles require more time to lay as indicated in Table 14–5. Reference is made to the discussion of Factors Affecting Labor in this chapter on pages 263–265.

Unit Costs for Asphalt Shingles

The unit cost for asphalt shingles is based on one square of material. The hours of labor are taken from Table 14–5.

Unit Cost of 3-Tab 235 Lb. Asphalt Shingles on Plain Roofs

1 square asphalt shingles	@ $ 20.00	=	$ 20.00
2 lbs nails	@ .60	=	1.20
2 hours labor	@ 12.00	=	24.00
	Unit cost per square =		$ 45.20
	Unit cost per sq ft =		$ .45

TABLE 14-5

Approximate Hours of Labor to Remove and to Lay One Square of
Roofing (Including Underlayment of Saturated Felt Where Required)

Type of Roofing	Pounds of Nails	Plain Roofs	Difficult or Cut-up Roofs	To Remove Old Roofing
Asphalt Strip Shingles	2	2	3	1
Asphalt Individual Shingles	4	4	6	1
Asbestos Rectangular Shingles	3	6	8	1½
Asbestos Hexagonal Shingles	3	4	6	1½
3/16"Slate Shingles	3	6	8	1½
Handsplit Shakes		1	2	1
Wood Shingles	2	4	6	1½
Corrugated Galvanized or Aluminum on Wood	2	2½	3½	1
Corrugated Galvanized Steel or Aluminum	2	6	8	1
3-Ply Built-up Gravel Roof	–	2	–	1½
4-Ply Built-up Gravel Roof	–	2½	–	1½
5-Ply Built-up Gravel Roof	–	3	–	1½
Roll Roofing	–	1½	2	1

The "Hour of Labor per Square" heading spans the "Plain Roofs", "Difficult or Cut-up Roofs", and "To Remove Old Roofing" columns.

FIBERGLASS SHINGLES

Fiberglass shingles are actually asphalt shingles with a *fiberglass mat* in the place of the *organic felt base*. This has resulted in a stronger shingle with considerably more asphalt for the same weight as the felt base shingle. Consequently the fiberglass shingle has better weather resistance and a longer life. Also, because the glass fiber base is noncombustible, the shingle is given a Class A label by the Underwriters Laboratory, whereas the organic felt base asphalt shingle is given a Class C label. Examples of fiberglass shingles to choose from are those of Johns-Manville,*

Fire-Glass III, a traditional square-cut, three-tab design, 12" × 36", 225 lb, 20-year prorated limited warranty.

Fire-King III, a heavyweight, three-tab shingle, 12" × 36", 265 lb, 25-year prorated limited warranty.

Woodlands, laminated shingle with a heavy shake-like appearance, 14" × 36", 300 lb, 25-year prorated limited warranty

Fiberglass shingles are furnished in the conventional three-tab size. Hip and ridge units are available per lineal foot in bundles. Application is much the same as for the organic-base asphalt shingle previously discussed.

---

*Information on Johns-Manville fiber glass shingles reprinted with permission of Manville Corporation.

A unit cost may be developed as follows using assumed prices and wages.

1 Square 225-lb, 12″ × 36″ fiberglass shingles	@ $ 60.00 =	$ 60.00
2 hours labor	@ 12.00 =	24.00
	Unit cost per square =	$ 84.00
	Unit cost per sq ft =	$ .84

SLATE SHINGLES

Slate is a laminated rock which is quarried and split into thin slabs by the manufacturer. The size of slate shingles varies between 6 and 8 inches in width, and between 12 and 24 inches in length. While the thickness of standard slate is about 3/16 inch, it is available in thicknesses of 1 inch and over, according to architectural requirements. A popular size slate shingle for dwellings is 10 inches wide and 20 inches long.

Slate is laid on solid roof decking with a 30-lb saturated felt underlay. They should not be laid over old existing roofing.

The holes in slate for nails are usually punched at the quarry. The *lap* of a slate is the distance by which the upper slate overlaps the head of the second slate below. The standard lap is 3 inches. The exposure of a slate is 1/2 of the length after deducting the lap. A 10-inch shingle with a 3-inch lap will have a 3 1/2-inch exposure ($\frac{10-3}{2}$ = 3 1/2 inches).

Hips and ridges are usually finished by alternately lapping shingles cut to proper size. The first course at the eaves is doubled for a starting course.

Waste from breakage and cutting, and allowance for the starter course, ridge and hip caps, varies between 10 and 20 percent as shown in Table 14–3.

Estimating Materials

The quantity of slate required is determined by adding the appropriate waste to the net roof area, and dividing by 100 to obtain the number of squares. Approximately 1/2 roll of 30-lb saturated felt is required for each square of slate.

Copper nails, 3d or 4d, are recommended. It requires about 3 pounds of nails to lay a square of slate shingles.

Estimating Labor

Experienced slaters are difficult to find in most areas. Two generally work together and occasionally, a helper is used to fill out a crew. The labor to lay a square of slate shingles varies with the size of the slate, the experience of the workers, and the shape of the roof. Table 14–5 shows the approximate number of hours per square under average conditions.

Unit Costs for Slate Shingles

A unit cost per square of slate shingles may be obtained by inserting the local material prices and hourly wage rates as shown in the following illustrations. The hours of labor may be taken from Table 14–5.

Unit Cost of 10″ × 20″ × 3/16″ Slate Shingles on Plain Roof

1 square 10″ × 20″ × 3/16″ slate	@ $ 120.00 =	$ 120.00
3 lbs 3d copper nails	@ 2.50 =	7.50
1/2 roll 30-lb felt	@ 4.00 =	2.00
6 hours labor	@ 12.00 =	72.00
	Unit cost per square =	$ 201.50
	Unit cost per sq ft =	$ 2.02

Unit Cost of 10″ × 20″× 3/16″Slate Shingles on Cut-up Roof

1 square 10″ × 20″ 3/16″ slate	@ $ 120.00 =	$ 120.00
3 lbs copper nails	@ 2.50 =	7.50
1/2 roll 30-lb felt	@ 4.00 =	2.00
8 hours labor	@ 12.00 =	96.00
	Unit cost per square =	$ 225.50
	Unit cost per sq ft =	$ 2.26

ASBESTOS-CEMENT SHINGLES

Asbestos-cement shingles are a rigid type, fire-resistive shingle made of a composition of asbestos fiber and portland cement under heavy pressure. They are available in the popular white or gray shingle textured to simulate wood, or in a variety of colors. The shingles are usually rectangular or hexagonal.

Asbestos-cement shingles are laid over solid roof decking with an underlay of 30-lb saturated felt paper. An advantage of asbestos-cement shingles is that they can be placed directly over old wood shingles if desired.

A shingle cutter is available for sale, or may be rented from a dealer to cut, punch, and notch asbestos-cement shingles.

Galvanized needle-point nails are recommended to lay these shingles. A 3d nail is used on new work, but when shingling over an old roof the 5d or 6d nail should be used. About 3 pounds of nails are required to lay one square.

Estimating Materials

Asbestos-cement shingles are sold by the square which is sufficient to cover 100 square feet when laid in accordance with the specified exposure. Exposures vary between 6 and 9 inches. Starter shingles for the first course, and also hip and ridge shingles are available.

The waste to be added to the net roof area for cutting and fitting, and for starters, hip and ridge shingles is shown in Table 14–3.

The number of squares of asbestos-cement shingles is obtained by adding the appropriate amount of waste to the net roof area and dividing by 100.

Approximately one-half roll of 30-lb asphalt-saturated felt and 3 pounds of nails are needed per square of shingles.

Estimating Labor

The number of hours required to lay asbestos-cement shingles varies with the type and size of the shingle. The 8″ × 16″ rectangular shingle, similar to the slate shingle, takes longer to lay than the larger hexagonal shingle, which is 14″ × 30″ wide. Table 14–5 shows the approximate number of hours of labor per square under average conditions.

Unit Costs for Asbestos-Cement Shingles

The following material prices and hourly wage rates are assumed; labor is taken from Table 14–5.

*Unit Cost of 8″ × 16″ × ¼″ Asbestos-Cement
Shingles on Plain Roof*

1 square asbestos-cement shingles	@ $ 100.00 =	$	100.00
3 lb 3d needle-point nails	@ .80 =		2.40
½ roll 30-lb sat. felt paper	@ 4.00 =		2.00
6 hours labor	@ 12.00 =		72.00
	Unit cost per square =	$	176.40
	Unit cost per sq ft =	$	1.76

*Unit Cost of 16″ × 16″ Hexagonal Asbestos-Cement
Shingles on Plain Roofs*

1 square asbestos-cement shingles	@ $ 60.00 =	$	60.00
3 lb 3d needle-point nails	@ .80 =		2.40
½ roll 30-lb sat. felt paper	@ 4.00 =		2.00
4 hours labor	@ 12.00 =		48.00
	Unit cost per square =	$	112.40
	Unit cost per sq ft =	$	1.12

ROLL ROOFING

Asphalt roll roofing comes in rolls that are 36 inches wide and 36 feet long. A roll contains 108 square feet and, making allowance for lapping, it is sufficient to cover 100 square feet. The 65-lb smooth surfaced, and the 90-lb mineral surfaced rolls are the most commonly used. An underlayment of 15-lb saturated felt is sometimes used, but not required. Lapping cement and 7/8 inch galvanized nails are provided inside each roll. When laid over old roofing longer nails should be used in order to penetrate the wood decking at least 3/4 of an inch.

Roll roofing is applied to roofs of every pitch. It is used on all types of structures with flat roofs, particularly where initial cost is a consideration. It is the least expensive roofing to apply.

Roll roofing may be laid parallel with the eaves or the ridge of a pitched roof. Caps for ridges and hips are usually made by splitting a roll down the center. For better quality of work, 9-inch starter strips are applied along the eaves and up the *rake* or gable ends.

The ends are lapped 6 inches and the edge laps are 2 inches. Roll roofing may be applied by nailing the upper edge of each strip and cementing the bottom edge. This is called the *concealed nail method.* The more common method of application is to nail the bottom edge of the overlapping strip which is called the exposed nail method.

Estimating Material

The quantity of roll roofing required for a particular roof is determined from the net roof area to be covered. The amount of waste in cutting, lapping at edges of eaves and gables, and for caps, varies with each job. Generally 10 percent is adequate to take care of most situations.

Estimating Labor

Labor factors which affect the rate of laying other types of roofing should be properly considered in estimating labor for roll roofing. The rolls are heavy and more difficult to manage as the pitch of the roof increases. On flat or slightly pitched roofs a roofer or carpenter should be able to cover a square in 1 1/2 hours. On steep-pitched or difficult roofs, the time should be increased to 2 hours per square.

Illustrative Example
(Wages and Prices Assumed)

1 roll 90-lb roofing	@	$	12.00 =	$	12.00
1 1/2 hours labor	@		12.00 =		18.00
		Unit price per square =		$	30.00
		Unit price per sq ft =		$	.30

BUILT-UP ROOFING

As its name indicates, built-up roofing is one that is applied by *building up* successive layers of saturated felt, lapped and cemented together with hot tar (pitch) or asphalt. The layers of felt are called *plies,* and this type of roof is identified by the number of plies, usually running from 3 to 5 thick. The top may be finished off with a mopping of tar or asphalt, and is referred to as a built-up smooth-top roof. Where slag or gravel is imbedded in the top coating, the roof is commonly spoken of as a tar and gravel, or a slag roof. When the incline of the roof exceeds 2 inches to the foot, the top ply is frequently mineral-surfaced 19-inch Selvage, Double Coverage Roll Roofing (see page 295).

The saturated felt used for built-up roofs is usually the 15-lb type. It is 36 inches wide, 144 feet long, and contains 432 square feet. One roll weighs 60 pounds. On many roofs, particularly bonded roofs, the first ply is a 30-lb felt which is available in rolls of 216 square feet weighing 60 pounds.

On a new roof over wood decking, a layer of rosin-sized building paper or one layer of 15-lb saturated felt is applied to prevent the pitch or asphalt from leaking through cracks, knot-holes, or joints in the boards. This layer of paper is not counted as one of the plies. Successive layers of 15-lb felt are then laid, lapping the sheets 19 inches and mopping between plies with hot pitch or asphalt. Approximately 30 pounds of asphalt or pitch per square is used for each mopping with at least double that quantity (60 pounds) applied to the surface if slag, gravel, or crushed marble is to be imbedded. About 300 pounds of slag or 400 pounds of gravel are required for each 100 square feet of surface covered. The quantity of pitch or asphalt applied to the surface when slag or gravel is not used is about 40 pounds per square.

In the application of lower quality built-up roofing, the first couple of layers of felt are laid dry, that is, without mopping between them. Other means of reducing both cost and quality include the use of less pitch or asphalt. An examination of a cross-section of a sample of the roofing that has been damaged will generally disclose the number of plies and quality of the roofing.

Estimating Materials

The number of squares of built-up roofing required is obtained from the net area to be covered. Waste is taken care of in the rolls of felt by the manufacturer. A roll of 15-lb felt contains 432 square feet and is intended to cover 400 square feet or 4 squares. A roll of 30-lb felt contains 216 square feet and will cover 200 square feet or 2 squares.

Measurements of the roof are taken from the *outside* of the walls. Allowance should be made for flashing against bulkheads and parapet walls. Normally, the flashing is carried up 1 foot, but in many older buildings the entire parapet may be flashed up to the coping.

Skylights, bulkheads, ventilators, and so forth, should not be deducted unless the area they occupy is in excess of 100 square feet, and then it is customary to deduct only one half of the area to allow for the additional work of cutting, flashing, and working around them.

The materials used in built-up roofing are saturated felt, asphalt, or coal-tar pitch, and slag or gravel when so surfaced. Different manufacturers and roofers vary the specifications, but for estimating purposes, the quantities shown in Table 14–6 are approximately correct.

TABLE 14–6

Materials Required to Apply 100 Square Feet Built-up Roofing Using Asphalt, and Surfacing with Slag or Gravel

Number of Plies	Rolls of 15-Lb Saturated Felt (432 Square Feet)	Pounds of Asphalt	Pounds of Slag	Pounds of Gravel
3	¾	150	300	400
4	1	180	300	400
5	1¼	230	300	400
3(1 dry ply)	¾	120	300	400
4(2 dry plies)	1	120	300	400
5(2 dry plies)	1¼	150	300	400

(For smooth-top finish without slag or gravel, deduct 20 pounds of asphalt per 100 square feet of roofing. When 30-lb saturated felt is used as a first ply, add ¼ of a roll of 15-lb felt.)

Estimating Labor

Built-up roofing is applied by a crew ranging from three to five workers, depending on the size of job and working conditions. All materials are raised with a rope and pulley hoist rather than carried up ladders. One worker generally tends the asphalt-heating equipment on the ground while the others lay the felt and mop.

Relatively flat roofs are easier to work on than those with slopes, or those built over crescent or bowstring trusses. The latter frequently require guardrails, and other safety devices to protect the workers. As the number of story heights increase, more labor is used in hoisting materials to the roof.

A roof with numerous skylights, ventilators, bulkheads and so forth, or one of irregular shape, requires more labor than a plain roof. Each job should be studied for its individual characteristics before deciding on the number of hours to allow per square. The approximate number of hours per square for uncomplicated roofs, including a superintendent, are:

> 2 hours for 3-ply
> 2½ hours for 4-ply
> 3 hours for 5-ply

Fuel, Mops, Supplies

The fuel required to heat asphalt or pitch, the mops to apply it, brushes for spreading gravel, pails, nails, and so forth, are all expendable supplies. In computing the cost of a roof, a flat allowance may be made to cover these items, or a percentage may be added to the *total cost of the labor and the material* for each job. The amount varies, but ranges between 3 and 5 percent.

Contractors Overhead and Profit

Because built-up roofing is subcontracted to roofers who are properly equipped, and who specialize in that work, it is important to add their overhead and profit to the cost of all material and labor. The amount to be added usually varies between 20 and 30 percent depending on the roofer who is doing the job.

Unit Costs for Built-up Roofing

The following formulas are used to show how to develop a unit cost. Overhead and profit should be added (Prices and wages assumed). Labor rates are from Table 14–5.

Unit Cost of 3-Ply Built-up Slag Surface Roofing
Over Wood Decking on a Flat Roof. One 15 lb
Felt Layed Dry, One 15 lb and one 30 lb Felt Mopped.

2 sq	rolls of 15 lb felt	@	$ 3.00 =	$	6.00
1 sq	rolls of 30 lb felt	@	6.00 =		6.00
140 lbs tar pitch		@	.14 =		19.60
400 lbs slag		@	.40 =		16.00
				$	47.60
			Mops and fuel 5 %		2.38
			Materials	$	49.98
Hours of labor, 2 @ $12					24.00
			Unit cost per square	$	73.98

$$\text{Unit cost per sq ft } \frac{\$73.98}{100} = \$ \ .74$$

*Unit Cost of 5-Ply Built-up Slag Surface Roofing
Over Flat Wood Deck. Five 15 lb Rolls of Felt,
Two layed Dry and Three Mopped.*

5 Rolls 15 lb Felt	. . . .	@	$ 3.00 =	$	15.00
200 lbs Tar pitch	. . . .	@	.14 =		28.00
400 lbs slag	. . . .	@	.40 =		16.00
				$	59.00
			Mops and fuel 5%		2.95
			Materials	$	61.95
Hours of labor, 3 @ $ 12					36.00
			Unit cost per sq	$	97.95

$$\text{Unit cost per sq ft } \frac{\$97.95}{100} = \quad \$ \quad .98$$

CORRUGATED FIBERGLASS ROOFING AND SIDING

Corrugated fiberglass roofing and siding sheets come in several widths and lengths, the 26″ × 12′-0″ being a popular size. The weight per square foot determines the price. Weights range from as little as 2 or 3 ounces to 8 ounces. The material is translucent and is furnished clear or in colors.

Application is similar to corrugated metal roofing and siding as to end and side laps, and to fittings (see below). A unit price may be developed in the following manner using assumed price and wages.

1 square 8 oz. fiberglass roofing	@	$ 100.00 =	$	100.00
2 1/2 hours labor	@	12.00 =		30.00
		Unit cost per square =	$	130.00
		Unit cost per sq ft =	$	1.30

CORRUGATED METAL ROOFING AND SIDING

Both corrugated aluminum and galvanized steel roofing and siding are used extensively on industrial-type buildings and on farm and secondary buildings. The sheets are available with either 1 1/4 or 2 1/2-inch corrugations, and come in different widths and lengths. The 26-inch wide sheets are most popular, and lengths ranging from about 6 to 12 feet should be selected according to the area to be covered to avoid as much cutting as possible.

Side laps should be a minimum of 1 1/2 corrugations, and end laps at least 2 to 6 inches depending on the pitch of the roof.

Ridge and valley rolls, rubber end-seals, and end-wall flashings may be obtained for both galvanized and aluminum corrugated sheets. Special rust-resistant nails and washers are used to apply this type of roofing. A 1 3/4 inch nail is used for new work on wood, and a 2-inch nail should be used when applied over an old roof; allow 2 pounds per square.

Corrugated sheets may be applied to solid wood decking, or over open framing. When applied over steel framing, special fittings are required. Add 5 to 10 percent to the cost of the material for rivets, clips, and so forth.

The price of corrugated roofing varies considerably with its thickness. Galvanized sheets are available in 12 gauge to 26 gauge. Aluminum sheet thickness is measured in inches and ranges from .019 inch (approximately 26 gauge) to .032 inch. The 26-gauge galvanized and .019 inch aluminum are frequently used on farm buildings, and the heavier types are used on industrial buildings.

Estimating Material

The quantity of corrugated roofing is obtained from the net roof area to be covered. The amount of *waste* to be added for lapping sides and ends, and for cutting and fitting varies between 15 and 25 percent depending on the amount of lap and cutting required. Generally 20 percent added to the net roof area will be adequate.

Two pounds of nails should be included for each square of roofing. Ridge rolls, valley rolls, and other accessories should be estimated by the number of lineal feet required.

Estimating Labor

The labor rate to apply corrugated roofing varies with the job conditions, to some extent the weight of the sheets, and whether or not it is being applied over wood or metal. Table 14–5 shows the approximate number of man-hours for applying corrugated roofing to wood and to steel.

Unit Costs for Corrugated Roofing
(Prices and Wages Assumed)

Unit Cost of .024 Inch Corrugated Aluminum
Roofing Applied Over Wood on Plain Roof

1 square .024 in. roofing	@	$ 75.00 =	$	75.00
2 lbs nails	@	.80 =		1.60
2½ hours labor	@	12.00 =		30.00
	Unit cost per square =		$	106.60
	Unit cost per sq ft =		$	1.07

Unit Cost of .032 Inch Corrugated Aluminum
Roofing Applied Over Wood on Plain Roof

1 square .032 in. roofing	@	$ 85.00 =	$	85.00
2 lbs nails	@	.80 =		1.60
2½ hours labor	@	12.00 =		30.00
	Unit cost per square =		$	116.60
	Unit cost per sq ft =		$	1.17

ALUMINUM SHINGLES AND SHAKES

Aluminum shingles or shakes have a heavily textured surface which resembles the traditional handsplit wood shingle or wood shake. They are furnished in 10″ × 60″, 12″ × 36″, and 12″ × 48″ sizes for easier handling.

While basically marketed for roofing they are frequently used on mansard roofs and as exterior siding. Aluminum shingles are relatively expensive, but have long durability, light weight, and require a minimum of maintenance.

The price of aluminum shingles depends on the gauge, .020 ga. or .030 ga., and on the finish. The price spread can range from $60.00 to $160.00 per square. Ridge and hip caps, starter strips, trim, etc., are available per lineal foot.

The labor to apply aluminum shingles, as with other kinds of roofing, will be governed by the usual considerations such as how cut-up the roof shape is; the pitch of the roof; the experience of the workers; whether scaffolding is needed, and how much preparation of the roof is required. A fair average time to apply one square will range between 4 and 6 hours.

Illustrative Example

1 Square .020 ga. colored aluminum shingles	@ $ 110.00 =	$ 110.00
Labor, 5 hours	@ 12.00 =	60.00
	Cost per square =	$ 170.00
	Cost per square foot =	$ 1.70

SQUARE FEET OF ROOFING DIRECT

FROM TABLES

The number of square feet of various kinds of roofing required for *gable, hip and intersecting roofs* may be read directly from Tables 14–8 through 14–13. Quantities shown are for all buildings with *ground floor areas* from 200 to 3,000 square feet and roof pitches of 1/8 to 1/2. They apply regardless of story height as long as the roof covers the measured *ground floor area*. Interpolation can be made for *ground floor areas* between the 100 square foot increments. (See pages 304–309).

Overhang

The quantities in the tables do not allow for overhang at the eaves or gables. For each 12″ horizontal overhang at eaves, add 5% to the table quantities.

Dormers

Roofing for dormers is approximately the quantity taken up in the main roof by the dormer. The only additional amount of decking is dormer overhang.

Hip and Gable Roofs

The surface area of a hip roof is identical to that of a gable roof of the same pitch and covering the same horizontal area.

Nails Required Per Square

Asphalt strip shingles	2 lbs	- 1¼ inch nail
Asphalt individual shingles	3 lbs	- ″ ″ ″

Asbestos cement shingles	3–4	- 3d nail
Slate 10″ × 20″, 3″ lap shingle	3	- 3d or 4d ″
Clay tile	3	- 1 ¾ inch nail
Wood, 16″–18″ shingle	2	- 3d
″ 24″ ″	2	- 4d
″ handsplit shakes	2	- 6d

FORMULAS FOR ESTIMATING ROOF AREAS AND
THE QUANTITY OF ROOFING MATERIAL

The areas of *gable and hip* roofs can be quickly obtained by multiplying the horizontal area to be *covered by the roof*, by one of the following factors corresponding to the particular roof pitch.

Roof Pitch	*Multiply Horizontal Area By:*
⅛ (3″ rise-per-foot-of-run)	1.03
⅙ (4″ rise-per-foot-of-run)	1.06
⁵⁄₂₄ (5″ rise-per-foot-of-run)	1.09
¼ (6″ rise-per-foot-of-run)	1.12
⅓ (8″ rise-per-foot-of-run)	1.20
½ (12″ rise-per-foot-of-run)	1.42

Multiply the roof area by the percentage of cutting and fitting waste shown in Table 14–7 on page 303.

Illustrative Example

A dwelling 30′ × 60′ has a gable roof with a pitch of ⁵⁄₂₄ and is surfaced with asphalt strip shingles.

The area of the roof with this pitch is 1,800 sq ft × 1.09 = 1,962 sq ft
The percentage to be added for cutting, fitting, etc., (page 303) is 10%.
The number of squares of shingles required is:

$$1,962 \times 1.10 = 2,158 \text{ sq ft}$$
$$= 21.58 \text{ squares rounded out to 22 squares.}$$

Roofing Materials Not Shown in Tables

The roof surface areas are shown in the second column of each table. For roof materials not covered by the tables, use the surface area and add the appropriate fitting and cutting waste.

DAMAGEABILITY OF ROOFING

No roofing material may be considered permanent inasmuch as all types are subject to normal weathering, or a shortened life because of improper application. All types of roofing are also subject to damage by fire, and by wind or hail. There are other miscellaneous causes of damage less frequently encountered such as lightning, collapse, falling trees or their limbs, and explosion.

TABLE 14–7

Cutting and Fitting Waste
The percentages that have been added to the areas covered in these
tables for cutting and fitting waste, starters, ridge caps, etc. are:

Roofing Material	Percentage Added	Multiply Area by
Asphalt strip and Asbestos Shingles	10%	1.10
18″ Wood Shingles _____ 4″ exposure	35	1.35
18″ ″ ″ _____ 5½″ ″	10	1.10
24″ ″ ″ _____ 6″ ″	38	1.38
24″ ″ ″ _____ 7½″ ″	10	1.10
Clay Tile Roofing	5	1.05
15 Lb Saturated Felt	10	1.10
30 ″ ″ ″	10	1.10

Normal Weathering

All roofing materials deteriorate from exposure to the weather. The rate of the deterioration depends to a large extent on the kind of material, the degree of maintenance, and the conditions of exposure. Rigid materials such as slate, metal, asbestos-cement, and tile are less vulnerable to the action of wind, rain and sun than the pliable asphalt roofing products.

Some roof materials in southern climates wear out faster than those in northern latitudes, and generally the south and west exposures of a roof will deteriorate quicker than the north and east exposures. As a rule, most roofing materials on steep slopes will outlast the same materials when applied to those of less pitch. Shaded roofs with wood shingles have a shorter life than those exposed to the sun and wind, which are necessary to keep them dry after wetting by rain, snow or dew.

Summer heat of extreme temperatures dries out the oils in asphalt roofing, and the shrinkage and expansion due to extreme ranges of temperature may cause cracking or alligatoring of the surface.

Maintenance

A roof that receives little or no maintenance weathers and wears out much more rapidly than one that receives proper attention. Wood shingle roofs that are periodically treated with creosote will lie flat and last much longer than untreated shingles. The painting of wood cedar shingles increases their life as the low rate of expansion and contraction does not disturb the film of paint. Heavy oil paints should not be used.

Built-up smooth surface roofs should be coated with bitumen periodically, particularly in areas of high temperatures, to replace the oils lost by exposure to the sun. Built-up roofs covered with crushed stone deteriorate in those places where the gravel, or slag, has been washed or blown away. Such areas should be cleaned off, recoated and the mineral should be replaced.

TABLE 14–8

Roofing for Gable and Hip Roofs, Including Intersecting Roofs, with Pitch of 1/8 (milling and cutting waste is included)

Ground Floor Area Sq. Ft.	Roof Surface Area Square Ft.	Asphalt Strip Shingles Square Ft.	Wood Shingles				Asbestos Cement Shingles Square Ft.	Clay Roofing Tile Square Ft.	Number of Rolls of Saturated Felt Roofing Paper	
			18" Long		24" Long				15-Lb.	30-Lb.
			4" Expo. Square Ft.	5½" Expo. Square Ft.	6" Expo. Square Ft.	7½" Expo. Square Ft.				
200	206	226	278	226	284	226	226	216	.57	1.14
300	309	339	417	339	426	339	339	324	.85	1.70
400	412	452	556	452	568	452	452	432	1.13	2.27
500	515	565	695	565	710	565	565	540	1.42	2.84
600	618	678	834	678	852	678	678	648	1.70	3.40
700	721	791	973	791	994	791	791	756	1.98	3.97
800	824	904	1112	904	1136	904	904	864	2.27	4.54
900	927	1010	1251	1010	1278	1010	1010	972	2.55	5.10
1000	1030	1130	1390	1130	1420	1130	1130	1080	2.83	5.66
1100	1133	1243	1529	1243	1562	1243	1243	1188	3.11	6.22
1200	1236	1356	1668	1356	1704	1356	1356	1296	3.40	6.80
1300	1339	1469	1810	1469	1846	1469	1469	1404	3.68	7.36
1400	1442	1582	1946	1582	1990	1582	1582	1512	3.96	7.92
1500	1545	1695	2085	1695	2130	1695	1695	1620	4.25	8.50
1600	1648	1808	2224	1808	2272	1808	1808	1728	4.53	9.06
1700	1751	1921	2363	1921	2414	1921	1921	1836	4.81	9.62
1800	1854	2034	2500	2034	2556	2034	2034	1944	5.09	10.18
1900	1957	2147	2640	2147	2700	2147	2147	2052	5.38	10.76
2000	2060	2260	2780	2260	2840	2260	2260	2160	5.66	11.32
2100	2163	2378	2920	2378	2982	2378	2378	2268	5.94	11.88
2200	2266	2486	3058	2486	3124	2486	2486	2376	6.23	12.46
2300	2369	2600	3200	2600	3266	2600	2600	2484	6.51	13.02
2400	2472	2712	3336	2712	3410	2712	2712	2592	6.80	13.58
2500	2575	2825	3475	2825	3550	2825	2825	2700	7.08	14.16
2600	2678	2938	3615	2938	3692	2938	2938	2808	7.36	14.72
2700	2781	3051	3753	3051	3834	3051	3051	2916	7.64	15.28
2800	2884	3164	3892	3164	3976	3164	3164	3024	7.92	15.84
2900	2987	3277	4030	3277	4118	3277	3277	3132	8.21	16.42
3000	3090	3390	4170	3390	4260	3390	3390	3240	8.49	16.98

Note: For types of roofing not shown, use Roof Surface Area column. Add the appropriate cutting and fitting waste.

TABLE 14–9

Roofing for Gale and Hip Roofs, Including Intersecting Roofs, with Pitch of 1/6 (milling and cutting waste is included)

Ground Floor Area Sq. Ft.	Roof Surface Area Square Ft.	Asphalt Strip Shingles Square Ft.	Wood Shingles 18" Long 4" Expo. Square Ft.	18" Long 5½" Expo. Square Ft.	24" Long 6" Expo. Square Ft.	24" Long 7½" Expo. Square Ft.	Asbestos Cement Shingles Square Ft.	Clay Roofing Tile Square Ft.	Number of Rolls of Saturated Felt Roofing Paper 15-Lb.	30-Lb.
200	212	234	286	234	292	234	234	224	.59	1.17
300	318	351	429	351	438	351	351	336	.88	1.76
400	424	468	472	468	584	468	468	448	1.17	2.34
500	530	585	715	585	730	585	585	560	1.46	2.93
600	636	702	858	702	876	702	702	672	1.76	3.51
700	742	819	1000	819	1022	819	819	784	2.05	4.10
800	848	936	1144	936	1170	936	936	896	2.34	4.68
900	954	1053	1287	1050	1314	1050	1050	1008	2.63	5.27
1000	1060	1170	1430	1170	1460	1170	1170	1120	2.93	5.85
1100	1166	1287	1573	1287	1610	1287	1287	1232	3.22	6.44
1200	1272	1404	1716	1404	1752	1404	1404	1344	3.51	7.02
1300	1378	1521	1859	1521	1900	1521	1521	1456	3.80	7.61
1400	1484	1638	2000	1638	2044	1638	1638	1568	4.10	8.19
1500	1590	1755	2145	1755	2190	1755	1755	1680	4.39	8.78
1600	1696	1872	2288	1872	2336	1872	1872	1792	4.68	9.36
1700	1802	1989	2430	1989	2480	1989	1989	1904	4.97	9.95
1800	1908	2106	2574	2106	2630	2106	2106	2016	5.27	10.53
1900	2014	2223	2717	2223	2774	2223	2223	2128	5.56	11.12
2000	2120	2340	2860	2340	2920	2340	2340	2240	5.85	11.70
2100	2226	2457	3000	2457	3066	2457	2457	2352	6.14	12.29
2200	2332	2574	3146	2574	3212	2574	2574	2464	6.44	12.87
2300	2438	2691	3290	2691	3358	2691	2691	2576	6.73	13.46
2400	2544	2808	3430	2808	3504	2808	2808	2688	7.02	14.04
2500	2650	2925	3575	2925	3650	2925	2925	2800	7.31	14.63
2600	2756	3042	3720	3042	3800	3042	3042	2912	7.61	15.21
2700	2862	3159	3860	3159	3942	3159	3159	3024	7.90	15.80
2800	2968	3276	4000	3276	4090	3276	3276	3136	8.19	16.38
2900	3074	3393	4150	3393	4234	3393	3393	3248	8.48	16.96
3000	3180	3510	4290	3510	4380	3510	3510	3360	8.78	17.55

Note: For types of roofing not shown, use Roof Surface Area column. Add the appropriate cutting and fitting waste.

TABLE 14–10

Roofing for Gable and Hip Roofs, Including Intersecting Roofs, with Pitch of 5/24 (milling and cutting waste is included)

Ground Floor Area Sq. Ft.	Roof Surface Area Square Ft.	Asphalt Strip Shingles Square Ft.	Wood Shingles 18" Long 4" Expo. Square Ft.	Wood Shingles 18" Long 5½" Expo. Square Ft.	Wood Shingles 24" Long 6" Expo. Square Ft.	Wood Shingles 24" Long 7½" Expo. Square Ft.	Asbestos Cement Shingles Square Ft.	Clay Roofing Tile Square Ft.	Number of Rolls of Saturated Felt Roofing Paper 15-Lb.	30-Lb.
200	218	240	294	240	300	240	240	228	.60	1.20
300	327	360	441	360	450	360	360	342	.90	1.80
400	426	480	588	480	600	480	480	456	1.20	2.40
500	545	600	735	600	750	600	600	570	1.50	3.00
600	654	720	882	720	900	720	720	684	1.80	3.60
700	763	840	1029	840	1050	840	840	798	2.10	4.20
800	872	960	1176	960	1200	960	960	912	2.40	4.80
900	981	1080	1323	1080	1350	1080	1080	1026	2.70	5.40
1000	1090	1200	1470	1200	1500	1200	1200	1140	3.00	6.00
1100	1199	1320	1617	1320	1650	1320	1320	1254	3.30	6.60
1200	1308	1440	1764	1440	1800	1440	1440	1368	3.60	7.20
1300	1417	1560	1911	1560	1950	1560	1560	1482	3.90	7.80
1400	1526	1680	2058	1680	2100	1680	1680	1596	4.20	8.40
1500	1635	1800	2205	1800	2250	1800	1800	1710	4.50	9.00
1600	1744	1920	2352	1920	2400	1920	1920	1824	4.80	9.60
1700	1853	2040	2500	2040	2550	2040	2040	1938	5.10	10.20
1800	1962	2160	2646	2160	2700	2160	2160	2052	5.40	10.80
1900	2071	2280	2793	2280	2850	2280	2280	2166	5.70	11.40
2000	2180	2400	2940	2400	3000	2400	2400	2280	6.00	12.00
2100	2289	2520	3087	2520	3150	2520	2520	2394	6.30	12.60
2200	2398	2640	3234	2640	3300	2640	2640	2508	6.60	13.20
2300	2507	2760	3381	2760	3450	2760	2760	2622	6.90	13.80
2400	2616	2880	3528	2880	3600	2880	2880	2736	7.20	14.40
2500	2725	3000	3675	3000	3750	3000	3000	2850	7.50	15.00
2600	2834	3120	3822	3120	3900	3120	3120	2964	7.80	15.60
2700	2943	3240	3969	3240	4050	3240	3240	3078	8.10	16.20
2800	3052	3360	4116	3360	4200	3360	3360	3192	8.40	16.80
2900	3161	3480	4263	3480	4350	3480	3480	3306	8.70	17.40
3000	3270	3600	4410	3600	4500	3600	3600	3420	9.00	18.00

Note: For types of roofing not shown, use Roof Surface Area column. Add the appropriate cutting and fitting waste.

TABLE 14–11

Roofing for Gable and Hip Roofs, Including Intersecting Roofs, with Pitch of 1/4 (milling and cutting waste is included)

Ground Floor Area Sq. Ft.	Roof Surface Area Square Ft.	Asphalt Strip Shingles Square Ft.	Wood Shingles 18" Long 4" Expo. Square Ft.	Wood Shingles 18" Long 5½" Expo. Square Ft.	Wood Shingles 24" Long 6" Expo. Square Ft.	Wood Shingles 24" Long 7½" Expo. Square Ft.	Asbestos Cement Shingles Square Ft.	Clay Roofing Tile Square Ft.	Number of Rolls of Saturated Felt Roofing Paper 15-Lb.	30-Lb.
200	224	246	300	246	310	246	246	236	.62	1.24
300	336	369	450	369	465	369	369	354	.93	1.86
400	448	492	600	492	620	492	492	472	1.24	2.48
500	560	615	750	615	775	615	615	590	1.55	3.10
600	672	738	900	738	930	738	738	708	1.86	3.72
700	784	861	1050	861	1085	861	861	826	2.17	4.34
800	896	984	1200	984	1240	984	984	944	2.48	4.96
900	1008	1107	1350	1107	1395	1107	1107	1062	2.79	5.58
1000	1120	1230	1500	1230	1550	1230	1230	1180	3.10	6.20
1100	1232	1353	1650	1353	1705	1353	1353	1298	3.41	6.82
1200	1344	1476	1800	1476	1860	1476	1476	1416	3.72	7.44
1300	1456	1600	1950	1600	2015	1600	1600	1534	4.03	8.06
1400	1568	1722	2100	1722	2170	1722	1722	1652	4.34	8.68
1500	1680	1845	2250	1845	2325	1845	1845	1770	4.65	9.30
1600	1793	1968	2400	1968	2480	1968	1968	1888	4.96	9.92
1700	1904	2091	2550	2091	2635	2091	2091	2006	5.27	10.54
1800	2016	2214	2700	2214	2790	2214	2214	2124	5.58	11.16
1900	2128	2337	2850	2337	2945	2337	2337	2242	5.89	1178
2000	2240	2460	3000	2460	3100	2460	2460	2360	6.20	12.40
2100	2352	2583	3150	2583	3255	2583	2583	2478	6.51	13.02
2200	2464	2706	3300	2706	3410	2706	2706	2596	6.82	13.64
2300	2576	2829	3450	2829	3565	2829	2829	2714	7.13	14.26
2400	2688	2952	3600	2952	3720	2952	2952	2832	7.44	14.88
2500	2800	3075	3750	3075	3875	3075	3075	2950	7.75	15.50
2600	2912	3198	3900	3198	4030	3198	3198	3068	8.06	16.12
2700	3024	3321	4050	3321	4185	3321	3321	3186	8.37	16.74
2800	3136	3440	4200	3440	4340	3440	3440	3304	8.68	17.36
2900	3248	3567	4350	3567	4495	3567	3567	3422	9.00	18.00
3000	3360	3690	4500	3690	4650	3690	3690	3540	9.30	18.60

Note: For types of roofing not shown, use Roof Surface Area column. Add the appropriate cutting and fitting waste.

TABLE 14-12

Roofing for Gable and Hip Roofs, Including Intersecting Roofs, with Pitch of 1/3 (milling and cutting waste is included)

Ground Floor Area Sq. Ft.	Roof Surface Area Square Ft.	Asphalt Strip Shingles Square Ft.	Wood Shingles 18" Long 4" Expo. Square Ft.	Wood Shingles 18" Long 5½" Expo. Square Ft.	Wood Shingles 24" Long 6" Expo. Square Ft.	Wood Shingles 24" Long 7½" Expo. Square Ft.	Asbestos Cement Shingles Square Ft.	Clay Roofing Tile Square Ft.	Number of Rolls of Saturated Felt Roofing Paper 15-Lb.	30-Lb.
200	240	264	324	264	332	264	264	252	.66	1.32
300	360	396	486	396	498	396	396	378	.99	1.98
400	480	528	648	528	664	528	528	504	1.32	2.64
500	600	660	810	660	830	660	660	630	1.65	3.30
600	720	792	972	792	1000	792	792	756	1.98	3.96
700	840	924	1134	924	1162	924	924	882	2.31	4.62
800	960	1056	1296	1056	1328	1056	1056	1008	2.64	5.28
900	1080	1188	1458	1188	1494	1188	1188	1134	2.97	5.94
1000	1200	1320	1620	1320	1660	1320	1320	1260	3.30	6.60
1100	1320	1452	1782	1452	1826	1452	1452	1386	3.63	7.26
1200	1440	1584	1944	1584	1992	1584	1584	1512	3.96	7.92
1300	1560	1716	2106	1716	2158	1716	1716	1638	4.29	8.58
1400	1680	1848	2268	1848	2324	1848	1848	1764	4.62	9.24
1500	1800	1980	2430	1980	2490	1980	1980	1890	4.95	9.90
1600	1920	2112	2592	2112	2656	2112	2112	2016	5.28	10.56
1700	2040	2244	2754	2244	2822	2244	2244	2142	5.61	11.22
1800	2160	2376	2916	2376	2990	2376	2376	2268	5.94	11.88
1900	2280	2508	3078	2508	3154	2508	2508	2394	6.27	12.54
2000	2400	2640	3240	2640	3320	2640	2640	2520	6.60	13.20
2100	2520	2772	3400	2772	3486	2772	2772	2646	6.93	13.86
2200	2640	2904	3564	2904	3652	2904	2904	2772	7.26	14.52
2300	2760	3036	2726	3036	3820	3036	3036	2898	7.59	15.18
2400	2880	3168	3888	3168	3984	3168	3168	3024	7.92	15.80
2500	3000	3300	4050	3300	4150	3300	3300	3150	8.25	16.50
2600	3120	3432	4212	3432	4316	3432	3432	3276	8.58	17.16
2700	3240	3564	4374	3564	4482	3564	3564	3402	8.91	17.82
2800	3360	3696	4536	3696	4648	3696	3696	3528	9.24	18.48
2900	3480	3828	4700	3828	4814	3828	3828	3654	9.57	19.14
3000	3600	3960	4860	3960	4980	3960	3960	3780	9.90	19.80

Note: For types of roofing not shown, use Roof Surface Area Column. Add the appropriate cutting and fitting waste.

TABLE 14–13

Roofing for Gable and Hip Roofs, Including Intersecting Roofs, with Pitch of 1/2 (milling and cutting waste is included)

Ground Floor Area Sq. Ft.	Roof Surface Area Square Ft.	Asphalt Strip Shingles Square Ft.	Wood Shingles				Asbestos Cement Shingles Square Ft.	Clay Roofing Tile Square Ft.	Number of Rolls of Saturated Felt Roofing Paper	
			18" Long		24" Long					
			4" Expo. Square Ft.	5½" Expo. Square Ft.	6" Expo. Square Ft.	7½" Expo. Square Ft.			15-Lb.	30-Lb.
200	284	312	384	312	392	312	312	298	.78	1.55
300	426	468	576	468	588	468	468	447	1.16	2.32
400	568	624	768	624	784	624	624	600	1.55	3.10
500	710	780	960	780	980	780	780	745	1.94	3.88
600	852	936	1152	936	1176	936	936	894	2.33	4.65
700	994	1092	1344	1092	1372	1092	1092	1043	2.71	5.43
800	1136	1248	1536	1248	1568	1248	1248	1192	3.10	6.20
900	1278	1404	1728	1404	1764	1404	1404	1340	3.49	6.98
1000	1420	1560	1920	1560	1960	1560	1560	1490	3.88	7.75
1100	1562	1716	2112	1716	2156	1716	1716	1639	4.26	8.53
1200	1704	1872	2304	1872	2352	1872	1872	1788	4.65	9.30
1300	1846	2028	2496	2028	2548	2028	2028	1937	5.04	10.08
1400	1990	2184	2688	2184	2744	2184	2184	2086	5.43	10.85
1500	2130	2340	2880	2340	2940	2340	2340	2235	5.81	11.63
1600	2272	2500	3072	2500	3136	2500	2500	2384	6.20	12.40
1700	2414	2652	3264	2652	3332	2652	2652	2533	6.59	13.18
1800	2556	2810	3456	2810	3528	2810	2810	2682	6.98	13.95
1900	2700	2964	3648	2964	3724	2964	2964	2830	7.36	14.93
2000	2840	3120	3840	3120	3920	3120	3120	2980	7.75	15.50
2100	2982	3276	4032	3276	4116	3276	3276	3130	8.14	16.28
2200	3124	3432	4224	3432	4312	3432	3432	3278	8.53	17.05
2300	3266	3590	4416	3590	4508	3590	3590	3427	8.91	17.83
2400	3410	3744	4608	3744	4704	3744	3744	3576	9.30	18.60
2500	3550	3900	4800	3900	4900	3900	3900	3725	9.69	19.38
2600	3692	4056	4992	4056	5096	4056	4056	3874	10.08	20.15
2700	3834	4212	5184	4212	5292	4212	4212	4023	10.46	20.93
2800	3976	4368	5376	5488	4368	4368	4172		10.85	21.70
2900	4118	4524	5568	4524	5684	4524	4524	4321	11.24	22.48
3000	4260	4680	5760	4680	5880	4680	4680	4470	11.63	23.25

Note: For types of roofing not shown, use Roof Surface Area column. Add the appropriate cutting and fitting waste.

When the top coating of bitumen is no longer heavy enough to imbed the mineral surfacing properly, the roof should be scraped. A heavy coating of bitumen should be applied and the slag or gravel replaced.

Flashings against parapet walls, bulkheads, and chimneys may crack from shrinkage, particularly if not properly applied. They require recovering or replacing. Low spots in flat-decked roofs accumulate water which freezes in winter and causes breaks in built-up or roll type roofing. Unless repairs are made promptly, water gets under the roofing surface, causing blisters, or rotting of the wood decking below.

Split wood and cracked slate, or asbestos-cement shingles, should be replaced or patched to prevent water from getting under the surface to cause further deterioration.

IMPROPER INSTALLATION

When roofing is not properly laid, it deteriorates more rapidly than when applied according to the manufacturer's specifications.

If the wood decking is laid with green lumber, the curling or cupping causes ridges which press upward to loosen nails and cut through the surface. Inadequate nailing of asphalt shingles or roll roofing exposes the laps to moderate wind action. Nailing asphalt shingles above the thick butt prevents them from lying flat. The use of nails which are too short, or are not rust-resistant, endangers the life of the roofing. The placing of more than two nails in wood shingles causes them to split when they expand and contract. Laying shingles with too great an exposure also exposes them to wind action and frequently causes curling.

Asphalt shingles should be laid over old wood shingles only after filling in the exposed butts with a feather-edge piece to give a solid base. Otherwise they will not lie flat and the surface will be uneven. When laid directly over old wood shingles, asphalt shingles are highly vulnerable to hail damage in the places that are not supported by a solid backing.

When a new asphalt shingle roof is laid over an old asphalt shingle roof, great care is necessary to obtain solid nailing and to select the proper length of nail which will penetrate the roof boards at least 3/4 of an inch.

Aluminum roofing placed in contact with steel or copper in the presence of moisture will corrode or pit from the electrolytic action. Most unprotected metal roofs that are exposed to spray from salt water will also deteriorate rapidly.

FIRE DAMAGE

Wood shingles and asphalt roofing products are combustible, although the degree of heat required to ignite the latter is relatively high. The principle value of asphalt roofing products in the presence of fire is their longer resistance to ignition.

Sparks and embers from exposure fires, although failing to ignite wood or asphalt roofing, can cause scorching and marring of the surface. In more serious exposure fires, the scarring due to embers is sufficient to require complete replacement. Rigid-type roofings such as tile, slate, asbestos-cement, or metal, are highly resistant to exposure fires.

Damage by fire to all types of roofing materials is the result, in most cases, of a fire which originates within the structure itself. It either chars or otherwise damages the roof framing and decking to the extent that replacement of the area is necessary, or the fire actually breaks through the roof. Frequently holes are chopped in roofs by firefighters to ventilate the fire or to extinguish it.

HAIL DAMAGE

The extent of damage to roofing by hailstones depends largely on the age of the roof, the type of material, and the size of the hailstones. Asphaltic roofing products are more vulnerable to the impact of hail than rigid materials. However, no roofing material is completely safe in storms involving hailstones as large or larger than golf balls, a phenomenon not uncommon in many parts of the country. Spanish tile and slate roofing surfaces have been pulverized in severe hailstorms.

The lighter-gauge metal roofs may be badly dented or pocked by pea-sized hail. Wood shingles can be dented and split. Asphalt shingles and roll or built-up roofing may also be dented and frequently hail perforates the materials.

Numerous instances are on record of hail 2 inches or more in diameter passing through the roof decking.

Damage to asphalt shingles applied over old wood shingles is more serious when no feather-edge inserts have been used to fill between the butts of the old shingles.

Built-up roofing surfaced with gravel or slag is much less susceptible to hail damage than smooth top roofing.

WIND DAMAGE

Damage to roofing during a windstorm, hurricane, or tornado may occur as a result of the direct force of the wind getting under the material; by suction; or by the collapse or destruction of the roof.

The effect of the direct action of wind on roofing depends on its velocity, the kind of material, its condition, and method of application. The Red Cedar Shingle Bureau states that: "To lift a shingle 8 inches wide away from a roof covered with No. 1 16-inch shingles laid with a 5-inch exposure requires a pull of 85 pounds—a force so much greater than a hurricane can exert that it can be conservatively stated that properly nailed shingles simply cannot be blown from a roof." Yet an old worn shingle roof or one laid with too much exposure can be seriously damaged by winds of considerably less velocity than those experienced in hurricanes. Asphalt shingles properly nailed and laid with a 4-inch exposure, will fare much better in high winds than shingles nailed too high, or with a 5- or 5 1/2-inch exposure. Shingles applied to steep-pitched roofs suffer less damage by wind than those laid on low-pitched roofs.

Both experience and experiments conducted in wind tunnels indicate that the roof shingles located 5 feet from the rake or edge of a roof, and four or five courses nearest the ridge, are subject to greater damage than other areas of the surface.

Asphalt shingles which are lifted during a strong wind may break along the nailing line. After the wind has subsided, the shingles return to their former position concealing the real damage.

Roll roofing, especially when laid with exposed nailing, is highly vulnerable to winds if the nails are too far apart or too close to the edge.

Slate, asbestos-cement, and tile roofs are extremely wind-resistant when properly laid. Usually they are not damaged except in the high winds of a hurricane or tornado.

Heavy-gauge corrugated metal or transite roofs properly applied will withstand winds of high velocities, but the light-gauge aluminum or galvanized corrugated roofs are subject to damage when wind gets underneath the edges or laps. Built-up roofing may be damaged by wind of high velocity if a break in the surface exists, or occurs during the storm, and gusts are able to get underneath. The force of hurricane or tornado winds may be sufficient to catch under a small break, and roll back large areas, or lift the roofing, loosening the entire surface.

In some cases, sections of built-up roofing are pulled away from the decking by suction, particularly if the first ply or layer is not adequately nailed. Frequently, after the storm, the surface of the roofing is loose and fluffed like a loose blanket. Damage of this kind usually cannot be repaired and the entire area involved must be removed and re-roofed.

Strong winds often blow the loose gravel or slag into piles on the roof, or blow it off the roof entirely.

METHODS OF REPAIR

The method of repairing or replacing of roofing that has been damaged by fire generally presents no unusual problems. The area damaged is plainly visible, unless destruction is complete, and repairs are determined by observation and measurement.

The extent of injury to roofing by hail or windstorm is not so readily determined without a careful survey. In all cases, when practical, the inspection should be made *on the roof* rather than from the ground. A roof that appears undamaged from ground level may show extensive injury upon close examination. Wind-broken shingles or small pocks or holes caused by hail are not always easy to observe unless seen at close range.

Repairing Asphalt Shingles

Replacing individual asphalt shingles is done by sliding a ripper underneath the shingle and cutting off the nails. A new shingle is then inserted and nailed. Cracked, torn, or otherwise damaged shingles should be removed and replaced. The cost of individual shingle replacement is primarily one of labor. Most roofers have a minimum flat price for a call, though only one shingle is involved. Where a number of them must be replaced, the charge may be based on a flat charge per shingle.

When the damage occurs over several square feet or more, ample allowance of material should be made for lacing in the shingles at the periphery of the damaged area. If the damaged portion of one side of a roof is so extensive or so scattered that the labor to patch and repair it equals or exceeds the labor required for complete replacement, it is advisable to estimate recovering the entire side of the roof.

Asphalt shingles that have been peppered by light hail, sufficient to mar the mineral surfacing, are not seriously injured except for appearance and possibly a slightly shortened life. However, when the shingle itself is perforated, even though immediate leaking is not apparent, it is advisable to replace or recover the roof with new shingles.

Matching asphalt shingles. The matching of asphalt shingles is sometimes a difficult problem. The inability to provide a suitable match may be due to the age and fading of the existing shingle, or it may be that the particular color or pattern is no longer available.

Where one or only a few individual shingles are required, it is sometimes possible to cut out a shingle from a piece of roll roofing of similar color. At other times a considerable amount of shopping around is necessary to locate proper matching shingles. Occasionally, where expensive shingles are involved, the manufacturer is able to produce enough to make repairs.

Where larger quantities of shingles are needed to patch a dwelling roof, many property owners are not adverse to having them removed from their garage. The latter is then re-roofed with a suitable shingle. On other occasions, matching shingles have been removed from the rear of a roof and placed on the front. The rear is then re-roofed with a shingle selected to satisfactorily blend or conform.

Patching or replacing small areas of asphalt shingles should be estimated on a time and material basis. The cost of laying single areas of one or more squares may be determined by the unit cost, provided ample consideration is given to the matching and lacing in.

Repairing Wood Shingles

A wood shingle roof surface that has been peppered by hail so that the painted or creosoted surface is unsightly, may be repainted or refinished with creosote to restore it. If the surface is only slightly marred, even by indentations, the same means of restoration are frequently satisfactory.

When wood shingles are split or broken by hail they must be replaced.

Shingles that have been loosened or dislodged by wind require replacement. When replacing individual shingles or a series of them to patch out a roof, the nails should be cut off with a ripper before the new shingles are slipped underneath the sound ones. The new shingles are then nailed with copper nails.

Patching of very old wood shingle roofs is sometimes done by use of tin pieces which are slipped under the split shingles. Loose nails should be drawn and new ones put in. Older wood shingle roofs are often satisfactorily patched by caulking or by cutting a piece of suitable size from a piece of 65-lb roll roofing, spreading roofing cement on it, and sliding it under the broken shingle. Copper nails are then driven to secure it.

As with asphalt shingles, all patch jobs should be estimated on a time and material basis. Solid areas of one or more squares may be figured by unit cost if proper consideration is given to lacing in the new shingles where they join the old.

Repairing Rigid Shingles

Tile, slate, or asbestos-cement shingle roofs may be treated very similarly to damaged wood shingle roofs. Individual replacements are made by use of a ripper to cut off the nails under the sound shingles. A new shingle is inserted and nailed with rust-resistant nails. Holes have to be drilled in asbestos-cement shingles for nailing. The nail head may then be covered with a copper strip made for that purpose. Lead plugs are sometimes used in place of the copper strip.

Broken clay tiles are removed. The joints are cemented and the new tile is wired in place with copper wire.

Repairing Built-up Roofing

The damage to built-up roofing, outside of replacement, when necessary, may be repaired in one of four ways: 1. Spot-patch. 2. Patch and mop. 3. Cap-sheet and mop. 4. Resurface with gravel or slag.

Before patching built-up roofing, all gravel or slag must be removed from the area, to be replaced after the repairs are complete.

Spot-patch. This method is used when small or medium-sized breaks occur in the surface involving one or more plies. The area around the break is cut out and removed. All moisture under the break should be thoroughly dried out. New felts are then laid and mopped between if more than one has been cut through. A coating of asphalt or pitch is then applied to the patch surface.

Patch and Mop. This method of repair is generally used where large patches or many small patches are necessary. The patching is done in much the same manner as in spot-patching. When the patches are complete, the entire surface of the roof is given a mop coat of asphalt or pitch.

Cap-Sheet and Mop. When hail has badly dented or has perforated a built-up roof, a method of repair is to clean off the surface of the roof area to be repaired, lay a 30-lb cap-sheet, and apply 25 or 30 pounds of asphalt or pitch per square.

Resurface with Gravel or Slag. In hurricanes and tornadoes, it frequently happens that no greater damage occurs to a built-up roof than having the loose gravel or slag blown around or off the roof. In some cases replacing the mineral surfacing is all that is required. At other times it is necessary to clean off the balded areas and mop coat with sufficient asphalt or pitch to hold new gravel or slag. Close inspection is the only way of determining which treatment is needed.

Repairing Metal Roofs

Whether or not corrugated aluminum or galvanized roofing has been injured by wind, hail, or fire is readily determined by examination. Badly twisted or dented sheets should be removed and replaced.

CHAPTER 15

Painting, Paperhanging and Glazing

PAINTING

In metropolitan New York, in New Jersey, and in several New England and Southern States, a unit cost for painting is applied to the *gross* wall and ceiling area of a room. In many of the midwestern and Pacific Coast states a unit cost is also applied to the gross area without deducting openings; but in the latter states a flat sum is added for each window and door opening, and the baseboard. These units are listed as *sides*. This practice tends to inflate the cost of painting, particularly in a room having several windows and doors. If openings are not deducted when applying the unit cost to the gross wall area, it amounts to doubling up on the cost of both labor and material where *sides* are added.

JOB CONSIDERATIONS

There are several different methods used to approximate the number of gallons of paint and the hours required to apply it. Some of these are described in the following pages. Whatever method is used, the main objective should be to compute as accurately as possible the amount of material and, considering all conditions that pertain to the particular job, to determine a reasonable number of hours for a painter to apply the materials. The following conditions are those which should be given particular attention as they most generally affect the cost of material and labor.

1. Shifting and protecting contents
2. Preparation of the surface
3. Kind of surface
4. Shape or form of surface
5. Type of paint
6. Number of colors
7. Number of coats
8. Method of application
9. Size and layout of area being worked in
10. Degree of worker's skill
11. Working from ladders or scaffold.

1. Shifting and Protecting Contents

A painter works a little faster in a room that is vacant than in one that requires shifting and covering the contents to conveniently reach the areas to be painted. It takes extra time to move furniture and take down pictures. In mercantile risks it should be determined who is to move stocks on shelves, and away from walls that are to be painted.

2. Preparation of the Surfaces

Walls that are badly smoked are customarily washed down to remove grease and dirt before applying paint. In many instances the washing down of smoked walls results in only one coat of paint being needed to cover the surface. Blistered paint from heat or water should be removed and sanded prior to applying paint. Frequently, minor cracks in plaster, or in the joints of wall board, have to be filled with patch plaster or spackle, and nail holes in woodwork must be puttied.

3. Kind of Surface

Both the labor and the quantity of paint vary with the kind of surface. Rough plastered brick, or cement walls require more paint than smooth plastered walls and woodwork. A painter using a brush can cover more square feet per hour on a smooth surface than on one that is porous. Exterior unpainted wood shingles and weather-worn siding require more paint and labor than surfaces where the old paint has been well maintained.

4. Shape or Form of Surface

Stairways, trims, cabinets, windows and doors take longer to paint per square foot than smooth uninterrupted areas such as walls and ceilings, bevel siding, and floors.

5. Type of Paint

The covering capacities of some types of paint are greater than others regardless of the kind of surface being painted. Cold water-mix paints and calcimine cover less square feet per gallon than oil, rubber or alkyd resin base paints, while shellac and oil stains have greater covering capacity than any of those named. In figuring the quantity of material it is important to know the probable covering capacity of the paint to be used. The type of paint also affects the cost of the material. See covering capacities in Table 15–1 on page 318.

Painters can apply different types of paint at varying rates. For example, while calcimine covers less area per gallon, a painter can apply it faster than oil paint.

6. Number of Colors

When doors, windows and trim in a room or on the exterior must be painted a different color from other areas, a painter will require a little more time than if only one color is used.

7. Number of Coats

The number of coats of paint required naturally affect the cost of the work. See Table 15–1 for covering capacity for each coat.

8. Method of Application

Brush work is a little slower than using a roller on walls and ceilings. In some jurisdictions a painter who uses a roller is paid a slightly higher rate of wages which may offset the advantage in speed. Where spray painting is permissible, the labor cost is considerably reduced although slightly more paint is usually required.

9. Layout of Area

Where a painter is working in a large area, as a loft building, the rate of applying paint on walls is higher than working in a confined area or a cut-up room. Areas such as stores, offices, or factories that are being painted during working hours have to be estimated at a slightly higher rate of application of paint because employees or customers moving about slow the progress of the painter.

10. Degree of Workman's Skill

Some painters have more experience and skill than others. A painting contractor will usually put workers where their talents and capabilities are most productive.

11. Working from Ladders or Scaffolds

A painter working from ladders or scaffolding generally works at a slightly slower rate of speed than when working at ground or floor level. It also requires extra time to set up and move ladders or scaffolds from one position to another.

COVERING CAPACITY OF PAINT

The number of square feet that one gallon of paint will cover varies with the condition and porosity of the surface, the type of paint, and the method used to apply it. Paint manufacturers usually specify on each container the approximate area that can be covered. A first coat generally covers less area per gallon than succeeding coats primarily because of the absorption. The first coat acts as a sealer. It is seldom that two painters will agree on the covering capacity of paint, but their differences of opinion should not be much more than 10 or 15 percent apart.

For estimating quantities, an average number of square feet per gallon of material is used. For practical purposes there is no advantage to be gained by attempting to use different spreading capacities for the first, second and third coats. One average figure should be used regardless of how many coats are to be applied. An exception is when applying paint to highly absorbent materials and surfaces like rough common brick and old unpainted wood that has weathered.

Table 15–1 shows the approximate number of square feet that a gallon of painting materials will cover on the most frequently encountered surfaces. When the surfaces are unusually rough or porous, the number of square feet should be reduced accordingly.

TABLE 15–1

Approximate Covering Capacity of Painting Materials Using Brush Application

USE OF PAINT MATERIALS	Sq Ft Per Gallon
Interior	
Flat paint on smooth walls	450 – 500
Flat paint on sand finish plaster	300 – 350
Flat paint on texture finish plaster	300 – 350
Gloss or semi-gloss on smooth plaster	400 – 450
Gloss or semi-gloss on sand plaster	300 – 400
Gloss or semi-gloss on texture plaster	250 – 350
Calcimine-Watersize	720
Concrete block - Resin Emulsion	200
Wood veneer-Lacquer	500 – 540
Wood veneer-Synthetic resin	600 – 675
Penetrating wax	600 – 675
Enamel on wood floors	400 – 450
Stainwax on floors or trim	500 – 600
Shellac (4 lb cut)	500 – 550
Varnish over shellac	500
Varnish remover	150
Exterior	
Wood siding and trim-flat or semi-gloss	450 – 550
Staining shingle roofs and siding	200 – 300
Oil Paint-asbestos shingles	200 – 300
Cedar shakes-Latex	200 – 300
Brick -Latex	200 – 300
Concrete floors and steps-enamel	300 – 400
Link fences	750
Gutters and leaders	200–250 lin ft per coat

Note: See pages 322 to 325, Measurement of Areas.

Wall and Ceiling Areas from Tables 15–2A, 15–2B, 15–2C

These Tables give the *total* square foot area of the walls and ceiling of rooms with lengths and widths shown and ceiling heights of 8, 9, and 10 feet. There has been no deduction for window or door openings and rooms are presumed to be rectangular in shape.

Interpolation can be made for room sizes to the nearest 6 inches.

TABLE 15–2A

Total Area . . . 4 Walls and Ceiling (in square feet) . . . Rooms with Ceilings 8 Feet

	3'	4'	5'	6'	7'	8'	9'	10'	11'	12'	13'	14'	15'	16'	17'	18'	19'	20'	21'	22'
3'	105	124	143	162	181	200	219	238	257	276	295	314	333	352	371	390	409	428	447	466
4'	124	144	164	184	204	224	244	264	284	304	324	344	364	384	404	424	444	464	484	504
5'	143	164	185	206	227	248	269	290	311	332	353	374	395	416	437	458	479	500	521	542
6'	162	184	206	228	250	272	294	316	338	360	382	404	426	448	470	492	514	536	558	580
7'	181	204	227	250	273	296	319	342	365	388	411	434	457	480	503	526	549	572	595	618
8'	200	224	248	272	296	320	344	368	392	416	440	464	488	512	536	560	584	608	632	656
9'	219	244	269	294	319	344	369	394	419	444	469	494	519	544	569	594	619	644	669	694
10'	238	264	290	316	342	368	394	420	446	472	498	524	550	576	602	628	654	680	706	732
11'	257	284	311	338	365	392	419	446	473	500	527	554	581	608	635	662	689	716	743	770
12'	276	304	332	360	388	416	444	472	500	528	556	584	612	640	668	696	724	752	780	808
13'	295	324	353	382	411	440	469	498	527	556	585	614	643	672	701	730	759	788	817	846
14'	314	344	374	404	434	464	494	524	554	584	614	644	674	704	734	764	794	824	854	884
15'	333	364	395	426	457	488	519	550	581	612	643	674	705	736	767	798	829	860	891	922
16'	352	384	416	448	480	512	544	576	608	640	672	704	736	768	800	832	864	896	928	960
17'	371	404	437	470	503	536	569	602	635	668	701	734	767	800	833	866	899	932	965	998
18'	390	424	458	492	526	560	594	628	662	696	730	764	798	832	866	900	934	968	1002	1036
19'	409	444	479	514	549	584	619	654	689	724	759	794	829	864	899	934	969	1004	1039	1074
20'	428	464	500	536	572	608	644	680	716	752	788	824	860	896	932	968	1004	1040	1076	1112
21'	447	484	521	558	595	632	669	706	743	780	817	854	891	928	965	1002	1039	1076	1113	1150
22'	466	504	542	580	618	656	694	732	770	808	846	884	922	960	998	1036	1074	1112	1150	1188
23'	485	524	563	602	641	680	719	758	797	836	875	914	953	992	1031	1070	1109	1148	1187	1226
24'	504	544	584	624	664	704	744	784	824	864	904	944	984	1024	1064	1104	1144	1184	1224	1264

TABLE 15-2B

Rooms with Ceilings 9 Feet

	3'	4'	5'	6'	7'	8'	9'	10'	11'	12'	13'	14'	15'	16'	17'	18'	19'	20'	21'	22'
3'	117	138	159	180	201	222	243	264	285	306	327	348	369	390	411	432	453	474	495	516
4'	138	160	182	204	226	248	270	292	314	336	358	380	402	424	446	468	490	512	534	556
5'	159	182	205	228	251	274	297	320	343	366	389	412	435	458	481	504	527	550	573	596
6'	180	204	228	252	276	300	324	348	372	396	420	444	468	492	516	540	564	588	612	636
7'	201	226	251	276	301	326	351	376	401	426	451	476	501	526	551	576	601	626	651	676
8'	222	248	274	300	326	352	378	404	430	456	482	508	534	560	586	612	638	664	690	716
9'	243	270	297	324	351	378	405	432	459	486	513	540	567	594	621	648	675	702	729	756
10'	264	292	320	348	376	404	432	460	488	516	544	572	600	628	656	684	712	740	768	796
11'	285	314	343	372	401	430	459	488	517	546	575	604	633	662	691	720	749	778	807	836
12'	306	336	366	396	426	456	486	516	546	576	606	636	666	696	726	756	786	816	846	876
13'	327	358	389	420	451	482	513	544	575	606	637	668	699	730	761	792	823	854	885	916
14'	348	380	412	444	476	508	540	572	604	636	668	700	732	764	796	828	860	892	924	956
15'	369	402	435	468	501	534	567	600	633	666	699	732	765	798	831	864	897	930	963	996
16'	390	424	458	492	526	560	594	628	662	696	730	764	798	832	866	900	934	968	1002	1036
17'	411	446	481	516	551	586	621	656	691	726	761	796	831	866	901	936	971	1006	1041	1076
18'	432	468	504	540	576	612	648	684	720	756	792	828	864	900	936	972	1008	1044	1080	1116
19'	453	490	527	564	601	638	675	712	749	786	823	860	897	934	971	1008	1045	1082	1119	1156
20'	474	512	550	588	626	664	702	740	778	816	854	892	930	968	1006	1044	1082	1120	1158	1196
21'	495	534	573	612	651	690	729	768	807	846	885	924	963	1002	1041	1080	1119	1158	1197	1236
22'	516	556	596	636	676	716	756	796	836	876	916	956	996	1036	1076	1116	1156	1196	1236	1276
23'	537	578	619	660	701	742	783	824	865	906	947	988	1029	1070	1111	1152	1193	1234	1275	1316
24'	558	600	642	684	726	768	810	852	894	936	978	1020	1062	1104	1146	1188	1230	1272	1314	1356

TABLE 15-2C

Rooms with Ceilings 10 Feet

	3'	4'	5'	6'	7'	8'	9'	10'	11'	12'	13'	14'	15'	16'	17'	18'	19'	20'	21'	22'
3'	129	152	175	198	221	244	267	290	313	336	359	382	405	428	451	474	497	520	543	566
4'	152	176	200	224	248	272	296	320	344	368	392	416	440	464	488	512	536	560	584	608
5'	175	200	225	250	275	300	325	350	375	400	425	450	475	500	525	550	575	600	625	650
6'	198	224	250	276	302	328	354	380	406	432	458	484	510	536	562	588	614	640	666	692
7'	221	248	275	302	329	356	383	410	437	464	491	518	545	572	599	626	653	680	707	734
8'	244	272	300	328	356	384	412	440	468	496	524	552	580	608	636	664	692	720	748	776
9'	267	296	325	354	383	412	441	470	499	528	557	586	615	644	673	702	731	760	789	818
10'	290	320	350	380	410	440	470	500	530	560	590	620	650	680	710	740	770	800	830	860
11'	313	344	375	406	437	468	499	530	561	592	623	654	685	716	747	778	809	840	871	902
12'	336	368	400	432	464	496	528	560	592	624	656	688	720	752	784	816	848	880	912	944
13'	359	392	425	458	491	524	557	590	623	656	689	722	755	788	821	854	887	920	953	986
14'	382	416	450	484	518	552	586	620	654	688	722	756	790	824	858	892	926	960	994	1028
15'	405	440	475	510	545	580	615	650	685	720	755	790	825	860	895	930	965	1000	1035	1070
16'	428	464	500	536	572	608	644	680	716	752	788	824	860	896	932	968	1004	1040	1076	1112
17'	451	488	525	562	599	636	673	710	747	784	821	858	895	932	969	1006	1043	1080	1117	1154
18'	474	512	550	588	626	664	702	740	778	816	854	892	930	968	1006	1044	1082	1120	1158	1196
19'	497	536	575	614	653	692	731	770	809	848	887	926	965	1004	1043	1082	1121	1160	1199	1238
20'	520	560	600	640	680	720	760	800	840	880	920	960	1000	1040	1080	1120	1160	1200	1240	1280
21'	543	584	625	666	707	748	789	830	871	912	953	994	1035	1076	1117	1158	1199	1240	1281	1322
22'	566	608	650	692	734	776	818	860	902	944	986	1028	1070	1112	1154	1196	1238	1280	1322	1364
23'	589	632	675	718	761	804	847	890	933	976	1019	1062	1105	1148	1191	1234	1277	1320	1363	1406
24'	612	656	700	744	788	832	876	920	964	1008	1052	1096	1140	1184	1228	1272	1316	1360	1404	1448

Illustrative Example

To find the total square foot area of a room 10'6" × 20'0" × 10'0":

Area of room 10' × 20' × 10'	= 800 sq. ft.
Area of room 11' × 20' × 10'	= 840 sq. ft.
Difference	40 sq. ft.
Area of room 10' × 20' × 10'	= 800 sq. ft.
Add ½ the difference	20 sq. ft.
Area of room 10'6" × 20' × 10'	= 820 sq. ft.

If the room size was 10'3" × 20' × 10', then 1/4 of the difference would be added or 10 sq. ft. giving that room an area of 810 sq. ft.

Interpolating ceiling heights. If the ceiling height is 8'6", add the area shown for 8' and 9' in the Table and divide by 2 to get the average. If the ceiling height is 9'6", add the area shown for 9' and 10' and divide by 2 to get the average. The answer is accurate.

Illustrative Example

Find the area of a room 10' × 20' with an 8'6" ceiling.

Area of a room 10' × 20' × 8'	= 680 sq ft
Area of a room 10' × 20' × 9'	= 740 " "
	1,420 " "
Area of a room 10' × 20' × 8'6"	= $\frac{1}{2}$ × 1420 sq ft
	= 710 sq ft

MEASUREMENT OF AREAS

The unit of measure for painting customarily is the square yard or square foot. In some parts of the country a *square* (100 square feet) is used as a unit.

All measuring is done in lineal feet, and the areas are first calculated in terms of *square* feet. Since this is true, and because the covering capacity of the materials and the rate of application are expressed in square feet, there would appear to be no reason to convert the figures to square yards or "squares." If it becomes necessary when analyzing or comparing estimates that express areas in terms of square yards or squares, the conversion can readily be done at that time. The unit that will be used in this chapter is the *square foot.*

When estimating the number of square feet of painting, the actual surface to be painted should be measured as accurately as possible because the cost of the material and the number of hours of labor will be determined from the surface area to be covered. The painting or decorating of flat wall surfaces such as ceiling and floor areas, presents no special problem in measurement, but questions frequently arise as to how to obtain the areas for windows, trimmed openings, stairs and balustrades, doors, exterior cornices, lattices and so forth. Rules-of-thumb are employed by most painters and, while there are differences in the methods used, each one strives to reasonably approximate the actual surface to be covered. The following methods for the measurement of flat and irregular surfaces are recommended.

Interior Walls and Ceilings

To obtain the area of the ceiling, multiply the length times the width. The area of the walls is obtained by measuring the distance around the room (the perimeter) and multiplying by the height from the floor to the ceiling. When walls are to be painted only above or below a wainscoting, the height measurement should be only for that portion to be painted.

Whether or not window and door openings are to be deducted will depend on the particular method used to estimate. When openings are deducted, the cost of material and labor to paint openings is added separately. If window and door openings are not deducted, then the additional time for cutting in windows and doors is included in the labor allowance for the room or per square foot area. Actually it requires no additional paint, even though different colors are used for the finish coat. It does require more time than painting a flat surface. (See "Methods of Estimating," this chapter.)

Windows and Doors

Measure the width and height from the outside of the trim of the opening. Add 2 feet to the width and height of a window. Add 2 feet to the width and 1 foot to the height of a door. This will allow for moldings, edges and frames. A window that measures $3' \times 6'$ would be figured as $5' \times 8'$ or 40 square feet. ($3' + 2' = 5'$ and $6' + 2' = 8'$.) A door measuring $3' \times 7'$ would be figured $5' \times 8' = 40$ square feet ($3' + 2' = 5'$ and $7' + 1' = 8'$). Many painters use the unit of 40 square feet for all average sized openings when figuring windows and doors.

Windows or doors with sash having more than one light of glass, are figured at 2 square feet for each additional light. A 6-light sash would add 10 square feet to the window area.

The proper measurement of window and door areas is very important when estimating brick, masonry, or other surfaces where the painting of such openings is strictly an individual operation.

Floors

The number of square feet in a floor is obtained by multiplying the length times the width. Openings less than $10' \times 10'$ in the floor are not deducted.

Wainscoting

If a wainscoting is painted on a flat surface such as plaster, the area is obtained by multiplying the length times the width. When the wainscoting is paneled, the area should be multiplied by 1 1/2 or 2, depending on the form of the surface.

Stairs and Balustrades

To obtain the surface area of stairs that are to be painted, allow 1 foot for the width of the tread, and 1 foot for the width of the riser to take care of the nosing and molding. Multiply this width by the length of the treads or risers, plus 2 feet for stair strings. A stairway with treads and risers 3'6" wide would have 3'6" + 2' = 5 1/2 square feet for each tread and riser. Multiply the area of the tread and the riser by the number of each.

A stairway with 18 risers 4 feet wide would have an area as follows:

$$
\begin{aligned}
\text{Risers} \quad &= 18 \times 1' \times (4' + 2') \\
&= 18 \times 1' \times 6' & &= 108 \text{ sq ft} \\
\text{Treads (1 less than number of risers)} \\
&= 17 \times 1' \times (4' + 2') \\
&= 17 \times 1' \times 6' & &= \underline{102} \text{ sq ft} \\
& & \text{Total} &= 210 \text{ sq ft}
\end{aligned}
$$

Some authorities count risers and multiply by 8 times the width.

The area calculation for a balustrade is obtained by measuring the height from the tread to the top of the handrail. Multiply this height by the length of the balustrade. The area of square edge and relatively simple designed balustrades should be multiplied by four. Turned and fancy balustrade areas should be multiplied by five. This takes care of both sides.

A balustrade that measures 2'6" $\times$ 18' has a plane area of 45 square feet. The paint area for a simple balustrade would be 45 square feet $\times$ 4 or 180 square feet. For a turned and fancy balustrade the paint area would be 45 square feet $\times$ 5 or 225 square feet.

Baseboard, Chair Rail, Picture Moldings, etc.

The paint area of baseboards, chair rails, plate rails, picture moldings and other individual trim members should be estimated on the basis of one square foot per lineal foot.

Trimmed Openings

The paint area of two sides of a *trimmed* opening (one without a door) is obtained by measuring the lineal feet around the two sides and top of the opening and multiplying by three.

A trimmed opening 8 feet wide and 7 feet high would have a paint area of 8' + 7' + 7' = 22' $\times$ 3 = 66 square feet. This area is for both sides of the opening. If it is computed in this manner for one room, care should be taken not to include it in the adjoining room also.

Built-in Shelving, Cupboards, and Cases

To obtain the paint area for built-in shelves, cupboards, book cases and cabinets, measure the front area (width $\times$ height). Multiply the front area by 3 for open-front units. Multiply the front area by 5 for units having doors. The total area obtained includes the complete finishing of the interior and exterior.

Exterior Shingles and Siding
(Wood, Metal and Composition)

To obtain the paint area of exterior shingles and siding, measure the actual wall area including gables, but do not deduct window or door openings less than $10' \times 10'$. For narrow clapboard exposures, and for all shingles, add 20 percent to the actual area. For wide clapboard exposures (over 5 inches) add 10 percent to the actual area.

Exterior Cornices and Eaves

Measure the length and width of the cornice to obtain the area. Multiply this area by 2 for relatively simple cornices, and multiply it by 3 for fancy or ornate cornices.

Where cornices are open, with exposed rafters, multiply the cornice area (length times overhang) by 3 to obtain the paint area.

For closed cornices or eaves that do not have the rafters exposed, add 50 percent to the area.

Exterior Masonry Walls

To obtain the paint area of brick, stucco, cement block, and other exterior masonry walls, measure the actual wall area including gables. Do not deduct window or door openings less than $10' \times 10'$.

Fences

To obtain the paint area of a solid board fence, measure the length and height. Multiply the length times the height to obtain the area of *one side*. Multiply by 2 if it is to be painted on two sides.

The paint area of *one side* of a picket fence is determined by multiplying the length times the height by 2. Two sides, multiply by 4. Chain link fences, measure one side, multiply by 3.

Lattice

To obtain the paint area of lattice work, multiply the length times the width. To paint *one side* multiply the actual area by 3.

ESTIMATING MATERIALS

Painting and finishing is one of the least difficult of building trades to estimate, because the quantities of material, and the numbers of hours of labor are both determined directly from the area to be covered.

When the paint area has been measured and calculated, the quantity of *material required per coat is obtained by dividing that area by the number of square feet a gallon of paint will cover.*

Illustrative Example No. 1

A room 14' × 18' × 9' is to be given *one* coat of flat paint on sheet rock walls and ceiling.

Ceiling area	14' × 18' = 252 sq ft
Wall area	2(14' + 18') × 9' = 576 sq ft
	Total = 828 sq ft

(*From Table 15–1*)

$$\text{Paint required for one coat} = \frac{828}{450} = 1.84 \text{ gallons}$$

If only one room is to be painted the quantity would be rounded out to 2 gallons. Where other rooms are to be painted, any adjustment to whole or half gallons is made at the end of the entire painting estimate.

Illustrative Example No. 2

A room 10' × 16' × 8'6" is to be painted *two* coats of flat on plastered walls and ceiling.

Ceiling area	10' × 16' = 160 sq ft
Wall area	2(10' + 16') × 8 ½ = 442 sq ft
	602
(Two coats)	× 2
	1,204 sq ft

(Tables 15–2A and 15–2B may also be used)
(*From Table 15–1*)

$$\text{Paint required} = \frac{1,204}{450} = 2.45 \text{ gallons}$$

ESTIMATING LABOR

The conditions which have an influence on painting labor specifically have been discussed in the first part of this chapter. Those factors which affect labor in general will be found under "Estimating Labor" in Chapter 2.

The labor-hours required to apply paint materials to 100 square feet of surface under a given set of conditions vary, but may be estimated reasonably closely on the basis of accepted average rates of applying paint. Table 15–3 shows the approximate number of square feet per hour required to apply paint materials for various types of interior work. They represent averages for two or three coats rather than individual first, second, or third coats. Included in the rates of application shown is the time needed for a normal amount of preparation such as sanding trim, patching minor plaster cracks, puttying nail holes, dusting, and so forth. The rates also contemplate working from step ladders to paint ceilings and working from extension ladders on exteriors. It includes time for getting ready and for cleaning up when the job is done.

Any special preliminary preparation such as having to move furnishings or merchandise, wash down walls, or burn off blistered paint, should be estimated separately and added to the over-all cost of the job.

The labor rates shown in Table 15–3 for "Interior" and in Table 15–4 for "Exterior" walls include the windows and doors found in the rooms of an average dwelling.

They also include the painting of baseboards, chair rails, and so forth. For example, the rate of 150 square feet an hour shown for smooth plastered walls including openings is an average between 200 square feet an hour for a flat surface with no openings and 125 square feet an hour for painting openings and trim. Whenever window and door openings and trim items are figured separately from the wall surfaces, the area of the openings should be deducted; otherwise there will be an overlapping of cost which will produce an excessive estimate. When it is desired to estimate openings and trim individually, the cost may be computed as outlined under "Measurement of Areas" or by using 1/16 of a gallon of material and 1/3 hour labor as shown in Table 15–3.

When unusual conditions are encountered, the labor rates shown in Tables 15–3 and 15–4 (page 328) may be modified according to judgment.

TABLE 15–3

Approximate Number of Hours to Apply Painting Materials for Type
of Interior Work Shown—Premises Unoccupied

TYPE OF WORK	Sq Ft Per Hour Per Coat	
	Brush	Roller
Washing down smoke-stained walls and ceilings	100–150	
Sanding woodwork, preparatory	150–250	
Sizing plastered walls	300–400	
Burning paint from trim	25–40	
Painting smooth plaster incl openings	150	
Painting smooth plaster excl openings	200–250	300–350
Painting sandfinish plaster excl openings	175–200	350
Painting sandfinish plaster incl openings	125	
Painting texture plaster-semi-gloss	100–125	350
Painting texture plaster-flat paint	180	350
Painting Cement block or Brick	150	250
Spray painting		350–400
Painting Concrete floors or walls	200–250	300–350
Floors-remove finish with liquid	50	
Floors shellac, paint, stain or varnish	200–300	400–500
Floors wax and polish	200	
Paneling-Staining, or finishing	250	
Trim-back priming-paint	350	
*Windows–one side	125–150	
**Doors–one side	125–250	
Moulding etc–1 to 6 inches wide	175 lin ft	
Note: No caffolding included in above hour rates.		

See pages 322 to 325, Measurement of Areas.
  *Average 12–14 openings per day, one coat, one side per 8 hours, 6/6 lites.
  **Flush-average 16–18 openings per 8 hours, one coat, one side.
   Panel–average 12–14 openings per 8 hours, one coat, one side.
There are some painters who estimate 1/16 gal. of paint per side per opening, and 1/3 hours labor per side per opening.

TABLE 15-4

Approximate Number of Hours to Apply Painting Materials for Type
of Exterior Work shown—Average from Ground and Ladders

TYPE OF WORK	Sq Ft Per Hour Per Coat	
	Brush	Roller
Sanding woodwork lightly	250	
Sanding blistered woodwork	100–200	
Burning off paint from woodwork	25–40	
Wood siding, incl window and door openings	100–125	
" " excl window and door openings	200–225	300–350
*Exterior trim only incl windows and doors	80–100	
Shingle staining-walls	150	200
Brick walls–Oil paint	125–150	200–250
Stucco walls–Oil Paint	100–150	200–250
Concrete block	100–125	250–300
Floors–wood	200	250
Floors–concrete	200–250	300–350
Roofs–Wood shingle stain	200–225	250–300
Spraying		500
Shutters	4–5 units	
Screens	5–6 "	
Storm sash	4–5 "	
Fences–solid board	250	400
" –picket	150–200	250–300
" –picket (spray)		450–500
" –link	125	300
Leaders and gutters	100 lin ft	

*Windows–painter will paint (one-coat) 12–14 openings per day, one side only.
  Flush doors–Painter will paint (one-coat) 16–18 openings per day, one side only.
  Panel doors–painter will paint (one-coat) 12–14 openings per day, one side only.
Note: See pages 322 to 325, Measurement of Areas.

METHODS OF ESTIMATING

There are five basic methods or systems of estimating painting and finishing which are in common use. In some sections of the country one particular method may become established by custom and usage to the exclusion of any other method. For that reason, and also for purposes of analyzing painting estimates, it is important to understand each system. When properly used, there should be little difference in the results obtained.

To demonstrate the application of each of the five methods, a single problem, which contemplates painting a typical room, is used.

Illustrative Problem

To paint the walls, ceiling and woodwork of a dwelling living room 16' × 20' × 9'. There are two doors with openings 3' × 7', and three windows with openings

3' × 6'. A 6" baseboard runs around the room. The walls and ceilings are finished in smooth plaster and no preparation of surfaces is required. The walls and ceiling are to be painted 2 coats of flat white. The woodwork and trim are to receive one coat of flat and one coat of semi-gloss enamel of a color different from the wall paint.

Areas

Ceiling	16' × 20' = 320 sq ft
Walls	2(16' + 20') × 9' = 648 sq ft
	Gross area = 968 sq ft

Openings (actual opening size)

2 Doors	3' × 7' =	42 sq ft
3 Windows	3' × 6' =	54 sq ft
Total		−96 sq ft

Net wall and ceiling area = 872 sq ft

Baseboard (openings not deducted)

$$2(16' + 20') = 72 \text{ lin ft}$$

For purposes of illustration painter's wages are assumed to be $12.00 an hour, and both flat and semi-gloss enamel to be $14.00 a gallon.

Method No. 1—Material and Labor

Separately Estimated

In this method the total number of gallons of paint, and the total hours of labor are determined from the *gross* wall and ceiling area. The number of gallons of paint is based on the average covering capacity of one gallon. The number of hours of labor is based on the number of square feet a painter can average in one hour.

Table 15–1 shows, for this type of paint and surface, an average covering capacity of 450 square feet for each gallon. By dividing the gross area of the room by the covering capacity per gallon, the number of gallons of paint required for one coat is obtained. For two coats it would require twice as much paint.

$$\frac{\text{Gross Area}}{\text{Coverage}} = \frac{968}{450} = 2.15 \text{ gallons per coat}$$

$$\text{Two coats} = \frac{\times 2}{4.3 \text{ gallons}}$$

Table 15–3 shows that a painter can apply an average of 150 square feet an hour on smooth plastered walls, ceilings and trim. The number of hours of labor required for each coat is obtained by dividing the gross area of the room by 150 square feet. The number of hours to apply two coats is double that required for one coat.

$$\frac{\text{Gross Area}}{\text{Rate per hr}} = \frac{968}{150} = 6.4 \text{ hours per coat}$$

$$\text{Two coats} = \frac{\times 2}{12.8 \text{ hours}}$$

Estimated Cost in Problem

4.3 gals paint	@	$ 14.00 =	$ 60.20
12.8 hours labor		12.00 =	153.60
		Total cost =	$ 213.80

When the gross area of a room is used to compute material and labor, it is contemplated that less paint is used on window openings. More labor, however, is needed to paint the window bars, and also the trim, particularly when a different color is used. The rate of 150 square feet an hour for applying paint takes into consideration the additional time for painting a normal number of openings and the trim in an average room. On a flat, smooth-plastered wall surface without openings an experienced painter would be able to cover between 200 and 300 square feet an hour with very little difficulty.

Method No. 2—A Unit Cost Applied to Gross Area

A popular method in some sections of the country is one in which a unit cost per square foot or square yard is developed and applied to the gross area without any deductions for openings. It employs the same principle as method No. 1 in that the rate of application makes allowances for the time required to cut in window and wood trim where different colors of paint are used.

Using a covering capacity of 450 square feet a gallon and a rate of application for a painter of 150 square feet an hour, the unit cost is obtained as follows:

1 gallon paint	@ $ 14.00 =	$	14.00
$\frac{450}{150} = 3$ hours labor	@ 12.00 =		36.00
Total cost of labor and material	=	$	50.00
Unit cost per sq ft $\frac{\$50.00}{450}$	=	$	.11 per coat
Unit cost per sq ft for 2 coats	=	$	.22

Estimated Cost in Problem

Gross area of room	=	968 sq ft
Unit cost per sq ft for 2 coats	=	× $.22
	Total cost =	$212.96

It should be noted that the covering capacity of a gallon of paint on smooth plastered walls is 450 square feet as taken from Table 15–1. The rate of application for a painter shown in Table 15–3 is 150 square feet an hour. The number of hours in this instance to apply one gallon of paint is obtained by dividing 450 by 150, which equals 3 hours.

The unit cost developed in this method is $.222, which in actual practice would be rounded off at $.22 per square foot.

Method No. 3—Separate Unit Costs Used for Flat Wall
and Ceiling Surfaces, and for Openings and Trim

A method that is quite common in many of the central, midwestern, and Pacific Coast states is one in which the walls and ceilings, the window and door openings, and the trim items are estimated on an individual unit cost. The unit cost that is used takes into consideration the conditions applicable to each type of surface, on the principle that paint coverage and the rate of application vary with the kind of paint and the form of the surface being painted. This method of estimating is taught in a number of trade schools for painters. Because of the refinement of the estimate, it is considered by many to be more accurate than any other system. It has particular merit where the treatment of the wall or ceiling surfaces and that given to openings and trim are substantially different in both character and in the cost per square foot. If, for example, the walls and ceilings are calcimined or painted with cold-water paints and all woodwork is painted a coat of flat and finished with semi-gloss, an average unit cost per square foot to be applied to the gross area would be difficult if not impossible to compute.

The deduction of openings from wall area is important where the cost of painting such openings is added as a separate item to the estimate because if a unit cost is applied to the gross wall area, the cost of the opening is figured twice. An average door opening, for example, is roughly 3' × 7', and if a unit cost of \$.20 per square foot is used, the cost of the opening is \$4.20. An average bedroom contains 2 doors and 2 windows, or 4 openings. Inclusion of the openings adds \$16.80 to the estimate. When 20 percent overhead and profit is added, the cost is further increased to \$20.16 for one room. There is no way to modify a unit cost which is to be applied to the gross area of a room so that it will take out openings. They must be physically deducted and a proper unit cost applied to the net wall area. These matters should be thoroughly understood as they become important when analyzing estimates on painting.

Under Method No. 3, it is customary to consider a baseboard as approximately equal to one opening. Because the cost of painting the openings is estimated separately from wall and ceiling surfaces, *all window and door openings are deducted* using the inside measurements of the openings. Different rates for applying paint are used for the walls and for the openings and trim. Table 15–3 shows a rate of 200 square feet an hour for plain walls and ceilings. It also shows a rate of 125 square feet an hour for painting windows, doors, and trim. Using these rates for applying paint, the following unit costs are developed:

Unit Costs for Walls and Ceilings

1 gallon paint	@ \$ 14.00 =	\$ 14.00
$\frac{450}{200}$ = 2 1/4 hours labor	@ ' 12.00 =	27.00
	=	\$ 41.00

$$\text{Unit cost per sq ft for 1 coat } \frac{\$41.00}{450} = \$.09$$

$$\text{Unit cost per sq ft for 2 coats} = \$ \quad .18$$

Unit Cost for Openings and Trim

1 gallon paint	@ $ 14.00 =	$ 14.00
$\frac{450}{125}$ = 3.6 hours	@ 12.00 =	43.20
		$ 57.20

Unit cost per sq ft for 1 coat $\frac{\$57.20}{450}$ = $.127

Unit cost per sq ft for 2 coats = $ .25

Estimated Cost in Problem

Net wall and ceiling area 872 sq ft × $ .18   = $ 156.96
Openings and Baseboard
3 windows   )
2 doors   )   6 openings @ 40 sq ft 240 sq ft
1 baseboard) $ .25   = $ 60.00
                                     Total cost $ 216.96

Method No. 4—Material and Labor Per Gallon
of Material

The cost of painting can be expressed in terms of the cost of labor and material per gallon. This unit cost per gallon can be multiplied by the number of gallons of paint required to obtain the total cost of labor and material.

Unit Cost Per Gallon

1 gallon paint	@ $ 14.00 =	$ 14.00
3 hours labor	@ 12.00 =	36.00
Cost of material		
and labor per gallon		$ 50.00

In the problem being used for illustration the total number of gallons of paint for two coats would be $\frac{968 \times 2}{450}$ = 4.3 gals.

Estimated Cost
4.3 gals. @ $50.00 = $215.00

Method No. 5—Material and Labor Per Hour

A method occasionally used by painters is to develop a unit cost per hour for both materials and labor. Using a covering capacity of 450 square feet per gallon and 150 square feet per hour, the unit cost would be developed as follows:

Unit Cost Per hour

$\frac{150}{450}$ = ⅓ gal of paint . . . . @ $ 14.00 = $ 4.66

1 hour labor . . . . @   12.00 =   12.00

Unit cost of material and labor per hour $ 16.66

Many estimators prefer this method, particularly for rough estimating, because they can look at a room, a house, or a complete job and approximate the number of hours of labor required to do the work.

In the illustrative problem the number of hours required at 150 square feet per hour of application would be $\dfrac{968 \times 2 \text{ (coats)}}{150} = 12.9$ hours

Estimated Cost. 12.9 hrs $\times$ $16.66 = $214.91

Summary of Methods

The choice of the method by which painting is estimated is one of individual preference, but whatever system is used, it should consider all of the conditions peculiar to the particular job. The five methods outlined show very little variation in cost when applied to the illustrative problem.

Methods	Total Cost
No. 1	$ 213.80
No. 2	212.96
No. 3	216.96
No. 4	215.00
No. 5	214.91

Unit Costs for Painting

A practical way to develop a unit cost is to use a gallon of material as a basis, because the covering capacity is expressed in terms of square feet. The rate at which painting material can be applied is expressed in square feet per hour. The following basic formula may be used in each instance:

Basic Formula

1 gal. paint	@ $ = $	
Hours labor = $\dfrac{\textbf{Coverage per gal.}}{\textbf{Rate of application per gal.}}$	@ $ = $_____	
Total cost of labor and material per gallon	= $	
Unit cost per sq ft = $\dfrac{\textbf{Total cost}}{\textbf{Coverage per gal.}}$	= $	

Illustrative Example No. 1

In Table 15–1 the covering capacity on exterior siding is given as 500 square feet per gallon. In Table 15–4 the rate of applying exterior paint on smooth surfaces including all openings is 125 square feet per hour. You may develop a unit cost for 1 coat painting on bevel siding as follows:

Exterior paint cost assumed @ $14.00 per gal.
Painter's wages assumed @ $12.00 per hour

Unit Cost

1 gallon paint	@ $ 14.00 =	$ 14.00
$\dfrac{500}{125}$ = 4 hours labor	@ 12.00 =	48.00
		$ 62.00

Unit co t per square foot $\dfrac{\$62.00}{500}$ = $.124 per coat

Unit cost per square foot for 2 coats $ .248 or $.25

Illustrative Example No. 2

To develop a unit cost for 1 coat of varnish on prepared floors. Table 15–1 shows the covering capacity of varnish as 400 square feet per gallon. Table 15–3 shows the rate of applying varnish to floors as 200 square feet per hour.

Floor varnish price assumed   @ $ 16.00 per gal.
Painter's wages assumed   @   12.00 per hour

Unit Cost

1 gallon varnish	@ $ 16.00 =	$ 16.00
$\dfrac{400}{200}$ = 2 hours labor	@ 12.00 =	24.00
		$ 40.00

Unit cost per square foot $\dfrac{\$40.00}{400}$ = $ .10 per coat

Illustrative Example No. 3

To develop a unit cost per opening for windows and doors using $1/2$ pint ($1/16$ gals.) material and allowing $1/3$ hour labor per opening per coat.

Paint cost assumed   @ $ 14.00
Painter's wages assumed   @   12.00

Unit Cost

1/16 gallons paint	@ $ 14.00 =	$ .88
1/3 hours labor	@ 12.00 =	4.00
Unit cost per opening, 1 coat	=	$ 4.88
Unit cost per opening, 2 coats	=	$ 9.76

APPLICATION OF UNIT COSTS FOR PAINTING

Illustrative Example No. 1

The walls and ceilings of the following four rooms are to be painted with two coats of flat, and the woodwork is to be painted with one coat of flat with a finish coat of semi-gloss.

Assuming that paint costs $14.00 per gallon and painter's wages are $12.00 per hour, obtain the total cost using Method No. 2.

1 gallon paint	Unit Cost @ $14.00 =	$ 14.00
$\dfrac{450}{150}$ = 3 hours labor	@ 12.00 =	36.00
		$ 50.00

Unit cost per sq ft for 2 coats $\dfrac{\$50.00}{450}$ × 2 = $ .22

Estimated Cost

Room size	Gross area of walls and ceiling		
10' × 14' × 9'	572 sq ft	@ 22¢	$ 125.84
12' × 18' × 9'	756 sq ft	@ 22¢	166.32
8' × 10' × 9'	404 sq ft	@ 22¢	88.88
11' × 12' × 9'	546 sq ft	@ 22¢	120.12
		Total cost	$ 501.16

Illustrative Example No. 2

The exterior bevel siding of a dwelling is to be painted with two coats. The size of the building is 45' × 30' with a 6' roof at the gable end. The height of the walls from foundation to the eaves is 12', and the cornices are finished flush.

Wall area	= 2(45' + 30') × 12'	= 1,800 sq ft
Gable area	= 6' × 30'	180 sq ft
		1,980 sq ft

Unit Cost
(Wages and Prices Assumed)

1 gallon paint	@ $ 14.00 =	$ 14.00
$\frac{500}{125}$ = 4 hours labor	@ 12.00 =	48.00
		$ 62.00
Unit cost per square foot 1 coat $\frac{\$62.00}{500}$	=	$ .124
Unit cost per square foot 2 coats	=	$ .25

Estimated Cost
1980 square feet @ $ .25 = $ 495.00

DAMAGEABILITY

Paints, varnishes and other decorative materials can be damaged during a fire by heat, smoke, water, and chemicals used to fight the fire, staining from contents, and by scratching or marring. Hailstones damage exterior paint, water from leaky plumbing or the backing up of sewers causes injury to interior paint, and wind-driven sand can scarify or completely remove exterior paint. Whatever the cause of the damage, it is visible and usually presents no problem of discernment with the exception of isolated cases of questionable damage by acid fumes, smoke, or other stains.

The main concern of an estimator is to determine the requirements and the specifications for restoring the surface to its original condition. The problem can be generally divided into two categories:

1. The preparation of the surface before finishing
2. The number of coats to be applied

Preparation of Surface

Before paint, varnish, or other materials can be applied, the surface must be cleaned to remove the dirt and grease. Blistered areas on wood must be burned off with a torch and sanded smooth. Scraping blackened and blistered woodwork with a knife will disclose whether or not the wood itself has been burned.

Walls and ceilings that are seriously coated by smoke should be washed before painting for two reasons. Washing removes dirt and grease and, on more occasions than are recognized, washing a surface will make restoration possible by the application of a single coat. Where the soilage is slight, washing may remove all evidence of injury eliminating the need for painting.

No painting should ever be done on wet wood or wet plaster surfaces. Nail holes, scratches, raised grain, and other surface defects and blemishes should be filled and sanded smooth. Cracks in plaster should be cut out and filled.

Many times steel members, masonry, metal ceilings, and so forth, must be wire brushed to remove rust, blisters or other loose material before painting. Varnished, enameled, and other glossy surfaces should be lightly sanded before applying new material.

Only a careful examination of the damaged property will make known the extent of necessary surface preparation. In most cases it will consist of washing or cleaning of the old surface, with possibly minor sanding or spackling. When there is a doubt that a surface requires, or can be, washed, brushed, sanded or scraped in order to prepare it, sample tests should be made. Tables 15–3 and 15–4 show the approximate number of square feet per hour required to prepare surfaces. The labor cost per square foot is obtained by dividing the local hourly wage rate by the number of square feet covered per hour. The cost of any materials such as washing powders, detergents, sponges, varnish remover and acids should be added.

Number of Coats to Be Applied

Exterior. New exterior woodwork should be given three coats of paint. On old painted surfaces, depending on their condition, one or two coats of the existing color are adequate. If the old paint is in good condition and the damage is slight, one coat of paint will generally give a satisfactory finish. If the old paint is in poor condition or has been seriously injured, it may require two coats.

Interior. Ordinarily new interior woodwork receives a prime coat and a finish coat of paint. Newly plastered walls should be given a coat of size, followed by two coats of paint. Calcimine or cold-water paints are applied over new plaster after applying a thin coat of shellac or sizing.

Old woodwork and plastered or dry walls may require one coat only if the soilage is slight, or if pre-washing will remove most of the stain. Otherwise, two coats are required to restore the surface.

PAPERHANGING

Wallpaper is sold by either the single or the double roll. A single roll is a strip 18 inches wide, 24 feet long, and contains 36 square feet. A double roll is 48 feet long and contains 72 square feet. Wallpaper borders are sold by the lineal yard. The quality and the price of wallpaper have wide ranges. It is, therefore, important to determine as closely as possible the quality of paper that was on the walls at the time of a loss.

Preparation of the Surface

Before walls are papered, they should be given a coat of glue size, the cost of which has been discussed under "Painting."

Old wallpaper should be removed, the walls washed, cracks patched and a coat of sizing applied. Wallpaper can be taken off by wetting it down with water, by brush or spray, and then scraping it off with a wide putty knife. An electric-steam appliance is available which speeds up the work of loosening the paper. A painter can remove old wall paper at the rate of 150 sq ft a single layer, and about 100 sq ft where there are two or more layers to be removed.

When permissible, new wallpaper can be applied directly over old paper provided the surface is smooth (not embossed) and the old paper is butt-jointed and tight on the wall. Under these conditions papering over old paper is entirely satisfactory.

TABLE 15–5

Estimating Quantities for Paperhanging Walls and Ceilings:

Measure straight area to secure number of square feet and divide by 30. (This allows 6 sq. ft. for waste.) Deduct ½ roll for every opening. Or, for the following standard-size rooms use this handy chart:

Size of Room	Single Rolls of Side Wall Height of Ceiling			Yards of Border	Rolls of Ceiling
	8 feet	9 feet	10 feet		
4 × 8	6	7	8	9	2
4 × 10	7	8	9	11	2
4 × 12	8	9	10	12	2
6 × 10	8	9	10	12	2
6 × 12	9	10	11	13	3
8 × 12	10	11	13	15	4
8 × 14	11	12	14	16	4
10 × 14	12	14	15	18	5
10 × 16	13	15	16	19	6
12 × 16	14	16	17	20	7
12 × 18	15	17	19	22	8
14 × 18	16	18	20	23	8
14 × 22	18	20	22	26	10
15 × 16	15	17	19	23	8
15 × 18	16	18	20	24	9
15 × 20	17	20	22	25	10
15 × 23	19	21	23	28	11
16 × 18	17	19	21	25	10
16 × 20	18	20	22	26	10
16 × 22	19	21	23	28	11
16 × 24	20	22	25	29	12
16 × 26	21	23	26	31	13
17 × 22	19	22	24	28	12
17 × 25	21	23	26	31	13
17 × 28	22	25	28	32	15
17 × 32	24	27	30	35	17
17 × 35	26	29	32	37	18
18 × 22	20	22	25	29	12

TABLE 15-5

Estimating Quantities for Paperhanging Walls and Ceilings:
Measure straight area to secure number of square feet and divide by 30.
(This allows 6 sq. ft. for waste.) Deduct ½ roll for every opening. Or, for the
following standard-size rooms use this handy chart:

Size of Room	Single Rolls of Side Wall Height of Ceiling			Yards of Border	Rolls of Ceiling
	8 feet	9 feet	10 feet		
18 × 25	21	24	27	31	14
18 × 28	23	26	28	33	16
20 × 26	23	28	28	33	17
20 × 28	24	27	30	34	18
20 × 34	27	30	33	39	21

Deduct one single roll of side wall for every two ordinary sized doors or windows or
every 36 square feet of opening. Yard goods with no match such as wide vinyls, measure area, take out for openings and allow 10″ for waste.
Courtesy of Painting and Decorating Contractors of America, Falls Church, Va.
22046

Estimating Materials

When hanging wallpaper, some allowance must be made for the waste which results from cutting, fitting around openings and built-in units, and for matching patterns. As a general rule the waste averages between 15 and 20 percent. A single roll of
paper contains 36 square feet. A practical and acceptable method for allowing waste is
to figure the rolls at 30 square feet. Double rolls are figured at 60 square feet to take
care of waste.

The cost of material and labor for papering is estimated by the roll. If the cost
per square foot or square yard is desired, it may be obtained in two ways: by dividing
the cost of papering a room by the wall area, and by computing the cost of hanging
one roll of paper and dividing by 30 square feet. (See "Methods of Estimating.")

To obtain the number of rolls of paper required, the distance around the room
(perimeter) is multiplied by the height from the top of the baseboard to the ceiling. *All
window and door openings are deducted.* The result is the wall area to be papered. That
area is divided by 30 to obtain the number of single rolls of paper. Paper is sold in full
rolls and the number of rolls needed must be carried out to the next whole roll. If double rolls are used, the area should be divided by 60.

Where walls are papered above a chair rail only, or above a wainscoting, the
height measurement should be taken from where the paper starts to the ceiling. When
a single wall is to be papered, the length of that wall is measured and multiplied by the
height.

Where ceilings are to be papered, the area is obtained by multiplying the length
of the ceiling by the width.

If a border is required, the distance around the room divided by 3 will give the
number of yards required.

Illustrative Example

Find the number of single rolls of paper and the number of yards of border required for the walls of a room 14' × 18' × 8' 6". The baseboard is 6" high. Openings consist of two windows 3' × 5' and two doors 3' 6" × 7'.

Area of Walls (above baseboard)
2(14' + 18') × 8'   = 512 sq ft
Deduct openings
2 windows 3' × 5' = 30 sq ft
2 doors 3½' × 7' = 49 sq ft
                                            $\underline{\hspace{1cm}}$ −79 sq ft
                        Area to be papered = 433 sq ft

$$\text{Rolls of paper} = \frac{433}{30} = 14.4 = 15 \text{ rolls}$$

$$\text{Yards of border} = \frac{2(14' + 18')}{3} = \frac{64}{3} \qquad = \quad 22 \text{ yards}$$

The amount of paste needed to hang paper varies with the type and weight of the paper. Heavy or rough textured papers require more than lightweight papers. For average estimating purposes a pound of paste makes 2 gallons, which will hang approximately 10 single rolls of paper.

Estimating Labor

A paperhanger will hang paper at the rate of 3 single rolls per hour under normal conditions. This rate is for butt-jointed work. If the joints are lapped, the rate should be increased to about 3 1/2 or 4 rolls per hour.

Where work and materials are of the highest grade, a paperhanger will hang paper at the rate of 2 rolls per hour.

Paper borders can be hung at the rate of approximately 30 yards per hour.

Table 15–6 shows the approximate labor hours required for various paperhanging operations (page 341).

METHODS OF ESTIMATING COST

Material and Labor Per Room

Some estimators prefer to obtain the cost of wall papering by estimating the material and labor required for each room rather than using a unit cost per square foot.

Illustrative Example

To find the cost of papering a room 12' × 16' × 8½'. It has two windows 3' × 5' and one door 3'6" × 7". Assume a baseboard 6" wide. The cost of the paper is $5.00 per roll, paste is $.30 per lb and paperhanger's wages are $12.00 per hour.

Area of Walls = 2(12' + 16') × 8½'   =   476 sq ft
Deduct openings
2 windows 3' × 5' = 30 sq ft
1 door 3½' × 7' = 25 sq ft
                                                          $\underline{\hspace{1cm}}$ −55 sq ft
Area to be papered   421 sq ft

Material cost

Rolls of paper needed $= \dfrac{421}{30} = 14$ @ $ 5.00 $=$ $ 70.00

Paste (1 lb for 10 rolls) 1.4 lbs   @   .30 $=$ .42

Total cost of material $=$ $ 70.42

Labor cost (Table 15–6)

Labor 3 rolls per hour $\dfrac{14}{3} = 4.7$ @   12.00 $=$ 56.04

Total cost of papering $=$ $126.46

A much simpler procedure is to develop a unit cost from a single roll of paper. The following basic formula may be used and local wage rates and material prices inserted. The illustration contemplates that a paperhanger can hang an average of 3 single rolls of paper an hour. When the rate of hanging is changed because of better grade work or an inferior grade, the formula may be adjusted accordingly. If 2 rolls are to be hung each hour, an allowance of 1/2 hour for labor is made in the formula. If 4 rolls per hour are contemplated, then the labor figure should be 1/4 hour. Paste is a relatively insignificant item and may be omitted if desired.

Unit Cost Per Square Foot

A unit cost per square foot may be obtained in two ways. The cost of the material and labor for a particular room may be computed and divided by the number of square feet to be papered. In the previous Illustrative Example the total cost of labor and material was $126.46. The area papered was 421 square feet. By dividing $126.46 by 421 a unit cost of $.30 is obtained. This method is useful where several rooms are to be papered. Once a unit cost is obtained for a typical room, it can be applied to other similar rooms.

Basic Formula

1 roll of wallpaper   @ $ = $

$\frac{1}{10}$ lb wallpaper paste   @   =

Hours labor   @   =   ____

Total cost per roll = $

Unit cost per sq ft $= \dfrac{\$}{30}$ = $

Illustrative Example

Using the material prices and labor wage rate in the previous example, a unit cost may be obtained.

1 roll of wallpaper   @ $5.00 = $5.00

$\frac{1}{10}$ lb wallpaper paste   @   .30 =   .03

$\frac{1}{3}$ hours labor   @   12.00 =   4.00

Total cost per roll = $9.03

Unit cost per sq ft $= \dfrac{\$9.03}{30} =$ $ .30

TABLE 15-6

Labor to Hang Paper

A. Ceilings (residential or commercial) rolls per 8 hr. day.

	Occupied	Vacant
a. Mica	18 to 24 rolls	26 to 30 rolls
b. Brush-tint	16 to 18 rolls	22 to 24 rolls
c. Lining (blank stock)	16 to 18 rolls	22 to 24 rolls

B. Sidewalls (residential or commercial) rolls per 8 hr. day.

	Occupied	Vacant
a. Lining (blank stock)	18 to 20 rolls	22 to 24 rolls
b. Ordinary pre-trim.	16 to 18 rolls	22 to 24 rolls
c. Hand-knifed	12 to 14 rolls	14 to 16 rolls
d. Cork	12 to 14 rolls	14 to 16 rolls
e. Grasscloth	14 to 16 rolls	18 to 20 rolls
f. Flock	8 to 10 rolls	. . . .
g. Burlap	12 rolls	. . . .
h. Silk	8 to 10 rolls	. . . .
i. Foil	8 rolls	. . . .
j. Scenics	10 panels per day	
k. Coated fabrics	14 to 16 rolls	18 to 20 rolls
l. Lining canvas	14 to 16 rolls	18 to 20 rolls

Vinyl or Coated Fabrics (commercial) rollage & yardage per 8 hr. day:

A. Light weight, 24" to 27", 18 to 22 rolls.

B. Heavy weight, 54", 30 lineal yards average; 45–65 lin. yds. corridor & volume.

Wood Veneer, Flexwood etc. sq. ft. basis

A. 105 to 145 sq. ft. per day.

Courtesy, Painting and Decorating Contractors of America

SANITAS AND WALLTEX

Sanitas and Walltex come in rolls containing 9 square yards. They are frequently used on the walls of bathrooms and kitchens because they are water-resistant and washable. The area is obtained in the same manner as for wallpaper. The quantity of Sanitas or Walltex required is equal to the area to be covered plus 10 percent for waste. Because they are expensive, measurements for Sanitas and Walltex should be taken with care.

A paperhanger can hang approximately 2 rolls of Sanitas and Walltex per hour.

DAMAGEABILITY

There is little to be said concerning the damageability of wallpaper that is not generally known. The problem that frequently confronts an estimator is deciding whether soiled wallpaper should be cleaned rather than replaced. Many times it is necessary to make sample tests, or call in wall cleaning firms for advice when they are available.

Most wallpapers that have been wet will have to be replaced. Unless perfect matching of the existing paper is possible, the entire room will have to be repapered even when small areas are stained.

Sanitas and Walltex that have been wet from the back side should be carefully examined to determine whether or not the glue has been injured.

Smoked or otherwise soiled walls can frequently be cleaned with commercial dough-type cleaners. A worker can clean wallpaper at the approximate rate of 100 square feet per hour by this method. Water-resistant fabric type materials, when subjected to cleaning, can usually be lightly washed with water and soap or detergents. Care should be taken not to remove any of the painted surface of the fabric.

GLAZING

The cost of glazing depends mainly on five conditions which should be considered in estimating:

1. Type and thickness of glass.
2. Quality of glass.
3. Size and shape of glass.
4. Type of sash.
5. Difficulties of installing.

Type and Thickness of Glass

There are many types of glass used in buildings. Clear window glass ranges in thickness from about 1/12 inch to 1/4 inch. It is available in two grades, or qualities, "A" and "B." Window glass is customarily identified in thickness as Single Strength (S.S.), Double Strength (D.S.), Semi Plate (S.P.), and Plate (Pl.) glass. The respective thicknesses are approximately 1/12 inch for Single Strength, 1/8 inch for Double Strength, 3/16 inch for Semi Plate, and 1/4 inch or 5/16 inch for Plate glass. Single Strength glass is not recommended for lights containing areas in excess of about 3 square feet. Double Strength may be used for lights up to 5 or 6 square feet. Crystal sheet glass (Semi Plate) is a type used where plate glass is too heavy and expensive, but where heavier than D.S. glass is desired. It comes in different thicknesses, but the 3/16 inch or 39 ounce per square foot is most commonly used.

In mercantile and industrial buildings wire glass is commonly used in doors and windows that are required to be fire resistive, and in areas exposed to possible fires in adjoining properties. The wire glass may be rough, ribbed, or polished. In specifying which type of wire glass is required, the thickness should also be indicated.

The ornamental types of glass include Florentine, Pressed, Ribbed, Flutex, Maze, Syenite, Frosted, Leaded and other art glass.

Size and Shape of Glass

The size of each light of glass should be measured carefully. In window glass the available sizes range in width from 6 to 12 inches by 1-inch intervals, and from 12 to 30 inches in 2-inch intervals. The sizes in length range from 8 to 40 inches in 2-inch intervals. Plate glass should be measured in even inches. Wire glass, Window glass, and Fancy glass lights should be measured exactly before obtaining prices from local dealers.

Glass is sold on discounted price lists. The amount of discount varies with the type, quality, size and quantity.

Specifications for glass in an estimate should include details of any special shape or finish of edges. Bent glass and beveled or polished edges all add to the price of the glass.

Quality of Glass

Single-strength glass, because of its thinness and slight variation in thickness, may have certain imperfections in the form of blisters or lines. There are three grades of quality in window glass. The most perfect glass is graded AA which has no objectionable imperfections. Grade A is the next highest and may contain a few flaws, none of which are prominent. Grade B is the lowest grade and generally contains noticeable imperfections.

Type of Sash

Wood sash with 1 3/8-inch frames require less putty than those with 1 3/4-inch frames. Steel sash generally require more putty than wood sash.

Difficulties of Installing

Allowances must be made for any difficulties a glazier will encounter when replacing lights of glass. If the putty is hard and adheres tightly to the frame, it may take more than normal time to cut it out. Glazing sash is easier from the floor or ground level than from exterior ladders or scaffolding. A glazier works outside faster in summer than in winter when the putty tends to stiffen with the cold.

Estimating Material

In estimating glass, the number of lights of each size should be noted with a description of type, thickness and quality. Prices should be obtained from a local dealer.

The amount of putty needed for glazing varies, and in relation to the total cost of glazing it is a minor item. For estimating the following amounts are recommended:

Wood sash (1⅜-inch frames)	1 lb per 8 lin feet of puttying
Wood sash (1¾-inch frames)	1 lb per 7 lin feet of puttying
Steel sash	1 lb per 5 lin feet of puttying

For steel sash ½ lb of putty per square foot is sometimes used.

Plate glass in doors, store fronts, transoms, and so forth should be measured to the next even inch. Individual sizes should be listed.

Estimating Labor

In estimating the time required to re-glaze sash, thoughtful consideration must be given to the physical problems that confront the person who will do the actual work. On new jobs, or those involving a large number of lights of glass in a concentrated area, the probable average number of hours for glazing can be estimated fairly closely. If, however, a windstorm has broken a sizable number of lights in a building, which requires moving constantly from one area to another, or from one room to another, much time will be taken up by that activity. Hardened putty in older wood frames is difficult to clean out without splitting away portions of the wood.

When the wood sash can be taken out of the frames, the glazing is easier and usually results in better work.

If a glazier must be hired to install one light of glass, the entire cost of coming from and returning to the shop may have to be charged against the cost of the one light.

The glazier's time when working from ladders or scaffolds is usually greater than when working from ground or floor level.

The best way to estimate the labor for reglazing is to separate the time for removing old putty and cleaning the frame, from the time required to install and putty in the

TABLE 15–7

Approximate Number of Hours to Remove Putty and Glaze Sash Per Light Under Favorable Working Conditions

Size of Glass (Inches) and Kind of Work	Hours to Clean, Remove Old Putty	Hours to Glaze	Total Hours
Glazing windows in wood sash:			
(S.S. and D.S.):			
8 × 12*	.21	.14	.35
10 × 16	.26	.17	.43
12 × 14	.26	.17	.43
14 × 20	.30	.20	.50
16 × 24	.40	.27	.67
20 × 28	.51	.34	.85
24 × 30	.55	.37	.92
Glazing windows in steel sash:			
12 × 16	.18	.12	.30
12 × 18	.25	.17	.42
16 × 20	.30	.20	.50
16 × 30	.37	.25	.62
16 × 48	.40	.33	.73
Monitor and skylight sash:			
24 × 36	.67	.50	1.17
24 × 48	.75	.67	1.42
24 × 60	.90	.75	1.65

*For glazing plate glass in wood sash, double the time shown for S.S. and D.S. For glass sizes other than those shown, interpolate between the size below and above it.

new glass. When it is not necessary to remove old putty, no allowance should be made. Table 15-7 shows the approximate number of hours of labor for each operation for various sizes and kinds of glass. These hourly rates should be used as a starting basis and modified according to the conditions found.

While glaziers or painters do most glazing, it is not uncommon in some localities to have carpenters or other mechanics install glass. If repairs in addition to glazing are being made, and one or even a few lights of glass are to be put in, the estimator should consider the possibility of having it done by workers on the job.

Methods of Estimating

Plate glass is estimated by the square foot, and local dealers should be consulted for guidance on prices. Estimators should avail themselves of the book of standard list prices, and current discounts on glass, so that they can approximate prices when necessary. There is frequently salvage in the larger plate glass windows. The amount of salvage will depend on the size of the remaining section. Glass companies customarily have standard practices in allowing for salvage.

Other types of glass are estimated by the individual kind, size and quality of each light. Material and labor costs should be figured for the locality. When a large number of lights other than wire glass are to be installed it is customary to allow approximately 3 percent for breakage. The following is an example of an estimate for glazing:

Illustrative Examples

1. Remove and replace 150 lights of wire glass in steel sash.
(Wages and Prices Assumed)
 Develop a unit cost:

1 light of glass 16″ × 20″ = 2.22 sq ft	@ $ 2.10	=	$ 4.67
(1/2 lb putty per sq ft) 1/2 × 2.22 =1.11	@ .60	=	.67
.50 hours labor (Table 15–7)	@ $12.00	=	6.00
	Unit cost per light =		$11.34

Cost of 150 lights = 150 × $11.34 = $1,701.00

Had it not been necessary to remove old glass the unit cost would be developed as follows:

1 light of glass 16″ × 20″		=	$ 4.67
1. 11 lbs putty		=	.67
.20 hours labor (Table 15–7)	@ $12.00	=	2.40
	Unit cost per light =		$ 7.74

Cost of 150 lights = 150×$7.74 = $1,161.00

2. Remove and replace one light d.t. glass 14″ × 20″ in 1 3/8″ wood sash.
 Develop a unit cost:

1 light of glass d.t. 14″ × 20″	@ $ 2.00	=	$ 2.00
Putty (8 lin ft per lb, page 343)			
$\dfrac{2(14″ + 20″)}{12} = \dfrac{5.7}{8} = .71$ lbs	@ .60	=	.43
.50 hours labor (Table 15–7)	@ 12.00	=	6.00
	Unit cost per light =		$8.43

CHAPTER 16

Lathing and Plastering

LATHING

Originally *plaster lath* denoted narrow strips of wood that were placed close together to provide a base or background for the application of plaster. It has come to include the numerous substitutes for that method of construction. Perforated sheets of thin steel and gypsum board, either solid or perforated, are known in the trade as metal lath and gypsum lath respectively. The cost of gypsum lath, in most places, is slightly less than wood lath, and wood lath is slightly less than metal lath. Lath of all kinds is applied by a lather where union regulations prevail. In other areas of the country lath is applied either by a lather or a carpenter.

MEASUREMENT OF AREAS

As in measuring areas for applying numerous other building materials, there are different practices and varying opinions on the question of deducting window and door openings. One person may deduct all openings in full, while another will deduct only one half of the area of the openings. Some estimators do not deduct any openings. Those who do not deduct openings, and those who deduct half of them, feel that their method will take care of the waste and the additional labor necessary to work around them. Those who deduct openings in full add a percentage allowance for waste, giving the waste a more realistic relationship to the actual quantity of material required to cover the surface. In the labor rate of application, a further allowance in time is then made for working around openings.

Although both methods have merit, the author recommends deducting all openings in full. An allowance of about 10 percent should be added to the "net" area for waste. This method of measuring areas is especially preferred in many of the modern industrial buildings and in house construction where large windows, glass walls, and trimmed openings are frequently encountered.

Whichever method is used, good judgment should be exercised to make certain that an adequate amount of material is allowed, and that sufficient labor is included for the physical layout of the surface to be covered.

WOOD LATH

Wood lath, becoming obsolete, are rough wood strips that measure 1 1/2" × 3/8" × 4' in length. They will reach from center to center on four studs, joists, or furring placed 16 inches apart. They are sold in bundles containing 100 lath, and are priced either by the bundle or per 1,000 lath. They are applied with 3d fine, No. 16 gauge, blued nails, and are spaced 3/8 inch apart for lime plaster and 1/4 inch for light gypsum plaster. This allows the plaster to be forced between the lath to form a hardened blob or "key" on the back side.

Estimating Quantities

The customary unit for computing quantities of wood lath is the square yard. It takes approximately 1,500 lath, 1 1/2" × 3/8", spaced 3/8" apart, to cover an area of 100 square yards. Approximately 10 pounds of nails are required for 100 square yards.

The *gross* area of the surface to be lathed is computed. *All openings are deducted*, and an allowance of 10 percent is added for waste (see Measurement of Areas).

Estimating Labor

Under average conditions such as might be found in dwellings with rooms that have one or two windows and doors, a lather can apply wood lath at the rate of about 10 square yards an hour, or 1,500 lath in 10 hours. On larger surfaces, with few or no openings, the rate of application can be increased to about 12.5 square yards an hour, or 1,500 lath in 8 hours.

Unit Costs

Illustrative Example
(*Material Prices and Wages Are Assumed*)

1500 spruce lath	@ $70.00 per M =	$105.00
10 lbs 3d lath nails	@ .60 per lb =	6.00
10 hours labor	@ 12.00 per hr =	120.00
	Unit cost per 100 sq yds =	$231.00
	Unit cost per sq yd =	$ 2.31

GYPSUM LATH

One of the most popular plaster bases is gypsum board. This is a core of gypsum between two layers of paper. It is relatively easy to apply, and the cutting and fitting around openings is quickly done by scoring the material with a knife or a hatchet. It is fire-resistive, offers desirable insulating qualities, and provides a better adhesive bond than wood or steel lath.

Gypsum lath is generally 16" × 48" and is 3/8" thick. Whether applied vertically or horizontally, the sheets will reach from center to center of studding that is spaced 16 inches apart. When studding is more than 16 inches on center but not more than

24 inches, gypsum lath of 1/2-inch thickness is used. It comes in solid sheets, or in sheets perforated with 3/4-inch holes every 16 square inches. The latter permits the plaster to form a key, or mechanical bond, on the back side; this type is frequently preferred for ceilings or is specified where a better fire-resistive rating is required. The mechanical bond tends to hold the plaster in place after the adhesive bond has been weakened by heat.

Estimating Quantities

The unit for computing quantities of gypsum lath is the square yard. Special flathead blued nails, 1 1/8-inches, long No. 13 gauge, are used in nailing gypsum lath. The nails are spaced about 5 inches apart. Approximately 15 pounds are required for 100 square yards. For gypsum lath 1/2-inch thick, nails 1 1/4 inches long are used and spaced 4 inches apart.

The gross area of the surface to be lathed is measured, and *all openings are deducted* (See Measurement of Areas). An allowance of about 10 percent is added for cutting and fitting.

Estimating Labor

A lather should be able to apply gypsum lath on frame construction at an average rate of 10 square yards an hour, or 100 square yards in 10 hours. When it is applied with staples, a lather can put on 100 square yards in about 8 hours. The difference in cost between nails and staples is not significant.

Unit Costs

Illustrative Example
(*Wages and Prices Assumed*)

100 sq yds gypsum lath	@ $80.00 =	$ 80.00
15 lbs gypsum board nails	@ .60 =	9.00
10 hours labor	@ 12.00 =	120.00
	Unit cost for 100 sq yds =	$209.00
	Unit cost per sq yd =	$ 2.09

Add for corner beads and other accessories

METAL LATH

Metal lath is made of steel. It comes either painted or galvanized to protect it against rusting. Metal lath gives the best mechanical bond of all laths. Portland cement mortar does not adhere well to gypsum lath; where that type mortar is required, metal lath is used.

Metal lath is made in a variety of patterns that may be generally classified as "expanded," "ribbed," "sheet," or "wire" laths. It comes in sheets measuring 27″ × 96″ with 10 sheets to the bundle. One bundle will cover 20 square yards. Metal lath is fastened to wood framing and furring by blued or galvanized nails. Where sheets are joined in between studding and joists, the metal lath is wired together with annealed wire.

Estimating Quantities

The unit for computing quantities of metal lath is also the square yard. Approximately 8 pounds of 1 1/2-inch No. 11 gauge, flat-head, blued nails are required for 100 square yards of lath. *In measuring areas for wire lath all openings are deducted in full.* (See "Measurement of Areas"). An allowance of 10 percent should be added for waste in cutting, lapping, and fitting.

Estimating Labor

Under average working conditions a lather can apply 100 square yards of metal lath to frame construction in 10 hours, or 10 square yards per hour.

Unit Costs

Illustrative Example
(*Wages and Prices Assumed*)

100 sq yds 3/8″ rib metal lath	@ $150.00 =	$150.00
8 lbs metal lath nails	@ .60 =	4.80
10 hours labor	@ 12.00 =	120.00
	Unit cost for 100 sq yds =	$274.80
	Unit cost per sq yd =	$ 2.75

GROUNDS AND SCREEDS

Grounds are wooden or metal strips that are nailed around rough interior door openings and some window openings to establish the thickness of the plaster when it is applied. They also act as a leveling guide for the mason. Grounds are also nailed along the base where the plaster is to stop. These may be nominal 1″ × 2″ furring. Usually they are left in place when they are flush with the plaster, and the baseboard is nailed over them.

Screeds are plaster strips placed by the plasterer along the outer edge of the ceiling to serve as a thickness and a leveling guide. Sometimes, they are also placed along the base of the wall.

The cost of screeds is included in the cost of the plastering labor. The cost of grounds at door or window openings may be estimated at approximately 1/2 hour per average opening. The cost of material is not significant. Baseboard grounds may be estimated at the rate of 100 lineal feet in 2 hours, and the material cost should be added.

CORNER BEAD

Galvanized steel corner bead is used at all corners for protection of the finished plaster edge. It also serves as a ground and leveling guide. Corner bead can be used with all types of lath.

The cost of corner bead is estimated on the basis of a lather's installation at the rate of 100 lineal feet in 2 hours. The cost of the corner head itself should be added.

Illustrative Example
(*Wages and Prices Assumed*)

100 lin ft corner bead	@ $ .10 =	$10.00
2 hours labor	@ 12.00 =	24.00
	Unit cost per 100 lin ft =	$34.00
	Unit cost per lin ft =	$ .34

PLASTERING

Plaster is actually mortar that is applied in successive coats to wall and ceiling surfaces. It is composed of a fine aggregate held together by either gypsum or portland cement as a cementatious material. Water is added to form a workable plastic mass that can be spread over the surface and smoothed out with a trowel or float. Gypsum is used for interior plaster, and portland cement is used for exterior plaster because it is weather resistant. Portland cement is also used for certain kinds of interior plastering work as basement walls or backing for ceramic tile.

A mechanic who applies plaster is known in the building trades as a *plasterer*. Because their work is somewhat related to masonry some plasterers are also masons and *vice versa.*

Plaster is applied in one, two, or three coats. Three-coat plaster consists of a *scratch coat*, a *brown coat*, and a *finish coat*. Three-coat work is considered to be the highest grade of plastering, but two-coat work is becoming increasingly common, particularly in house building. One reason for the popularity of the latter is the lower cost in labor. Another reason is that the solid gypsum board lath enables a plasterer to obtain a true surface with two coats rather than three, which is generally necessary on wood and metal lath. One-coat plaster is generally limited to secondary structures and out-buildings, where a rough or uneven surface offers no objection and where the plaster is basically functional.

Both gypsum and portland cement plaster may be applied to masonry walls. In such cases two-coat work is considered adequate. Local building codes may, in certain instances, require three coats.

INTERIOR PLASTERING

Three-coat Plastering

Three-coat plaster is applied over metal or wood lath. It consists of two base coats and one finish coat. The first base coat in three-coat work is called a *scratch coat* because it is scored, or scratched, while still wet to provide a better bond for the second base coat. Much of the plaster applied on the first coat is forced through the spaces in wood and metal lath to provide a *key* effect behind the lath, giving the plaster a mechanical bond. Because first-coat plaster is applied about 1/4 inch thick, it leaves the ridge marks of the lath. A second, or *brown coat*, is applied after the scratch coat has set firm but not dried out. This coat is troweled level and smooth in preparation for the finish coat. The finish coat usually is a mortar of gypsum and hydrated lime. It is called a putty or *white coat*. It is applied approximately 1/16 inch thick.

Two-coat Plastering

Two-coat plaster is applied over a solid plaster base such as gypsum board or masonry. The first coat is put on either in one operation, or by the *double back* method. In the latter method a thin coat is first applied. Before it has set, the plasterer goes back over the area filling in and leveling the surface to the full thickness of the grounds. A finish coat is then applied in the same manner as in three-coat work after the base coat has set.

White Putty Finish

The plaster finish in most common use is the lime-putty finish. It is applied over the base plaster in a thickness of about 1/16 of an inch. It produces a smooth hard white surface, which may be painted or papered.

Lime that has been partially hydrated by the manufacturer is known as "quick-lime." It must be further hydrated on the job by soaking (slaking) from 12 to 24 hours before it can be used. Double-hydrated finishing lime such as USG Ivory Finishing Lime, requires no slaking on the job because it is 92 percent hydrated during manufacture. When mixed with water, the lime immediately develops high plasticity, eliminating overnight soaking.

Neat gypsum plaster is added to the lime putty as a setting agent to produce hardness, and also to resist shrinking which is a characteristic of hydrated lime. The proportions specified by the ASA are:

1 (100-lb) bag of gypsum gauging plaster to not more than
4 (50-lb) bags of hydrated lime

Keene's Cement Finish

During the calcination of natural gypsum, if all of the chemically combined water is removed, instead of 75 percent as in the case of plaster of Paris, the resultant product is called Keene's cement. It is a form of gauging plaster which, when mixed with lime putty, produces a very hard and durable surface.

The ASA (American Standards Association) specifications for mixing proportions of *medium hard finish* Keene's cement are:

1 (100-lb) bag of Keene's cement to not more than
1 (50-lb) bag of hydrated lime.

The ASA specifications for *hard finish* Keene's cement are:

1 (100-lb) bag of Keene's cement to not more than
1/2 (50-lb) bag of hydrated lime

Skimcoat Plastering

Where economy is a factor in less expensive dwelling construction, a thin coat of lime putty plaster finish is applied *directly* to gypsum board after the joints have been taped. It is found in many of the low-cost homes in the southern states where extreme changes in temperature are rare.

The thickness of this skimcoat is similar to the 1/16-inch finish coat applied over regular two- and three-coat plastering. The cost may be obtained as follows:

Illustrative Example
(*Material Prices and Wages Assumed*)

2 (100-lb) sacks gysum plaster	@ $	4.00 =	$ 8.00
8 (50-lb) sacks hydrated fin. lime	@	2.50 =	20.00
12 hours plasterer	@	12.00 =	144.00
12 hours laborer	@	9.00 =	108.00
	Cost per 100 sq yds =		$280.00
	Unit cost per sq yd =		$ 2.80

Sand Finish

A *sand finish* employs mortar similar to the scratch and brown coats except that a very fine sand (sometimes white sand) is used. A cork float is employed to obtain a sandpaper-like surface. Occasionally, colors are used to tint the sand finish making it unnecessary to paint or decorate.

MATERIALS

Gypsum is a gray or white, calcium, rocklike mineral that is found abundantly in nature in most parts of the world. It contains, by weight, approximately 21 percent of chemically combined water. When the gypsum rock is pulverized, and subjected to heat, water is driven off. This process is called *calcination*. If 75 percent of the water is removed, the resulting product is *plaster of Paris*. This calcined gypsum when remixed with water, sets and returns to gypsum rock *with its original chemical composition*.

Calcined gypsum is the cementatious base material for most of the interior plaster mortars. Mixed with sand, perlite, or vermiculite, it is used as a rough or base plaster for scratch, brown, or sand-finish coats. When mixed with hydrated lime, it is used as a putty or white-finish coat.

PROPORTIONS

Job-mixed Plaster

Gypsum plaster for base coats may be mixed on the job by machine or by hand in a mortar box. It may also be purchased ready-mixed with mineral aggregate already added.

When job-mixed, the following proportions of calcined gypsum to aggregate are recommended by the Gypsum Association:

For two-coat or "double back" plastering: 100 pounds of calcined gypsum plaster to 250 pounds of damp, loose sand, or 2½ cubic feet of perlite* or vermiculite.**

For three-coat application of plasters: The first (scratch) coat is 100 pounds of calcined gypsum plaster to 200 pounds of damp, loose sand, or 2 cubic feet of perlite or vermiculite.

The second (brown) coat is 100 pounds of gypsum to 300 pounds of damp, loose sand, or 3 cubic feet of perlite or vermiculite.

In lieu of the above, the scratch coat may consist of 100 pounds of gypsum to 250 pounds of sand, or 2½ cubic feet of perlite or vermiculite providing the brown coat is of similar proportioning.

It is common practice to designate proportioning by symbols such as 1:2 meaning 100 pounds of gypsum to 200 pounds of sand, or 2 cubic feet of perlite or vermiculite depending upon the aggregate indicated. Similarly, 1:2½ means 100 pounds of gypsum to 250 pounds of sand, or 2½ cubic feet of perlite or vermiculite.

Where double symbols occur as "1:2, 1:3," the first symbol refers to the scratch coat proportioning and the second symbol to the brown coat. A single designation "1:2½" indicates that proportioning is the same for both coats.

The use of sand, perlite, or vermiculite, as an aggregate in job-mixed plaster is a matter of choice and also dependent upon several conditions. For ordinary plastering, sand is generally used where an ample supply of clean, sharp sand is available at low cost. Because of the better insulating and fire-resistive qualities of the lightweight aggregates, they may be preferred over sand. Where it is necessary to comply with specific regulations on fire ratings, perlite and vermiculite are used. When plaster thicknesses are specified in excess of those normally used, lightweight aggregates may be preferred.

Neat gypsum is the term applied to gypsum with no aggregate added. It comes in 100-pound sacks. Perlite and vermiculite aggregate are sold in sacks containing 3 to 4 cubic feet. Their weight varies between 7 1/2 and 10 pounds per cubic foot. Sand is sold by the ton or cubic yard. A cubic yard contains 27 cubic feet and weighs between 2,600 and 2,800 pounds.

Ready-mixed Plasters

Much of the gypsum for scratch and brown coats is mill-mixed, and delivered to the job in sacks. It requires only the addition of water for use.

While at one time sand was the principal aggregate used in mill-mixed plasters, it has almost been superseded by the lightweight perlite and vermiculite aggregates. These lightweight aggregates weigh about 10 percent as much as sand. They have superior sound and heat insulating qualities. They also have a better fire-resistive rating than sand. Plasters with the lightweight aggregates handle with greater ease and, to some ex-

---

Perlite is a siliceous volcanic glass expanded by heat.

**Vermiculite* is a micaceous mineral expanded by heat. Both are lightweight aggregates with fire resistive properties superior to sanded plasters.

tent, can be applied more rapidly by a plasterer. An important advantage of mill-mixed plasters is their uniformity of mixture, which is not always easy to control with mixing on the job.

Ready-mixed plasters are favored by many builders for winter plastering. Mixing gypsum and sand on the job in wet and freezing weather can be awkward and costly.

ESTIMATING QUANTITIES

The unit of measurement in plastering is 1 square yard. The gross area of a room is computed and *all window and door openings are deducted in full as in lathing.* (See Measurement of Quantities under Lathing.)

When estimating the quantity of materials required for a square yard of one-, two-, or three-coat work, it is important to determine the thickness of the plaster that existed in the damaged building. This can be done usually by measuring with a rule *from the face of the lath* to the outer surface of the finished plaster. It is general practice in most parts of the country to apply plaster over the face of the plaster base in approximately the following thicknesses including the finish coat:

```
Wood lath . . . . . . . . . . . . . . . . . . . . . . . . . . . . . . . . . . . . . . . . . . . . . . . . . . . . . . .5/8"
Metal lath . . . . . . . . . . . . . . . . . . . . . . . . . . . . . . . . . . . . . . . . . . . . . . . . . . . . . . . .5/8"
Gypsum lath . . . . . . . . . . . . . . . . . . . . . . . . . . . . . . . . . . . . . . . . . . . . . 3/8" and 1/2"
Masonry walls  . . . . . . . . . . . . . . . . . . . . . . . . . . . . . . . . . . . . . . . . . . . . . . . . . . . . .5/8"
```

Since wood and metal lath have openings through which the plaster is forced to provide a key, or mechanical bond, much more plaster is required than is indicated for either gypsum or masonry bases. It is more difficult to control the amount of plaster on wood and metal lath. Too much pressure on the trowel forces more plaster than is necessary through the openings and it drops behind the lath. Table 16–1 shows the approximate number of sacks of prepared perlite gypsum plaster required for 100 square yards, applied on different kinds of plaster base. The thickness shown includes a 1/16-inch white-coat finish.

TABLE 16–1

Approximate Number of Sacks of Prepared Perlite
Gypsum Plaster Required to Cover 100 Square Yards

Kind of Base	Total Plaster Thickness (Inches)	Number of Sacks
Wood lath	5/8	22(80 lbs per sack)
Metal lath	5/8	35(80 lbs per sack)
Gypsum lath	1/2	20(80 lbs per sack)
Masonry walls	5/8	24(67 lbs per sack)

The approximate quantities of materials for job-mixed, sanded, or perlite gypsum plaster for 100 square yards are shown in Table 16–2.

TABLE 16-2

Approximate Quantities of Materials Required for 100 Square Yards
of Job Mixed Plaster

Kind of Base	Total Plaster Thickness Inches	100 Lb Sacks of Neat Gypsum	Cu Yds Sand 1:2½ mix	Aggregate Perlite		
				Cu Ft	4 Cu Ft Sacks	Mix
Wood lath	5/8	11	1.1	27.5	7	1:2½
Metal lath	5/8	20	2.0	50.0	12.5	1:2½
Gypsum lath	1/2	10	1.0	25	6.5	1:2½
Masonry walls	5/8	12	1.2	30	7.5	1:3

ASA permit 250 lbs damp loose sand or 2½ cu ft of vermiculite or perlite provided this proportioning is used for both scratch and brown coats on three-coat work. The number of 4 cu ft sacks of perlite are shown to nearest ½ sack.

Table 16–3 shows the approximate quantities of gypsum and hydrated lime required to cover 100 square yards with white coat finish 1/16 inch thick.

The quantities indicated in Tables 16–2 and 16–3 are within the coverage ranges published by the U.S. Gypsum Company.

TABLE 16-3

Approximate Quantities of Materials Required for 100 Square Yards
White Coat Finish 1/16" Thick

Kind of Finish	100 Lb Sacks Neat Gypsum	100 Lb Sacks Keene's Cement	50 Lb Sacks Hydrated Lime
Lime putty	2	—	8
Keene's cement	—	4	4

Based on ASA mix of 1(100-lb) sack neat gypsum gauging plaster to 4(50-lb) sacks hydrated lime, and 1(100-lb) sack of Keene's cement to 1(50-lb) sack hydrated lime for medium hard finish.

ESTIMATING LABOR

When estimating labor costs for plastering, just as is done in the case of painting or other operations which involve the interior of a building, particular attention should be given to the physical circumstances of each job. Where patching of walls or ceilings is required, or where certain ceilings and walls require complete replacement, and the premises are furnished and occupied, the labor factor is considerably more than when the plasterer is working in a vacant and unoccupied building. Where the workers are free to move about, and are unhampered by furnishings that need to be moved or protected, their rate of applying plaster approaches that for new construction.

The cost of plaster per square yard on small areas is usually much higher than the cost on large areas. The cost of plastering ceilings only in most instances is greater per square yard than it is for side walls. Closets and cupboards slow a plasterer down. All factors affecting labor, as discussed in Chapter 2, should be reviewed from the standpoint of their influence on plastering work.

When estimating the cost of completely plastering rooms, several rooms, or an entire building, the use of average labor rates of application is entirely in order.

It was formerly a custom to allow the same number of hours for a plasterer's helper, or laborer, as for a plasterer, when most plastering was done on wood lath, and the aggregate was mainly sand and was mixed by hand. The plaster was heavy, and much time was required to mix materials, carry them to a plasterer and arrange scaffold horses and plank. With present day lightweight aggregates, and machines for mixing plaster as it is required, the laborer's time is reduced. A rough ratio of a laborer's hours to a plasterer's is approximately 75 percent. There will be situations, however, where a laborer is required for the full time that the plasterer is working. Proper consideration should be given in each instance to the probable number of hours a laborer will be needed to serve the plasterer.

Table 16–4 shows the approximate number of hours required for a plasterer and a laborer, or plasterer's helper, to apply 100 square yards of various kinds of plaster on different types of plaster base. It is assumed that the plaster base has been prepared for the plasterer.

TABLE 16–4

Approximate Hours Required to Apply 100 Square Yards Gypsum
on Various Kinds of Plaster Base

Kind of Base	Plaster Thickness Inches	No. of Coats Excluding White Coat	Plasterer	Laborer
Wood lath	5/8	2	14	10
Metal lath	5/8	2	15	12
Gypsum lath	1/2	1	10	8
Gypsum lath	3/4	2	14	10
Masonry walls	1/2	1	10	8
Masonry walls	5/8	2	12	9
White Finish Coat				
Lime putty	1/16	1	10	8
Keene's cement	1/16	1	13	9

ESTIMATING COSTS

The cost of plaster applied to a surface depends on the type of plaster base or lath, the thickness of the plaster, and the number of coats to be applied. In most sections of the country the cost of plaster is figured on a square yard basis, as opposed to estimating the materials and labor for each room or for the area to be covered. However, the cost per square yard must be determined first, and this is done by obtaining the cost for 100 square yards, because most coverage tables for material, and also for labor, are in terms of 100 square yards. These unit costs are developed for each type of plastering on different types of bases.

Before attempting to compute a unit cost for plastering on any loss, it is important to determine by inspection the kind of plaster that existed. Insurance policies that cover property damage provide for replacing with like kind and quality. The kind of plaster that is originally applied in a building is controlled by several considerations, among which are local custom and practice, climate, economy, and building code regulations. The quality ranges from the application of *skimcoat* plaster on gypsum board, two-coat plaster on gypsum lath, to three-coat work on wire lath. It is also significant to base unit costs on the exact specifications necessary to restore a plastered wall to its original condition. If only the white finish coat is affected, or if the lath does not require replacing, the unit cost should be confined to those operations. As an illustration, while wood lath is used infrequently on new construction, much of it is found in existing older buildings. Unless wood lath is burned or destroyed, the old plaster frequently can be removed and new plaster applied without replacing lath.

Unit Costs

The costs for common kinds of plastering may be computed as shown in the following examples: By referring to Tables 16–1 through 16–4 unit prices may be obtained for types of plastering bases and finishes not shown in the illustrations. Where a different finish coat is used, the cost of "lime putty finish" should be deducted and the cost of the finish coat to be applied should be added.

The following illustrative unit costs with material prices and hourly labor rates assumed.

Prepared Perlite Gypsum Plaster-Unit Cost of Two-Coat "Double Back" Plaster on Gypsum Lath (Grounds 7/8"-Plaster Thickness 1/2" Including 1/16" finish)

Table	Base Coat			
16–1	20(80-lb) sacks prepared plaster	@ $	4.00 =	$ 80.00
16–4	10 hours plasterer	@	12.00 =	120.00
16–4	8 hours laborer	@	9.00 =	72.00
	Lime Putty Finish			
16–3	2(100-lb) sacks gauging plaster	@	3.00 =	6.00
16–3	8(50-lb) sacks hydrated finish lime	@	2.50 =	20.00
16–4	10 hours plasterer	@	12.00 =	120.00
16–4	8 hours laborer	@	9.00 =	72.00
		Cost per 100 sq yds =		$490.00
		Cost per sq yd =		$ 4.90

Unit cost of Three-Coat Plaster on Metal Lath (Grounds 3/4"—Plaster Thickness 5/8" Including 1/16" Finish)

Table	Base Coat			
16–1	35(80-lb) sacks prepared plaster	@ $	4.00 =	$140.00
16–4	15 hours plasterer	@	12.00 =	180.00
16–4	12 hours laborer	@	9.00 =	108.00

Lime Putty Finish

Table					
16–3	2(100-lb) sacks gauging plaster	@	3.00 =		6.00
16–3	8(50-lb) sacks hydrated finish lime	@	2.50 =		20.00
16–4	10 hours plasterer	@	12.00 =		120.00
16–4	8 hours laborer	@	9.00 =		72.00

Cost per 100 sq yds = $646.00
Cost per sq yd = $ 6.46

Unit Cost of Three-Coat Plaster on Wood Lath (Grounds 7/8"-Plaster Thickness 5/8" Including 1/16" Finish)

Table	Base Coat				
16–1	22(80-lb) sacks prepared plaster	@ $	4.00 =	$	88.00
16–4	14 hours plasterer	@	12.00 =		168.00
16–4	10 hours laborer	@	9.00 =		90.00

Lime Putty Finish

16–3	2(100-lb) sacks gauging plaster	@	3.00 =		6.00
16–3	8(50-lb) sacks hydrated finish lime	@	2.50 =		20.00
16–4	10 hours plasterer	@	12.00 =		120.00
16–4	8 hours laborer	@	9.00 =		72.00

Cost per 100 sq yds = $564.00
Cost per sq yd = $ 5.64

Unit Cost of Two-Coat Plaster on Masonry (Grounds 5/8"-Plaster Thickness 5/8" Including 1/16" Finish)

Table	Base Coat				
16–1	24(80-lb) sacks prepared plaster	@ $	4.00 =	$	96.00
16–4	12 hours plasterer	@	12.00 =		144.00
16–4	9 hours laborer	@	9.00 =		81.00

Lime Putty Finish

16–3	2(100-lb) sacks gauging plaster	@	3.00 =		6.00
16–3	8(50-lb) sacks hydrated finish lime	@	2.50 =		20.00
16–4	10 hours plasterer	@	12.00 =		120.00
16–4	8 hours laborer	@	9.00 =		72.00

Cost per 100 sq yds = $539.00
Cost per sq yd = $ 5.39

JOB-MIXED GYPSUM PLASTER USING

SAND AGGREGATE

The following basic formulas and illustrative examples use sand as the aggregate to be mixed with gypsum plaster. Where specifications call for perlite aggregate refer to Table 16–1 for quantities required.

Unit Cost of Two-Coat "Double Back" Gypsum Plaster on Gypsum Lath (Grounds 7/8" — Plaster Thickness ½" Including 1/16" Finish)

Illustrative Example

Table	Base Coat				
16–2	10(100-lb) sacks gypsum plaster	@ $	4.00 =	$	40.00
16–2	1 cu yd sand	@	7.00 =		7.00

16–4	10 hours plasterer	@	12.00 =	120.00
16–4	8 hours laborer	@	9.00 =	72.00
	Lime Putty Finish Coat			
16–3	2(100-lb) sacks gauging plaster	@	3.00 =	6.00
16–3	8(50-lb) sacks hydrated finish lime	@	2.50 =	20.00
16–4	10 hours plasterer	@	12.00 =	120.00
16–4	8 hours laborer	@	9.00 =	72.00

$$\text{Unit cost for 100 sq yds} = \$457.00$$

$$\text{Unit cost per sq yd } \frac{\$457.00}{100} = \$\ 4.57$$

Unit cost of Three-Coat Plaster on Metal Lath (Grounds ¾"— Plaster Thickness 5/8" Including 1/16" Finish)

Illustrative Example

Table	*Base Coat*			
16–2	20(100-lb) sacks gypsum plaster	@ $	4.00 =	$ 80.00
16–2	2 cu yds sand	@	7.00 =	14.00
16–4	15 hours plasterer	@	12.00 =	180.00
16–4	12 hours laborer	@	9.00 =	108.00
	Lime Putty Finish Coat			
16–3	2(100-lb) sacks gauging plaster	@	3.00 =	6.00
16–3	8(50-lb) sacks hydrated finish lime	@	2.50 =	20.00
16–4	10 hours plasterer	@	12.00 =	120.00
16–4	8 hours laborer	@	9.00 =	72.00

$$\text{Unit cost for 100 sq yds} = \$600.00$$

$$\text{Unit cost per sq yd } \frac{\$600.00}{100} = \$\ 6.00$$

Unit Cost of Two-Coat Plaster on Masonry (Grounds 5/8"— Plaster Thickness 5/8" Including 1/16" Finish)

Illustrative Example

Table	*Base Coat*			
16–2	12(100-lb) sacks gypsum	@ $	4.00 =	$ 48.00
16–2	1.2 cu yds sand	@	7.00 =	8.40
16–4	12 hours plasterer	@	12.00 =	144.00
16–4	9 hours laborer	@	9.00 =	81.00
	Lime Putty Finish Coat			
16–3	2(100-lb) sacks gauging plaster	@	3.00 =	6.00
16–3	8(50-lb) sacks hydrated finish lime	@	2.50 =	20.00
16–4	10 hours plasterer	@	12.00 =	120.00
16–4	8 hours laborer	@	9.00 =	72.00

$$\text{Unit cost for 100 sq yds} = \$499.40$$

$$\text{Unit cost per sq yd } \frac{\$499.00}{100} = \$\ 5.00$$

Unit Costs Including Lath

Because of the numerous combinations of the different kinds of lath or plaster base, and the various kinds of plaster applied, it is recommended that a unit cost per square yard be obtained for the particular *lath* involved and added to the cost per square yard for the type of plaster being considered.

MACHINE PLASTERING

Plastering machines have been developed for applying rough or basecoat plasters. They are being improved continually and manufacturers of plastering materials are revising their formulas to adapt them to machine application. There are some who predict that much, if not most, of the future plastering will be applied by machine. They are portable units operated by electric or gasoline motors.

These machines actually spray mortar on the walls and ceilings. Where a true and smooth surface is required, a plasterer must still darby and trowel the surface after the machine has deposited the mortar.

Machine application uses more mortar, but requires considerably less labor, which is the more expensive factor in plastering. Because it is a specialized operation and used mainly where large quantities are needed, no attempt should be made to estimate the unit cost of machine plastering without obtaining the aid of a subcontractor familiar with the work.

EXTERIOR PLASTERING

Stucco, which is the term often used for exterior plaster, is composed of a mortar made with portland cement rather than gypsum because it is exposed to the weather. The aggregate in the mortar is sand, although a fine pebble gravel, marble dust, white sand, and colors are sometimes used in the finish coat. Hydrated lime is added as a plasticizer to make the mortar workable when being applied.

Exterior plaster can be applied directly to brick, concrete, concrete block, hollow terra cotta tile, and other kinds of masonry. It can also be applied on wire lath nailed directly to studding, or over sheathing. Three-coat exterior plaster is considered the best grade of work; under many building codes it is a requirement. However, there is some two-coat work being done, both on masonry and lath bases in low-priced homes and on secondary buildings. As with interior plastering the successive coats are scratch, brown, and finish coats. The customary thickness of exterior plaster is about 3/4 inch but may be found in thicknesses that vary between 5/8 and 1 inch. Before estimating, the actual thickness of the plaster should be scaled at the loss as it determines the quantity of material required, and also the labor.

Measurement of Areas

Under *Lathing* the different customs for measuring areas were discussed, and the reader is referred to that section as the comments apply also to exterior plaster work. It is recommended that gross areas be computed and *all window and door openings be deducted in full.* An allowance of 10 percent should be added to the net area for waste.

Metal Lath Base

The metal lath applied to frame buildings as an exterior plaster base comes in many forms. They may be classified into three kinds: the expanded metal lath, the welded wire lath, and the woven wire lath. The latter, because of its similarity, is frequently called *chicken wire*, and is used extensively. Because it is not good practice to nail lath solidly to studding or sheathing, a special 1 1/2-inch self-furring nail is used which holds the lath 1/4 to 3/8 of an inch away from the surface. This allows the mortar to flow in back of the lath and form a proper key at all points.

Whether the metal lath is applied over open studding, or solid wood sheathing, a layer of 15-lb saturated felt paper should first be applied. In the case of open studding it acts as a backing behind the lath for the plaster. When applied over solid wood sheathing, it prevents the wood from soaking up the water in the plaster. When asphalt-coated composition sheathing is used, it is not necessary to apply saturated felt paper, as the coating prevents absorption of the water from the mortar.

Before applying saturated felt paper on open studding, 18-gauge steel wires are stretched horizontally, spaced about 6 or 8 inches apart. In some sections of the country the wires are drawn tightly in the horizontal position; in other sections it is the practice to staple the wires on every other stud. At the alternate studs the wires are then drawn upward until taut and stapled to the stud. The effect is the same: The wires provide rigidity back of the wire lath and paper against pressure from the plastering trowel when the mortar is being applied.

To estimate the quantity of wire lath and the labor for applying it, refer to that subject under "Lathing" in this chapter. Costs may be obtained by using the formula shown for metal lath. The labor cost of stringing 18 gauge wire every 8 inches should be estimated at an average of 6 hours per 100 square yards of surface area. The number of lineal feet of wire is obtained by multiplying the area by 1.5. It requires about 8 pounds of 18 gauge wire per 100 square yards.

Another type of metal lath used on exterior open-stud stucco work is a paper-backed product of 2" × 2" mesh 16 gauge steel wire. The trade name is *Steeltex*. When this type of metal lath base is applied over open studding, neither the wire back nor 15-lb saturated felt paper is required.

The labor required to apply 100 square yards of paper-backed stucco lath would be an average of 10 hours.

Masonry Base

All masonry bases must be completely cleaned to remove dirt and grease before plastering; otherwise, the mortar may not adhere. In some cases chipping, scratching or roughing the surface is necessary to obtain a good mechanical bond.

Exterior Plaster Mortar

For estimating the cost of portland cement mortar, the proportions of materials for a cubic yard are identical to those used for masonry cement mortar (Chapter 9) employing a 1:3 mix with hydrated lime equal to 10 percent of the weight of cement added.

Cement	10 sacks	@ $4.00 =	$40.00
Sand	1 cu yd	@ 7.00 =	7.00
Hydrated lime	2 (50-lb) sacks	@ 2.00 =	4.00
		Cost per cu yd =	$51.00

Estimating Quantities

The unit used for quantities of portland cement plastering is 1 square yard. The area to be covered is obtained by measurement and converted to square yards by dividing the number of square feet by nine. Table 16–5 shows the number of cubic yards of cement mortar required for 100 square yards of plastering of the thickness indicated.

TABLE 16–5

Cubic Yards Portland Cement Plaster Required for
100 Square Yards for the Various Thicknesses
Shown (1:3 Mix)

Plaster Thickness (Inches)	Cubic Yards on Masonry Base	Cubic Yards on Wire Lath Base
1/4	0.75	0.90
1/2	1.50	1.80
5/8	1.37	1.64
3/4	2.25	2.70
1	3.00	3.60

TABLE 16–6

Approximate Hours Required to Apply 100 Square Yards
Portland Cement Plaster on Wire Lath and
Masonry Bases (Float Finish)

Base	Plaster Thickness (Inches)	Number of Coats	Plasterer	Labor
Masonry	1/2	1	13	6
	5/8	2	21	10
	3/4	3	28	20
Wire lath	5/8	2	30	20
	3/4	3	40	30

Add approximately 4 hours to plasterer's time for a troweled finish.

In computing the quantity of cement plaster mortar, no regard need be given to the number of coats as the thickness of the plaster controls the quantity. When estimating labor, the number of coats is important because each coat is separately applied.

Estimating Labor

A plasterer and a laborer work together on exterior work in the same manner as interior plastering. The laborer mixes the mortar, arranges horse and plank scaffold, and serves the plasterer with necessary materials. On small to medium-sized jobs a laborer's time will be practically equal to the plasterer's. When larger areas are being plastered and the mortar is machine mixed, the laborer's time is reduced considerably. The ratio of a laborer's time to that of the plasterer should be judged for each particular job. Table 16–6 shows the approximate number of hours required for a mason and laborer to apply 100 square yards of plaster of various thickness and number of coats on both masonry and wire lath base.

Scaffolding and Hoisting

When structural scaffolding is to be erected by carpenters, the cost should be added as a separate item. The cost of hoisting materials or equipment to floors above grade should also be added when such operations are indicated.

Estimating Costs

It is customary to compute the cost of exterior plastering for 100 square yards, and then reduce this to a cost per square yard as is done in the case of interior plastering. (Refer to "Estimating Costs" under "Interior Plastering" for considerations prior to estimating).

The following examples are given for different kinds of exterior plaster work (prices and wages assumed):

Unit Cost of 100 Sq Yds Two-Coat Stucco on Masonry Base (1:3 Mix-5/8" Thick—Float Finish)

Illustrative Example

Table				
16–5	1.37 cu yds plaster mortar	@ $ 51.00 =	$	69.87
16–6	21 hours plasterer	@ 12.00 =		252.00
16–6	10 hours laborer	@ 9.00 =		90.00

$$\text{Unit cost for 100 sq yds} = \$ \ 411.87$$
$$\text{Unit cost per sq yd} \ \frac{\$411.87}{100} = \$ \ 4.12$$

Unit Cost of 100 Sq Yds One-Coat Stucco on Masonry Base (1:3 Mix—½" Thick-Float Finish)

Illustrative Example

Table				
16–5	1.5 cu yds plaster mortar	@ $ 51.00 =	$	76.50
16–6	13 hours plasterer	@ 12.00 =		156.00
16–6	6 hours laborer	@ 9.00 =		54.00

$$\text{Unit cost for 100 sq yds} = \$ \ 286.50$$
$$\text{Unit cost per sq yd} \ \frac{\$286.50}{100} = \$ \ 2.87$$

Unit Cost of 100 Sq Yds Two-Coat Stucco on Woven Wire Mesh Lath on Open Studs (1:3 Mix— 5/8" Thick-Float Finish)

Table

	8 lbs 18 gauge wire	@ $ 1.00 =	$	8.00
	100 sq yds wire mesh lath	@ .90 =		90.00
	2½ rolls 15-lb saturated felt paper	@ 3.20 =		8.00
	8 lbs self-furring lath nails	@ .60 =		4.80
	6 hours lather applying wire	@ 12.00 =		72.00
	10 hours lather applying lath	@ 12.00 =		120.00
16–5	1.64 cu yds plaster mortar	@ 51.00 =		83.64
16–6	30 hours plasterer	@ 12.00 =		360.00
16–6	24 hours laborer	@ 9.00 =		216.00

$$\text{Unit cost for 100 sq yds} = \$ \ 962.44$$

$$\text{Unit cost per sq yd} \ \frac{\$962.44}{100} = \$ \quad \$ \ 9.62$$

DAMAGEABILITY OF GYPSUM PLASTER

The degree of damage that gypsum plaster has suffered from fire, water, explosion, blasting, settlement, or other perils can best be determined only by careful inspection. To assume that plaster must be removed and replaced simply because it has been exposed to fire or has been wet is an unsound approach. Much will depend on the type of lath or plaster base, the extent and duration of a fire, whether it is sidewall or ceiling plaster, the age of the plaster, its composition and its physical condition at the time the damage took place.

Damage by Fire

When gypsum plaster is directly exposed to the heat of fire, the water of crystallization is slowly driven off. Long-continued heat causes the surface to powder which, although increasing its insulating properties, weakens it. The thin white putty coat finish becomes chalky and may spall or scale off the rough plaster base. In extremely hot fires of long duration the plaster may completely crumble and disintegrate.

When the back of plaster on metal lath is exposed to hot fire for any length of time, the key can be weakened seriously. Wood lath will char and burn when directly exposed to a fire so that, even though the plaster appears sound on the room side, replacement may be required. Gypsum lath base will withstand the effects of heat and fire in back of it for a longer period than either wood or metal lath without serious injury to the plaster. Where any question arises concerning the necessity of replacing plaster, proper examination and tests should be made to determine the condition and firmness of the plaster. Scraping the surface with a knife will disclose whether it has powdered or is chalky. Tapping firmly with a stout rod, broom handle, or any other convenient object will indicate by the sound, and by any movement, whether or not the plaster has been loosened.

Damage by Water

One of the characteristics of gypsum is that it can be wet down to the point of being plastic, but upon drying out, it returns to its original hardness. This is the reason it makes such an excellent cementatious material for plaster.

Plaster that has been wet by water from a leaky roof, a fire hose, or even inundation will show no harmful effects, provided it is not continuously or periodically exposed to water and is allowed to dry out. Probably there is no better example of the ability of gypsum plaster to withstand water effects than that contained in the Gypsum Association's report on the Topeka flood of 1951. Two motels were submerged under 8 feet of water for five days. Less than six weeks after the waters receded, they both were back in business with no repairs being needed to the walls.

There are innumerable instances of sidewall plaster being so saturated with water used to extinguish a fire that has occurred above, that a lead pencil could be pushed through the plaster. After several days the plaster returns to its original hardness with no evidence of damage. The drying out process should, however, be controlled. Usually a moderate amount of heat, with the windows opened to permit air circulation, will be satisfactory. It is important, of course, in winter months, to protect the wet plaster from freezing.

When ceilings have been seriously wet down and the plaster is on a wood lath base, there is always a strong possibility that the wood has swelled and this, together with the added weight, will break the key. The same thing may also happen in the case of certain types of metal lath. The adhesive bond of gypsum lath will generally hold, but frequently the weight of the wet plaster and gypsum board may pull the nails through.

It is always better, if possible, to allow wet plaster to dry out for several days before estimating. At that time appropriate tests may be made to ascertain its true condition.

Repeated wetting of plaster such as might take place when a roof continues to leak over a period of time, will eventually rot the plaster.

Damage by Cracking

The causes of cracks in plaster are numerous aside from inherent cracking due to poor materials, improper proportioning, and inferior work. Cracks most commonly found within insurance coverage are caused by settlement of the ground, blasting, jarring or twisting of a structure during strong or violent winds, land or mud slides, earthquake, lightning, explosion, and heat from a fire. They may occur in both ceiling and sidewalls. One of the difficulties encountered is that of trying to distinguish old cracks in plaster from those attributed to the immediate casualty. When the plaster immediately adjacent to the cracks is sound, an experienced plasterer can perform repairs which cannot be detected after repainting or repapering.

METHODS OF REPAIRING GYPSUM PLASTER

Estimating the cost to repair damaged plaster will be governed by the specifications taken in the field. These specifications may call for patching an area of white coat that has scaled off or they may call for patching a section of three-coat work including the lath.

Lath

When lath in a patch has to be replaced before new plaster is applied, gypsum lath, when permitted, can be inserted regardless of the type of existing lath. Where metal lath is required to comply with building regulations, however, it must be used.

Wood lath, if not burned or broken, can be renailed wherever loose after all old plaster has been knocked from between the lath. Prior to replastering on wood lath, it is wet down to prevent the water in the plaster from being absorbed before it has set.

Masonry Base

All that is required to prepare a masonry base for replastering is to clean and brush it thoroughly, then wet it so that it will not absorb the water in the plaster too rapidly.

Patch Plastering

When preparing a section to be patched, it is important that all loose plaster be removed to be sure that the remaining plaster adjacent to the patch area is attached firmly to the base. This presents a difficult problem in older buildings where over a long period the plaster has sagged or pulled away from the lath, particularly wood lath. In some cases both the lath and plaster have become loosened.

If only the white-finish coat has been damaged, this is removed wherever loose, and the old brown-coat base is wet at the edges of the patch only. The brown coat is scored or scratched to provide a proper bond.

When finish coat and rough base coat, or coats, are to be replaced, the old plaster is removed and the edges are wet down. New coats are applied successively just as in new plastering. The old plaster edges serve as grounds for the final coat.

Patching Cracks

Large cracks in plaster should be cut out removing all loose plaster along the edges. Sometimes it is advisable to widen the area around the crack as much as 4 or 6 inches, and nail a strip of wire lath in the opening. In such instances rough coats and a finish coat are applied rather than attempting to fill so wide an opening in one step.

Smaller cracks are easily repaired by a skilled plasterer. They are opened slightly with a sharp tool, wet, and then filled with plaster of Paris.

Large Area Repairs

When an entire ceiling or a whole wall has to be replastered, it may be estimated either at a unit price or on a material and labor basis. If a unit price is developed and used, it must be remembered to add all operations necessary in addition to the labor of direct application of the plaster.

Estimating Repair Costs

Except where large areas are involved, all patch jobs in plastering should be estimated on a material and time basis. Consideration should be given to all factors that the worker will be confronted with on the particular job. It should be remembered that in many areas there is a minimum charge by plasterers for patching regardless of whether it takes six hours, a whole day, or thirty minutes.

DAMAGEABILITY OF STUCCO PLASTER
AND REPAIR METHODS

Portland cement plaster is used on exterior work because it is water-resistant. Therefore, its damageability by the effects of water, even submersion, is negligible.

In the presence of fire it is less damageable than gypsum plaster but, in severe fires of long duration, it will suffer some injury depending upon its thickness. Only careful examination will determine the extent of damage.

Patching Stucco

The procedures followed in patching gypsum plaster apply generally to portland cement plaster. Areas to be repaired are cut out back to a firm base, and new plaster is applied in successive coats. It is then finished to conform with the original surface.

Cracks are cut out and filled with mortar.

Large Areas

When large areas of cement plaster have been affected, it is frequently more economical, when permitted, to prepare the old surface by scratching and roughing, then to apply new stucco over the entire area.

CHAPTER 17

Sheet Metal Work

Sheet metal work is a specialized building trade encompassing the fabricating and installation of building units made of sheet metal, including but not limited to such things as gutters and downspouts, roofing, flashing and counter flashing, ventilators, skylights, and metal doors.

Duct systems for heating, air conditioning, and manufacturing processes, while falling into the classification of sheet metal work, will not be taken up here. Estimating these items should be referred to those in the business who are experts in the field.

FLASHING AND COUNTER FLASHING

Flashing is used mainly in building construction to prevent water from entering where there are joints created by intersecting roofs (valleys); chimneys, stack pipes, ventilators, etc., which project through a roof; over windows and doors; at skylight curbings; and where roofs (flat or pitched) join a vertical wall. Another and more limited use for flashing is as a shield at foundation sills and post bases against insect damage, particularly by termites.

The metals used most commonly for flashing are galvanized sheet metal, copper, tin, and aluminum. In areas where the atmosphere tends to corrode more rapidly (near oceans, paper mills or certain chemical plants), copper is preferable to galvanized metal. Occasionally lead or zinc flashing is used.

Copper flashing is commonly 16-ounce but is also available in 20 and 24-ounce rolls and sheets. *Galvanized sheet metal* is usually recommended to be 26-gauge. It comes in 8″, 14″ and 20″ widths. A common thickness of *aluminum flashing* is .019 inches and is furnished in rolls 6″ to 28″ wide and 50 lineal feet long.

Estimating Costs

Flashing materials should be estimated by the lineal foot specifying the kind of material, thickness and width. Installation costs including material should be estimated by the square foot.

Labor should be estimated by the individual job considering the location of the work, the type, and working conditions. As a guide, a worker can place roll flashing in a roof valley at the average rate of 10–20 lineal feet per hour depending on the slope. Placing flashing over windows and doors where narrow widths are used, a worker should average 15 lineal feet per hour. Ridge and hip flashing can be applied at the average rate of 15–20 lineal feet per hour.

Illustrative Example

Assuming prices and wages shown, the unit cost to flash valleys on a medium pitched roof would be:

100 sq ft. .019 aluminum flashing	@ $ .35 = $	35.00
12 hours labor	@ 12.00 =	144.00
	Unit cost per 100 sq ft = $	179.00
	Unit cost per sq ft = $	1.79

GUTTERS AND DOWNSPOUTS

A gutter installed at the cornice line of a roof to collect water runoff, is sometimes called an eave trough. The vertical downspout is sometimes called a leader pipe or conductor. Fittings include hangers, end caps, slip joints, funnels, elbows, shoes, leader straps or hooks, and wire strainers. Most of these are self descriptive. Elbows are used to connect the gutter outlet to the downspout which is located against the building. Shoes attach to the bottom of the downspout to divert the water away from the foundation.

Gutters and leaders are made in several styles such as half-round and box type. Downspouts are shaped round and rectangular, both of which may be smooth or corrugated.

The materials may be galvanized steel, copper, vinyl, fiberglass and, less frequently, wood.

In estimating material costs, the shape of the gutter, the material and its thickness or gauge, are important considerations.

Illustrative Example No. 1

A unit cost for "Ogee" shape 26-gauge white enamel finish aluminum gutters would be as follows assuming prices and wages shown:

100 lin ft 5" enamel "ogee" 26-gauge aluminum gutters	@ $ .70 = $	70.00
Add for fittings (as required)		
4 hours labor	@ 12.00 =	48.00
4 hours labor for helper	@ 9.00 =	36.00
	Unit cost per 100 lin ft = $	154.00
(excludes fittings)	Unit cost per sq ft = $	1.54

Illustrative Example No. 2

A unit cost for 2¼" × 3¼" aluminum downspout with enamel finish including straps would be as follows assuming prices and wages shown:

100 lin ft downspout and hangers	@ $ .75 = $	75.00
4 hours labor	@ 12.00 =	48.00
4 hours labor for helper	@ 9.00 =	36.00
	Unit cost per 100 lin ft = $	159.00
	Unit cost per lin ft = $	1.59

VALLEY FLASHING

FLASHING

SHINGLE FLASHING

DORMER

A

STANDING CENTER

B

CHIMNEY

CRICKET

COUNTER-
FLASHING

SHINGLE
FLASHING

C

M 87657 F

Flashing at center of rooflines. A, Valley and shingle flashing;
B, standing-center valley flashing; C, chimney and shingle flashing.

Figure 17–1

Wood-Frame House Construction . . . Agriculture Handbook No. 73.

STUD

STUCCO

EXPANDED-TYPE
METAL LATH

FLASHING

DRIP CAP

FRIEZE

WATERPROOF PAPER

SIDING

SHEATHING

A

SIDING

BUILDING PAPER

FLASHING

CLEARANCE

ROOFING

STARTER
STRIP

PORCH
RAFTER

B

Location of flashing. *A*, at material change; *B*, at roof deck.

Figure 17–2
Wood Frame House Construction . . . Agriculture Handbook No. 73.

Illustrative Example No. 3

A unit cost for copper, half round, 5" gutter assuming the following price and wages shown:

100 lin ft 5" copper gutter	@ $ 3.00 =	$	300.00
4 hours labor	@ 12.00 =		48.00
4 hours labor for helper	@ 9.00 =		36.00
	Unit cost per 100 lin ft =	$	384.00
	Unit cost per lin ft =	$	3.84

Illustrative Example No. 4

A unit cost for vinyl 5" gutters assuming the following price and wages would be:

100 lin ft 5" vinyl gutter	@ $ 1.00 =	$	100.00
4 hours labor	@ 12.00 =		48.00
4 hours labor for helper	@ 9.00 =		36.00
	Unit cost per 100 lin ft =	$	184.00
	Unit cost per lin ft =	$	1.84

METAL SKYLIGHTS

Prices of skylights installed should be obtained from subcontractors. Approximate estimates may be used based on the cost per square foot. The cost depends on whether it is a hip skylight or a gable skylight. Also, the cost varies according to whether the unit has ventilators. When estimating, all of these features should be determined as well as the size. Curbing is figured separately.

METAL VENTILATORS

Mushroom, spinner, gable, and other types of ventilators should be measured, fully described and priced by a metal worker.

RIDGE ROLLS

Ridge rolls can be placed on ridges and hips at the average rate of 100 lineal feet per 8 hour day.

CHAPTER 18

Kitchen Cabinets and Built-ins

Kitchen cabinets have been generally standardized as to dimensions and to a large extent as to construction. There are many designs and sizes to adapt them to particular areas into which they will be installed.

Basically, there are three classifications of kitchen cabinets, *base, wall* and *tall* or *utility* cabinets.

Base cabinets are those fitted with counter-tops to provide working space or into which such things as sinks and table-top ranges are recessed. They have shelves and drawers. Base cabinets may be installed directly against the walls of the kitchen or out in the open in which event they are *island counters.*

Wall cabinets, half as deep as base cabinets, are attached to the wall above the counter tops, about 18 inches, unless they are above a range, refrigerator, wall oven, or similar built-in appliance. They have shelves but no drawers. Figure 18–1 shows a profile view of base and wall cabinets with typical measurements indicated. More recent kitchen designs provide for a drop ceiling over the wall cabinets. Where there is none, the wall cabinets frequently extend to the ceiling or have a useless space of several inches from the top of them to the ceiling. (See page 374.)

Tall cabinets, sometimes called utility cabinets, are 12″ deep or the same depth (24 inches) as the base cabinet. A popular height is 7 feet with a shelf or two at the top, and open below for storage of brooms, mops and similar paraphernalia. In some instances the tall cabinets have only shelves which may be used as storage for foodstuffs and cleaning materials. They have one or two doors.

MATERIALS OF CONSTRUCTION

Plywood and particleboard (flake board) are used for sides, backs, bottoms, doors and partitions. Particleboard is becoming very popular even in high grade cabinets as it does not warp. It is less costly than plywood or solid wood boards.

Framing members for fronts, sides, backs, and so forth, may also be plywood or particleboard, although pine is also used. Drawers are made of similar materials.

Finish of the cabinet fronts may be wood or vinyl veneer; stained, or painted.

The usual hardware consists of door hinges, handles or drawer pulls. Also, there are patented drawer guides and glides to take the place of wood guides. Metal, adjustable (slotted) shelf standards have a distinct advantage over fixed wood shelf supports.

Counter tops will be discussed in a later section.

Figure 18–1

Base and wall cabinets in profile and in perspective with typical measurements shown.

METHODS OF CONSTRUCTION

There are various ways and means of cutting and joining the members in cabinet construction. Much depends on the tools available, the quality of workmanship and the cabinetmaker's preference, experience and background.

Some frames are doweled or mortise and tenon, others may employ corrugated nails or similar fasteners. Many quality-made cabinets have no backs in them as the walls serve as backs. Where cabinet backs are necessary, 1/4 inch plywood or hardboard is used.

Drawer construction varies. Sometimes the sides are butt-jointed but a better method is an end lap or rabbet joint, or better yet, a dado box corner. A more expensive joint, found in only the highest-quality cabinets, is a box joint or dovetail. See Figure 18–2 for details.

In less expensive drawer construction the bottom is nailed flush to the sides. In better cabinets the sides and front of the drawers are rabbetted just above the bottom and the 1/4 inch panel is inserted into the rabbet to form the drawer bottom.

Figure 18–3 shows the dimensions and construction of an actual base and wall cabinet that was made in a cabinetmaker's shop. Particleboard and plywood were used

END LAP JOINT

BOX JOINT

BUTT JOINT

Figure 18-2
Drawer joint construction.

and a facing of vinyl sheet was applied to the front edges and back of the doors, and also to the face only of the front frames and side panels of the cabinets. (See page 376.)

The materials, excluding the counter-top, are listed below. It is important to note the order in which materials are listed and, also, that the method of listing would be the same in these cabinets regardless of whether they were made with plywood veneer, plywood to be painted or solid wood to be stained and varnished. The difference would be in the cost of the material, type of hardware and perhaps some minor variations in construction details.

Base Cabinet

Platform frame	14 lin ft 1" × 3"	= 14 lin ft	@ $ .25 =	$	3.50
Front frame—top	1-1" × 2" 4'-0"	= 4 " "	@ .20 =		.80
bottom	1-1" × 1" × 4'-0"	= 4 " "	@ .15 =		.60
stiles	3-1" × 2" × 3'-0"	= 9 " "	@ .20 =		1.80
Counter supports	3-1" × 2" × 2'-0"	= 6 " "	@ .20 =		1.20
Rear top ledger	1-1" × 8" × 4'-0"	= 4 " "	@ .75 =		3.00
Panels, ¾ " particle board					
bottom	1-2' × 4'-0" =	8 sq ft			
ends and partition	3-2' × 3'-0" =	18 " "			
shelves	2-2' × 1'-6" =	6 " "			
	Total panels	32 sq ft	@ $ .40 =	$12.80	

Figure 18–3

Base cabinet and wall cabinet

Doors, 5/8" particle board	2-2' × 2'-6" = 10 sq ft		@ .35 =	3.50
Hardware—2 pr door hinges (self closing)	....		@ 3.50 =	7.00
2 door handles			@ 1.50 =	3.00
8 metal shelf standards			@ 2.00 =	16.00

Total material $53.20

Wall Cabinet

Front frame—top	1-1" × 2" × 4'-0" = 4 lin ft	@ $ .20 = $	.80
bottom	1-1" × 1" × 4'-0" = 4 " "	@ .15 =	.60
stiles	3-1" × 2" × 2'-6" = 8 " "	@ .20 =	1.60
Rear top ledger	1-1" × 2" × 4'-0" = 4 " "	@ .20 =	.80

Panels, ¾" particle board

bottom	1-1' × 4'-0" = 4—sq ft	
sides and partition	3-1' × 2'-6" = 7½ sq ft	
shelves	2-1' × 2'-0" = 4—sq ft	

Total panels 15-½ sq ft   @ $ .40 = $ 6.20

Doors, 5/8" particle board	2-2' × 2'-6" = 10 sq ft	@ .35 =	3.50
Hardware—2 pr door hinges		@ 3.50 =	7.00
2 door handles		@ 1.50 =	3.00
8 metal shelf standards		@ 2.00 =	16.00
	Total material		$39.50
	Total material cost, both cabinets		$92.70

Vinyl Veneer

Vinyl sheet veneer, 1/32" thick was glued to the units in shop. It required 63 square feet of material and epoxy glue as shown below.

Base cabinet
Glue			$ 2.60	
2-end panels	2' × 3'-0" =	12 sq ft		
2 doors (sides and edges)	2' × 2'-6" =	20 " "		
face of front frame and base		4 " "		
		36 sq ft	36.00	$ 38.60

Wall cabinet
Glue			2.60	
2-end panels (sides and edges)	1' × 2'-6" =	5 sq ft		
2 doors	2' × 2'-6" =	20 " "		
face of front frame		2 " "		
		27 sq ft	27.00	29.60
	Total veneer material			$ 68.20

Labor

It took an experienced cabinet maker 8 hours to cut out and assemble the two cabinets and another 6 hours to apply the vinyl veneer or:

14 hours @ $12.00 per hour	$ 168.00
When the cabinets were delivered it took 4 hours to install them in the kitchen	48.00
Total labor cost	$ 216.00
Summary of cost excluding counter top	
Material for cabinets	92.70
Vinyl veneer material	68.20
Labor to make	168.00
Labor to install	48.00
Subtotal	$ 376.90
Overhead 20%	75.38
	$ 452.28
Profit 10%	45.23
Grand total	$ 497.51

COUNTER TOPS

The materials for counter tops on base cabinets range from the inexpensive linoleum type with metal edging, to the more costly molded melamine (Formica) with rounded edges and splashback or stainless steel. Most tops are laid over 3/4" particleboard or plywood. Occasionally ceramic tile is used.

The present trend is toward laminated plastics that are heat and stain-proof. They come in widths of 25 inches which permits an overhang of 1 inch, and they have a cove-molded front and a backsplash, so that it is actually a one-piece top. Lengths are available from 4 to 12 feet in the straight sections, and 6 to 12 feet in L-shape corner sections. Prices range from $10 to $15 a running foot.

There is also a popular kitchen counter top of laminated maple 1 1/2″ thick and 25 inches wide with a 4 inch splashback. These cost $25 to $30 a running foot. This type top is sometimes referred to as a chopping-block counter.

COST OF CABINET INCLUDING COUNTER TOP

Cost of cabinets				$ 497.51
4′ Formica counter top	@ $10 per ft =	$	40.00	
Installation, 1 hour	@ 12.00 =		12.00	
		$	52.00	
	Overhead 20% =		10.40	
		$	62.40	
	Profit 10% =		6.24	68.64
				$ 566.15

Developing Unit Costs

An accepted method of estimating kitchen cabinets on an approximate basis is to use a cost-per-lineal-foot. The unit is applied to the base cabinet and the wall cabinet separately inasmuch as the construction cost for each is considerably different.

In the foregoing example the cost of the cabinets was estimated as follows assuming that the assembly and installation cost is divided equally.

Base cabinet material	$ 53.20			
vinyl veneer	38.60			
Counter top	52.00			
Labor assembling and installing	108.00			
	$ 251.80			
Overhead 20%	50.36			
	302.16			
Profit 10%	30.22			$ 332.38
Wall cabinet material	$ 26.43	$ 39.50		
vinyl veneer	17.30	29.60		
Labor assembling and installing	45.00	108.00		
	$ 88.73	$ 177.10		
Overhead 20%	17.75	35.42		
	$ 106.48	$ 212.52		
Profit 10%	10.65	21.25		
			$ 233.77	
			$ 566.15	

The unit cost for the base cabinet is $\dfrac{\$332.38}{4} = \$83.10$ per lin ft

The unit cost for the wall cabinet is $\dfrac{\$233.77}{4} = \$58.44$ per ″ ″ ″

Modification of Unit Costs

The unit prices developed for the cabinets on the previous pages are for a specific type of construction, finish and design. The cost-per-lineal-foot will vary up or down as the design and construction change. If the vinyl finish is not put on, and a less expensive counter top is laid, then the unit cost will be much less. But if drawers are added the cost will be a little more.

It is also true that in identical type cabinets the greater the length, the lower the cost-per-lineal-foot, and conversely the shorter the cabinet, the higher the unit cost-per-lineal foot. This can be readily demonstrated by taking the catalogue of a lumber supply company, or home furnishing company, or a mail order company and check unit costs-per-lineal-foot for identical kitchen cabinets. These sources are also excellent to develop a sense of values for cabinets of various types and quality.

BASIC CONSTRUCTION TYPES

Figure 18–4 shows exploded views of the two basic types of cabinet construction. At the top is the *box or panel type* which has no frame. Plywood, particleboard or flake board 3/4-inch thick, are used for sides, bottom and back. The top may be of the same material or one of the laminated or molded types. (See page 380.)

The panels are assembled by gluing and nailing with finishing nails. Doors are usually 1/2 or 5/8-inch stock and may have a lip or may close flush on the face. Drawer and shelving arrangement is optional. This type cabinet is used for base, wall and tall utility cabinets. The finish may be paint, stain or laminated.

At the bottom of Figure 18–4 is shown an exploded view of the *frame type* of cabinet. As its name indicates, a frame for the cabinet is made using nominal 1″ × 2″ stock which is actually 3/4″ × 1 1/2″. The front or face of the frame should be clear stock unless it is to be laminated or otherwise concealed. Other members can be almost any stock since it will be covered.

Jointing of the frame varies from doweling to mortise and tenon construction depending on the quality of the product.

The ends are usually a good grade of plywood and may be prefinished. A back is optional and if used is either 1/4 inch plywood or hardboard. Tops are also optional as are drawer and shelving arrangements. Doors are generally the same as in the box or panel type cabinets.

While we have illustrated the two basic types of construction in Figure 18–4 there are minor variations. Sometimes a certain amount of miscellaneous framing is used in the box or panel type as shown earlier in Figure 18–1. The important thing is to establish exactly how the cabinet you are estimating is constructed, then proceed.

ESTIMATING A BOX-TYPE CABINET

If we assume that the box-type base cabinet in Figure 18–4 is 48 inches long, 24 inches wide and 34-1/2 inches high, the material including doors but excluding the counter top would be as follows, keeping in mind the *nearest practical size* the material

BASE CABINETS

Figure 18–4

Exploded view of both "box" and "frame" construction of cabinets.

will have to be cut from. For example the height while 34-1/2 inches requires end panels cut from a 36-inch board.

¾" Plywood
Ends . . . . 2 pcs 24" × 36" = 12 sq ft
Partition . . . . 1 " 24" × 30" = 5 " "
Bottom . . . . 1 " 24" × 48" = 8 " "
Kick board . . . . 1 " 4" × 48" = $\frac{2}{3}$" "
                                          25$\frac{2}{3}$ sq ft   @ $.42 = $ 10.78
¼" Plywood back panel   36" × 48" = 12 sq ft   @ .20 =   2.40
2½" " doors   24" × 30" = 10 " "   @ .40 =   4.00
Glue and nails   @ 2.00 =   2.00
2 pr hinges   @ 3.50 =   7.00
2 magnetic door catches and knobs   4.00
                                                            _____
                                                           $ 30.18

If we added a plain particleboard counter top and splash board we include the following:

Splashboard	¾" × 6" × 48" = 4 lin ft @ $.60 =	$ 2.40	
Counter top ¾" particleboard	24" × 48" = 8 sq ft @ .90 =	7.20	
			9.60

If we add 2 shelves we include the following:

2 Plywood shelves	¾" × 24" × 24" = 8 sq ft @ $.45 =	$ 3.60	
8 Adjustable shelf standards 24"	@ 2.00 =	16.00	
			19.60
	Total material =		$ 59.38

If we assume the labor to build this cabinet would be 8 hours, then in an area where carpenters are paid $12.00 per hour, the labor charge would be $96.00. The total cost to make the cabinet would be $155.38 or about $38.85 a running foot. This does not include finishing or hanging and only has a particleboard top; also, this cabinet has no drawers and is basically quite plain.

As the quality increases, so does the cost.

ESTIMATING A FRAME CONSTRUCTED CABINET

If we assume the same size cabinet but follow the frame construction shown in Figure 18–4, we develop the following cost assuming the prices shown.

Front frame			
2-1" × 2" × 4'-0" bottom and top rails	= 8 lin ft		
3-1" × 2" × 2'-6" stiles	= 7½ " "		
	15½" "		
Back frame identical	×2		
	31 " "		
6 Stretchers 1" × 2" × 2'-0'	12 " "		
Total framing (1" × 2")	43 lin ft	$ 8.60	
2-5/16" dowels		1.50	
-Glue		1.00	
Kickboard 1-pc ¾" × 4" × 4'-0" = 4 lin ft		1.20	
Plywood			
Ends	2-24" × 36" × ¼" = 12 sq ft	2.40	
Back	1-36" × 48" × ¼" × 12 " "	2.40	
Bottom	1-24" × 48" × 3/8" = 8 " "	3.04	
Doors	2-24" × 20" × ½" = 10 " "	4.00	
2 pr hinges		7.00	
2 magnetic door catches and knobs		4.00	
		$ 35.14	

As in the example of the box type cabinet if we add a plain particleboard counter top we include the following:

Splashboard	¾″ × 6″ × 48″	= 4 lin ft	$ 2.40
Counter top ¾″ particle board 24″ × 48″		= 8 sq ft	7.20
			$ 9.60

If we add 2 shelves we include the following:

2 – Plywood shelves ¾″ × 24″ × 24″	= 8 sq ft	3.36
8 Adjustable shelf standards 24″		16.00
		$ 19.36

	$ 35.14
	9.60
	19.36
Total material	$ 64.10

The labor to make the frame by doweling will be more than nailing the box type cabinet together depending on the tools used. If we assume it will take 3 hours longer or 11 hours at the rate of $12.00 per hour, the cost to assemble the cabinet will be $132. The total cost, not including any finish or surfacing of the counter top is $196.10 or $49 per running foot.

ESTIMATING TALL CABINETS

Figure 18–5 shows an exploded drawing of a *tall cabinet* of box or panel construction. Using the measurements shown we can estimate the cost, assuming prices and wages, as shown below.

¾″ Plywood

2 sides	12″ × 7′-0″ = 14	sq ft	
1-bottom	12″ × 3′-0″ = 3	″ ″	
1-top	12″ × 3′-0″ = 3	″ ″	
2-doors	18″ × 7′-0″ = 21	″ ″	
3-shelves	10″ × 3′-0″ = 7½ sq ft		
	48.5 ″ ″		$ 20.37
¼″ Plywood back 3′ × 7′	= 21.0 sq ft		4.20
Kickboard 1-pc ¾″ × 4″ × 4′ = 2/3 ″ ″			1.20
2 pr hinges			7.00
2 door catches and knobs			4.00
Glue and nails			1.50
4 adjustable shelf standards 4′ – 0″			8.00
			$ 46.27

The labor to build this box-type tall cabinet would be about 6 to 8 hours. Assuming 6 hours at $12 per hour the labor cost is $72.00. The total cost of labor and material is $118.27. The cost per lin ft would be $39.42.

POINTS TO REMEMBER WHEN ESTIMATING CABINETS

The design, construction and finish of cabinets vary greatly.

1. Draw a free-hand sketch of the cabinet. If part or all is still in existence, look inside and see how it is constructed.
2. Note the kind of material, its thickness and quality. Take all measurements necessary to develop the cost.

3/4" sides, bottom
and top

3/4" or 5/8" doors

1/4" back

TALL   CABINET

Figure 18–5

Exploded view of a "tall" cabinet.

3. Note the type and quality of hardware, drawer glides and shelf brackets. *This is important; note type of counter top.*

4. Note how any drawers are constructed, look at the corners.

5. Carefully list the bill of materials on the premises or if your notes are complete do it in your office or home.

6. Consider very carefully the hours of labor you think it will take to build or repair the cabinet.

7. Add the cost of trucking if the cabinet is to be made away from the premises.

8. Add the cost of installation.

9. Add any subcontractor's overhead and profit if warranted.

10. Go back over your estimate and double-check the details.

CHAPTER 19

Electrical Wiring

Electrical power reaches a building through service wires extending from the power company's distribution system to the building, passing through a meter to the service equipment. This generally consists of a meter, a main entrance switch, followed by a distribution panel with fuses, circuit breakers or other protective devices for each *branch* circuit. Branch circuits bring current to points within the premises by way of junction boxes, to switches, outlet receptacles, lights, motors and other equipment.

Adequate wiring contemplates:

1. Service entrance of sufficient capacity.
2. Wires of sufficient capacity throughout.
3. Sufficient number of circuits.
4. Adequate number of outlets, switches, etc., for livable flexibility.

ELECTRICAL TERMS

To understand the language of electrical wiring, whether in manuals, on plans, or in a discussion with electrical contractors one should be familiar with the various terms used. Those terms most frequently encountered, and their definitions, are the following:

Alternating current (A.C.). Electrical current that reverses direction in a circuit at regular intervals.

Ampacity. The maximum number of amperes a wire of specified diameter can safely carry continuously is called the ampacity of that wire.

Ampere. The rate of flow of electric current through a conductor. More technically, the amount of current that will flow through one ohm under pressure of one volt.

Circuit. A continuous path for current to travel over two or more conductors (wires) from source to point of use and back to source.

Circuit breaker. A device designed to open automatically on a predetermined overload of current, and close, either automatically or manually, all without injury to the device itself.

Conductor. Any material that will permit current to flow through it. Wire in a circuit is a conductor.

Current. A flow of electrical charge.

Cycle. The flow of alternating current in one direction, then in the other direction, is one cycle. In a 60-cycle circuit this occurs 60 times a second.

Direct current (D.C.). Current in which electricity flows in only one direction. Current from any type of battery is *always* direct current.

Fuse. A protective device inserted in a circuit, containing a short length of alloy wire with a low melting point and which will carry a given amperage indefinitely. Any overload of current melts the alloy wire and opens the circuit.

(Plug fuses, when burned out, cannot be re-used.)

Fusetron. A time-delay fuse designed to tolerate temporary overloads for several seconds without blowing out.

Fustat. A nontamperable fuse designed to prevent changing fuses to a higher amperage. For example, No. 14 wire has an ampacity of 15 amps. If larger amp fuses are used there is danger of the wires in the circuit heating sufficiently to cause a fire on the premises before the fuse blows.

Grounding. A connection from the wiring system to the ground.

Horsepower. Watts measure the total energy flowing in a circuit at a given moment; 746 watts equal one horsepower.

Insulation. Any material which will *not* permit current to flow, is an insulator. When such material is used to isolate a charged conductor, it is called *insulation*.

Kilowatt. One thousand watts.

Kilowatt hours (k.w.h.). A kilowatt of power used for one hour.

National Electrical Code (N.E.C.). A set of standards for the design and manufacture of electrical materials and devices and the manner of their installation.

Ohm. The unit of electrical resistance. It is the resistance through which one volt will force one ampere.

Outlets. Any point in a circuit where electrical power can be taken and consumed.

Parallel wiring. This type of circuit wiring is commonly used in electrical construction wiring. Outlets and lights are wired across the two source wires rather than in series along one of the source wires.

Phase. The type of power available, single-phase or three-phase. The former is used in residence wiring, while three-phase is used in commercial and manufacturing properties where heavy power is required.

Resistance. Different materials offer resistance to the flow of current in different degrees. Aluminum wire, the same diameter as copper wire, has a greater resistance than the copper wire. Iron wire has a greater resistance to the flow of current than aluminum wire. Resistance depends on the kind of material, the size of the conductor and its length.

Receptacle. An outlet into which appliances can be plugged.

Series wiring. A scheme of wiring, seldom used, where the current path is along one conductor or wire. Example: Christmas tree lights, when one lamp is removed all others go out because the circuit is broken.

Source. Wherever the current comes from; may be a generator, battery, or where it enters the premises at the meter.

Switch. A device to open and close an electrical circuit or divert current from one conductor to another.

Service drop. The overhead service conductors (wires) at the building.

Transformer. An apparatus to increase or decrease voltage. The former are *step-up* and the latter *step-down* transformers.

Underwriters Laboratories. (UL) A "not-for-profit" organization whose cost of operation is supported by manufacturers who submit merchandise for testing and labels. Many cities require UL listing on all electrical materials used or sold in their jurisdiction. Underwriters Laboratories, Inc. was created in 1894 by the insurance industry but today is independent and self-supporting.

Volt. The pressure to force one ampere through a resistance of one ohm. Voltage is electrical pressure.

Voltage drop. The voltage loss due to overcoming resistance in wires and devices. The Code recommends wire sizes be used so that voltage drop will not exceed three percent in any branch circuit, measured at the furthest outlet.

Watt. The unit of electrical power. Amperes $\times$ Volts = Watts.

ELECTRICAL SYMBOLS

Electrical installation plans, whether for new work or repairs, consist of sketches or outline drawings of the room, or area involved. On these drawings are shown, by *symbols*, the location of equipment, fixtures, outlets, etc. It is as important to understand these symbols as it is to understand the language or terms used in electrical construction wiring. In Figure 19–2 are shown some of the most commonly used symbols.

Figure 19–1

A few of the labels of the Underwriters' Laboratories, Inc. (UL). The labels are affixed to electrical wires, conduit devices, etc. when they have UL approval.

LIGHTING OUTLETS

CEILING   WALL

SURFACE OR PENDANT FIXTURE

RECESSED FIXTURE

SURFACE OR PENDANT EXIT LIGHT

BLANKED OUTLET

JUNCTION BOX

RECEPTACLE OUTLETS

GROUNDED   NOT GR'D'D

DUPLEX RECEPTACLE OUTLET

RANGE OUTLET

SINGLE SPECIAL PURPOSE RECEPTACLE

SPECIAL PURPOSE CONNECTION
USE SUBSCRIPT LETTERS TO INDICATE FUNCTION*

SWITCH OUTLETS

S   SINGLE POLE SWITCH

S2   DOUBLE " "

S3   THREE WAY " "

SWITCH AND SINGLE RECEPTACLE

CEILING PULL SWITCH

CIRCUITING

WIRING CONCEALED IN CEILING OR WALL

WIRING CONCEALED IN FLOOR

WIRING EXPOSED

*DW - dishwasher

Figure 19-2

A few of the graphic symbols for electrical wiring and layout diagrams used in architecture and building construction. Courtesy The Institute of Electrical and Electronics Engineers.

Figure 19–3 is a floor plan of a residence showing how symbols are used in actual practice. The lines, dotted and solid, show the location of the wiring system.

SIZES OF WIRE

Electrical wire sizes are referred to by *number* in accord with the gauge of the American wire gauge (AWG) which is the same as that of Brown and Sharpe (B & S gauge) for non-ferrous metals. For example, No. 14 wire, a minimum size for general building wiring, is 0.064 inches (64 mils) in diameter. Smaller wires range from No. 16 to No. 36 on the standard wire gauge. Larger wires range from No. 12 to No. 0. There are larger and also smaller wires made but we are not concerned with them. Note that as the number gets larger the wire gets smaller.

Many years ago No. 14 wire was considered adequate for wiring residential and farm buildings. With the increase in appliances that have motors and heating elements, this size wire is not approved in most cities of any size. No. 12 wire is specified and in some cases No. 10. Table 19–1 shows the ampacity (see electrical terms page 384) of various sized copper wires.

TABLE 19–1
Ampacity of Various Size Wires

No. of Wire	Ampacity	No. of Wire	Ampacity
14	15	4	70
12	20	2	95
10	30	1/0	125
8	40	2/0	145
6	50	3/0	165

Note also that these are the maximum ratings for protective devices (fuses) for the size wire shown.

COLORS OF WIRE

The insulation on building wires is colored for a specific purpose—to assure that "hot" wires will be connected only to "hot" wires and neutral wires will run uninterrupted back to ground terminal. The white may be used as a grounding neutral wire only.

The colors of wires, established by the N.E.C. are:

2 - wire circuit . . . . . white, black
3 - wire circuit . . . . . white, black, red
4 - wire circuit . . . . . white, black, red, blue
5 - wire circuit . . . . . white, black, red, blue, yellow

Sometimes another grounding wire is used but it must be green, with a yellow stripe, or completely bare.

Figure 19–3

Use of symbols in the wiring layout for a residence.

METHODS OF WIRING

There are many different methods of wiring but the most commonly used systems for residential, farm and non-commercial or non-manufacturing buildings are the following:

Rigid conduit	Non-metallic sheathed cable
Thin-wall conduit	Armored cable
Flexible conduit	Surface raceways

Rigid Conduit

Much like water-pipe but annealed for ease of bending and treated to prevent corrosion, this conduit comes in lengths of 10 feet with Underwriters Label. It is sold in nominal inside diameter sizes from 1/2 to 6 inches although actual measurements are slightly larger. The pipe is threaded for coupling and connecting to boxes. Conduit systems are installed before the wire is pulled through.

Galvanized rigid conduit can be used inside, outside, and also underground. The *black enamel* conduit may only be used indoors away from moisture.

The N.E.C. limits the number of wires permitted in the conduit depending on the size of the pipe and the size and type of wire. Both rigid and thin-wall conduit are more costly to install than other systems.

Thin-Wall Conduit

This is the common name for *electrical metallic tubing* or EMT as designated by the N.E.C. In many respects this conduit is similar to rigid conduit; it is sold in 10 ft lengths, carries the Underwriters Label, the internal diameters are the same, and the method of installation is similar. Its advantage over rigid conduit, obviously, is its ease of cutting, bending and handling. Fittings are compression types as it cannot be threaded. EMT must be supported within 3 ft of a box, fitting or cabinet, and at least every 10 ft.

Flexible Conduit

This is to be distinguished from *armored cable* (BX) which contains wires. The wires have to be pulled through *flexible conduit* in the same manner as rigid conduit or EMT. Its use is somewhat limited to situations requiring flexibility, or in conjunction with rigid conduit where freakish bends are required.

Non-metallic Sheathed Cable (originally called Romex)

This system of wiring is used where economy is a major influence due to lower material costs and ease of installation. The N.E.C. recognizes two types. In Type NM the wires are enclosed in an outer fabric braid or plastic sheath which is flame-retardant and moisture resistant. It may be used for exposed and concealed work where excessive moisture is not present. There should be a strap every 4 1/2 feet and within 12 inches of every outlet box.

Type NMC has the wires imbedded in solid plastic, and may be used in damp locations but not in water or underground.

Both Types NM and NMC are available with an uninsulated ground wire. Connections may only be made in junction boxes.

Local codes will govern the use of this system of wiring.

Armored Cable

This type of wiring was originally called BX which was a trade name. It is still referred to as BX regardless of brand. It consists of two or more insulated wires enclosed in a flexible, spiral, galvanized steel sheath to protect the wires from mechanical injury. It is not a watertight armor and nails can be driven through it. Its use should be confined to interiors, never underground, and in relatively dry surroundings. It is the second easiest system to install, non-metallic sheathed cable being first. It has a small bond wire assuring a continuous circuit through the metal covering. Only steel junction boxes, switches and outlet boxes should be used.

Surface Raceways

These are channels of metal or plastic for holding and actually concealing wires on the surface of a wall or ceiling. Special elbows, junction boxes, switches, receptacles, fittings and connectors are available. Usually the raceways and fittings are mounted and then the wires are pulled through. This system of wiring is often used in making repairs where it is undesirable to break through walls, ceilings and decorations.

NUMBER OF BRANCH CIRCUITS

In determining the number of circuits required there are four considerations to be explored:

1. Lighting circuits
2. Small appliance circuits
3. Circuits serving individual equipment
4. Special laundry circuits

Lighting Circuits

The Code recommends, in Section 220–3(a), one 20 ampere circuit for every 500 square feet of livable floor area for *lighting* (or a 15 ampere circuit for every 375 square feet). A dwelling, for example, with 1500 square feet of livable area would require 3 circuits; one with 2500 square feet would need 5 circuits. The minimum required is slightly less and is based on 3 watts for every square foot of area under Section 220–2. Practically, one circuit for each 400 feet of floor area would be better. Open basements that are not used for living area, such as a recreation room, may be ignored in computing floor area.

TABLE 19–2

Watts Consumed by Electrical Units

Lighter Units

Blankets (electric)	150–	200
Blender (food)	250–	1,000
Fans (portable)	50–	200
Heat lamp (infrared)	250–	up
Knife	100–	
Lights, incandescent - noted on lamp	10–	up
Lights, fluorescent	20–	up
Mixer (food)	150–	
Radio (variable)	30–	150
Refrigerator	200–	300
Shaver	10–	
Stereo, Hi-Fi	300–	
Sump pump	300–	
Sun lamp (ultraviolet)	250–	400
T-V	300–	
Water pump	300–	

Intermediate Units

Electric iron	800–	1,200
Freezer	300–	500
Power saw (depends on motor)	300–	600
Vacuum cleaner	300–	600
Waffle iron	500–	1,000

Heavier Units

Air-conditioner (window type)	600–	1,500
Coffeemaker	600–	1,000
Deep fryer	1,300–	1,500
Dishwasher	1,000–	1,500
Clothes dryer	4,000–	8,000
Fry pan	1,000–	1,200
Garbage disposal	600–	1,000
Grill (outdoor type)	1,300–	
Heater (portable)	1,000–	1,600
Range-burners and oven on	8,000–	14,000
" -oven on only	4,000–	6,000
" -burners on only	4,000–	8,000
Rotisserie	1,200–	1,500
Washing machine	600–	800
Water heater	2,000–	5,000

Small Appliance Circuits

Section 220–3(b) of the Code requires, in addition to the previous lighting recommendation, that two or more 20-ampere *appliance branch circuits* shall be provided for the outlets servicing small appliances in the kitchen, pantry, family room, dining room, and breakfast room of dwelling occupancies. Both circuits must run to the kitchen; other rooms may be served by either of the circuits.

Special Laundry Circuit

At least one 20-ampere branch circuit should be provided for *laundry* appliance receptacle(s).

No lighting outlets may be connected to the foregoing circuits.

A three-wire 115/230-volt branch circuit is equal to two 115-volt receptable branch circuits.

Circuits Serving Individual Equipment

Appliances or equipment that use large amounts of current are usually provided with an individual circuit. While the Code does not require that all of the following appliances have a separate circuit, it is good practice to provide one.

Electric range
Water heater
Electric clothes dryer
Automatic clothes washer
Dishwasher
Disposal
Appliances rated over 1,000 watts
Electric motors rated over 1/8 hp

These circuits may be 115 or 230-volt depending on their rating. Table 19–2 shows the watts consumed by many electrical units.

From the foregoing three considerations, the total number of circuits may be determined. Assume a dwelling with a total of 1,500 square feet of livable floor area. Also assume it has the individual equipment noted.

Illustrative Example

1. Lighting Circuits 1,500 sq ft ÷ 500   3 circuits
2. Small Appliances (minimum)   2   "
   Laundry circuit outlets for small appliances 1   "
3. Individual Equipment
   ¼ hp motor, oilburner   1   "
   Range   1   "
   Water heater   1   "
   ½ hp motor water pump   1   "
   Clothes washer   1   "
   Clothes dryer   1   "
   Bathroom heater   1   "
                                      Total 13 circuits

SERVICE ENTRANCE

Electrical current from the power company comes into the premises through lead-in cables called a *service drop* if it is from overhead. It passes through a watt-hour meter to measure consumption and to the main power switch or *service disconnect*, and into the *distribution panel*. This panel (*service panel*), *panel board* contains the protective devices for each branch circuit. These may be fuses or circuit breakers. The branch circuits are designed to handle, safely, subdivisions of the total anticipated electrical requirements.

60 Ampere Service. Standard for 1930–1940, but considered minimum in 1975 by NEC to "get by." Provides for lights, appliances including range, washer-dryer, *or* hot water heater. Considered adequate where gas is used to heat water and for clothes dryers. Entrance service is over three No. 6 gauge wires.

100 Ampere Service. Standard for 1950–1960 when all-electric homes became popular. Now considered *minimum* for lighting, hot water, dryer, freezer, irons, refrigeration, air conditioner, electric range and small appliances. Entrance service is over three No. 2 or No. 3 gauge wires of RH-RW insulated type.

200 Ampere Service. The coming of electric heat, plus 12,000 watt ranges, dryer, central air conditioning, roasters and rotisseries, and other appliances with heating elements, and numerous small appliances made the 100-ampere service inadequate. Entrance service is over three No. 1/0 or No. 3/0 wires of RH-RW insulated type. Some local codes now require 200-ampere minimum service.

Determining Total Service Entrance Load

The Code requires a minimum service of 100 amperes, 115/230 volts, if there are over five 2-wire branch circuits.

It is possible to fairly well approximate the size service entrance needed by totaling up the load carried. Returning to the four considerations in determining the number of branch circuits, the wattage (load) is computed.

1. The Code permits an alternate method of computing lighting and general purpose circuits. That is to figure 3 watts for each square foot area, the same area used to determine the number of circuits on pages 392–393.
2. Small appliance load under Section 220-3(b) was two 20-ampere 115-volt circuits. This works out to $20 \times 115 = 2,300$ watts ($A \times V = W$); but it is recommended 3,000 watts be used.
3. The special laundry outlet required by the Code in 1968 is rounded out at 1,500 watts.
4. Individual equipment (heavy appliances) are rated individually. In Table 19–2 there is a listing of the most common units and the range of wattage for each. Include each at full rating in watts.

By adding the wattage from these four considerations and dividing by 230 volts, service entrance amperage is determined.

Special Notes: The Code has provided a "demand factor" for safety. If the total for the above *items 1, 2 and 3* exceeds 3,000 watts 35 percent of the excess over 3,000 is added, *not the entire excess.*

Electric ranges get special attention. Since not all burners will be in use at once with the oven turned on, the range's full rating in watts is not used.

If the range is rated at 12,000 watts, use an arbitrary 8,000 watts.

If the range is rated over 12,000 watts use 8,000 watts plus 400 for each one thousand watts in excess of 12,000. Add wall-mounted ovens, counter-mounted (table-top) burners together to get the equivalent range rating.

Electric heating treatment under the Code can be complicated but in general the *total* wattage consumed by all heating elements is used.

Air-conditioning requires that the rating in watts of the motor or motors be used.

Illustrative Example

(Dwelling 1,500 sq ft)

	Estimated Watts	Demand Factor %	Net Watts
1. Lighting 1,500 sq ft @ 3 watts	4,500		
2. Small appliances	3,000		
3. Special laundry outlet	1,500		
First 3,000 watts Total	9,000	100	3,000
Balance, 6,000 watts		35	2,100
4. ¼ hp motor Oil burner			300
½ hp motor Water pump			500
Range rated 15,000 watts (arbitrary)			8,000
Balance 7 × 400			2,800
Water heater			3,500
Clothes washer			600
Clothes dryer			6,000
Bathroom heater			1,000
		Total	27,800

The total watts, 27,800, divided by 230 volts equals 120.87-amp entrance service needed. A 100-amp service would not be adequate but a 150-amp service would be ideal as it would have a good safety factor for any expansion.

ESTIMATING ELECTRICAL WIRING

All of the basic principles discussed in Chapter 2, "Fundamental Considerations," are applicable to estimating the cost of electrical wiring. The major factors in forming judgment and making decisions will be *specifications, material, labor, overhead* and *profit.*

SPECIFICATIONS

In drawing up specifications for restoring electric wiring to its original condition prior to the fire, windstorm, flood or whatever peril was involved, one must be acquainted with the local ordinances, as well as the National Electrical Code. It is important that the person making the survey be knowledgeable about the damageability of wiring, appliances and devices. Frequently, circuits must be tested and motors and equipment checked.

Damageability of electrical wiring. Electrical wiring, fixtures and receptacles, when exposed to heat from a fire, generally begin to deteriorate and the extent of damage is related to the temperature, its duration, and the type of material and cable. The tendency to damage by heat of the new types of cables that have a fiberglass inner insulation and polyvinyl sheathing is somewhat less than the steel armored cable or cable in metal conduit.

Water from sprinklers in operation during a fire, and from fire hoses, which wet down the wiring system, may cause rusting of armored cable sheathing if it is not dried out quickly. Water that gets into conduit in fixtures and in receptacles will start corrosive action or possible rust unless promptly dried out. There is also the danger of the water causing short circuits if the current is turned back on before the system is fully dried out.

The possibilities of concealed damage or the development of future trouble due to serious exposure to fire and water frequently prompt municipal authorities to require a complete rewiring of the system involved.

However, the mere exposure to fire or water does not mean that the wiring or its appurtenances have suffered permanent injury. Many electrical installations, including motors and the lighting fixtures, have been exposed to moderate heat of a fire, or have been actually submerged by flooding, yet with prompt attention, with drying out and replacing of minor units, they have continued to function satisfactorily.

A circuit that is partially damaged or destroyed can sometimes be cut, a junction box installed, and the system restored. Other times an entire circuit, back to the distribution box, is necessary. Only a thoughtful inspection will determine procedure.

Sketches are an aid in determining what work has to be done and what materials will be needed, lengths of cable or other conductors, especially where major repairs are contemplated. A line drawing of the floor plan, with room sizes, symbols and notations is usually sufficient (see Figure 19–3.)

In some cities and towns, when the electrical wiring has been damaged a specified percentage or extent, and the wiring is substandard in the sense it is not up to the existing Code, the local ordinance requires the entire premises to be rewired to meet the new code. This may be a situation where there is knob and tube wiring or perhaps No. 14 wire is used. The Code since that installation may require No. 12 wire throughout. Outlet boxes, devices, fuses, even the distribution panel, may have to be replaced. Most insurance policies exclude any increased cost of repairs or reconstruction made necessary by the local building ordinances. Therefore the specifications for repair and replacement should be confined to what existed at the time of loss.

It is not always necessary to install the wiring in precisely the original location. Old cable may be left intact if concealed and new cable installed by lifting floor boards, boring through partition plates, or running cable down to the basement and up through a partition to the new outlet or switch. In older buildings, where cosmetic considerations are not a factor, surface wiring through raceways may be acceptable to the owners. Good judgment and common sense are necessary in writing up specifications of electrical repair in damaged buildings.

Where the property is completely destroyed and must be rebuilt, working from the new floor plan makes the job easier.

MATERIALS

It is best to make up a *branch circuit schedule* in which the location of lighting outlets, switch outlets, and convenience receptacles is shown. Figure 19–4 is a sample of a type used.

BRANCH CIRCUIT SCHEDULE

Circuit No.	Location on Premises	Light Outlets		Switch Outlets			Receptacles Outlets	Special Circuit
		Clg	Wall	S_1	S_3	S_4	General	
1	Kitchen	4			2			
2	Laundry	1			1		1	
3	Kitchen						3	Oven & Range
	Dining Rm.		2		1		2	
4	Living Rm.		2	2			6	
5	Master B.R.	1		1			4	
	Master Bath	2		1			1	Heater
6	Guest Rm.	1		1			4	
	Guest Bath	1		1			1	
7	Garage	2			1			Furnace

Figure 19–4

Branch Circuit Schedule

From this schedule the number and type fixtures are listed for later pricing.

The next step for estimating materials can be quite complicated on jobs other than minor repair and replacement. It will require meticulous attention to detail. A typical *Circuit Material Schedule* is shown in Figure 19–5. There are other methods of take-off of materials that are less detailed and used by estimators of long experience. Some use a notebook in which they write down the materials as they inspect and examine the premises, or study a floor plan. (See page 398.)

Since most building losses are less than total, the material listing will be rather short.

LABOR

The number of hours for an electrician to install new electrical wiring, outlets, boxes and so forth, calls for some experience and judgment. Much will depend on the type of wiring as to whether it is exposed, concealed or in raceways; also, how much cutting or boring through partitions, floors or joists is needed to "fish" wires through. Much thought and consideration must be given to those factors that affect labor as outlined in the tables.

The *labor hours* given in Table 19–3 for some of the operations encountered in run-of-the-mill electrical wiring should be used as a starting point in estimating the hours required for the particular job being considered. In many cases, pure judgment will supplement or even supersede the hours shown.

PRICING THE ESTIMATE

The principles for pricing the labor and material in an electrical wiring estimate follow those outlined in Chapter 4.

BRANCH CIRCUIT MATERIAL SCHEDULE

No.	Cable No. 12 2-wire	Cable No. 12 3-wire	Boxes 4" Oct.	Boxes 4" Sq	Boxes SR	Plaster Rings 4" R	Plaster Rings 4" Sq	Plaster Rings 4" SS	Switches S₁	Switches S₃	Switches S₄	Plug Recep	Plates SS	Plates 2G	Plates PR	W.N	C	H	G.S.
1	120'	20'	2	2	6	2	2		2			4	2		4	6		4	
2	160'	35'	1	3	8		1	1	2	4		3	2	4	3	10	4	2	2
3	40'	14'			5	1	1	1			1	8		1	8	4	2	6	4
4	18'	70'		3	2				1	3		4	1	3	4	4	2	2	
ETC																			
Total																			

SR Standard receptacle box   2G Two gang switch plate
SS Single pole switch plate   PR Plug receptacle plate

Figure 19-5

Branch Circuit Material Schedule

Table 19–3

Approximate Hours to Install Various Electric Wiring and Units

Type of Work		Hours
Conduit, Cable and Wire		Hours Per 100 Lin Ft
Rigid conduit–galvanized	½" to 1"	10–14
" " "	1¼" to 2"	16–20
Rigid conduit–Thin wall	½"	6–10
" " " "	1"	10–12
" " " "	1½"	12–16
" " " "	2"	14–18
Flexible conduit (Greenfield)	½" to 1"	6– 8
" " "	1½" to 2"	10–14
Non-metallic sheathed cable	#12 and #14	3– 4
Armored cable (BX)	#10 wire	5– 6
" "	#12 and #14 wire	4– 5
Pulling wire through conduit	#10, #12 and #14	2– 4

Service Entrance Installation

Meter	1–2 each
Main switch	1½–2 "
Panel or circuit-breaker box–200 amp	2–3 "

Setting Outlet Boxes

4" × 3" and octagonal	0.5 "
2 gang	0.6 "
4 gang	1.0 "

Installing Switches and Receptacles	.3 to .5 "

Overhead and Profit

An electrician is entitled to a reasonable charge for his overhead and profit. The same factors will apply in arriving at these charges as are outlined in detail under Fundamental Considerations (Chapter 2). As previously pointed out, the general contractor or builder adds only a supervisory charge to the electrical contractor's estimate.

Alternate Estimating Method

Frequently electrical contractors are able to estimate the cost of wiring in buildings on a unit cost for labor and material per outlet. This method is based on knowledge and experience in numerous and similar type jobs.

Under this method each lighting outlet, and each convenience outlet is counted and multiplied by the unit cost. To the total must be added any special circuits to heavy duty appliances or equipment, and also any lighting fixtures, door chimes and so forth.

CHAPTER 20

Thermal Insulation and Vapor Barriers

Many materials used in building construction possess some degree of insulating quality. Even "dead air space" in the walls, floors and ceilings provides a certain amount of insulation. However, the term "dead air space" can be misleading because temperature differentials on the opposite sides of the air space can cause air movement. For example, on a very cold day one finds a current of air being drawn in at a rapid rate around loose-fitting window sash or doors due to the inside and outside temperature differences.

Table 20–1 lists various insulating materials and the type marketed for building insulation. Their relative conductivity or insulating qualities are shown. The lower the K factor, the higher the insulating value of the material based on the rate of heat transmission or heat loss.

INSULATING MATERIALS

The kinds of materials that are used for building insulation are numerous, ranging from those made from minerals like glass and slag or rock, to processed vegetable fibers such as wood, cotton and bark (cork). *Polystyrene* and *urethane*, products of organic chemistry, have become popular because of their good insulating qualities and the fact they are available as foam-in-place, and in board form.

Mineral wool (rock wool) or silicate cotton, consists of thread-like filaments produced by the action of steam or air under pressure upon slag in a molten state.

Fiberglas is made from molten glass which is blown by machine into strands finer than human hairs. These are then spun into wool-like material for blanket and batt insulation.

Vermiculite and perlite (see pages 454 and 448).

TYPES OF INSULATION

The physical forms or types that various insulating materials are sold in, fall into five general categories.

1. Loose fill
2. Flexible (blanket or batt)
3. Rigid
4. Foamed-in-place (sprayed)
5. Vapor barriers

TABLE 20-1

Thermal Conductivity Values of Insulating Materials

Type	Material	K Factor (Conductivity)
Flexible	Rock wool	.27
	Glass fiber	.25
Loose fill	Standard materials	.28–.30
	Vermiculite	.45–.48
Rigid	Insulating fiberboard	.35–.36
	Sheathing "	.42–.55
	Corkboard	.26
Foam	Polystyrene	.25–.29
	Urethane	.15–.17
Wood	Low density	.60–.65
Reflective (2-sides)	(vapor barrier)	(a)

(a) Insulating value is equal to slightly more than 1-inch of flexible insulation.

Loose Fill Insulation

This type of insulation is available in bulk form, in bales or bags. It can be put in place by pouring, or packing by hand in around windows and doors. The material for loose fill insulation is usually Fiberglas or mineral wool. Vermiculite, pelletized polystyrene, and ground cork sometimes are used as loose fill. When large quantities are required, loose fill can be blown into the walls or ceilings with special portable equipment. In existing buildings that were not insulated when built, this method is the most economical.

This form of insulation is best when sidewalls are filled between studs and at least 4 inches is placed between joists or rafters.

Estimating material. The quantity of loose fill needed is dependent on the type, how well it is packed down (density), and the depth. A density of 6 to 8 lbs per cu ft is recommended.

The material comes in bags weighing 40 lbs and contains 4 cubic feet. If the material is packed to a density of 6 lbs for each cubic foot, then 40 lbs would cover $\frac{40}{6}$ or 6.67 cubic feet.

This would cover an area of 80 square feet 1 inch deep ($6.67 \times 12 = 80$). The area taken up by the joists or studs has to be added to determine the actual area of the insulation. When the area a bag will cover 1 inch deep is determined, the area a bag will cover 2, 3, 4 or 6-inches deep is obtained by dividing by the depth in inches. In other words if a bag covers 80 square feet, 1 inch deep, at a density of 6, it will cover 20 square feet $\frac{(80)}{4}$ 4 inches deep, and 40 square feet 2 inches deep.

The following formula (A) may be used to obtain the area any bag will cover, 1 inch deep, for a specific density.

(A) $\dfrac{\text{lbs per bag} \times 12}{\text{density}}$ = **Square foot area covered 1 inch deep**

The following formula (B) may be used to obtain the area to be added which is taken up by joists (or studs).

(B) $\dfrac{\text{Area covered 1 inch deep} \times .125}{\text{Spacing of joists or studs in feet}}$ = **Square foot area of joists.**

Illustrative Example

At a density of 8, determine the area, 1 inch deep, a 40 lb bag of loose fill will cover if the joists are 16 inches on center.

Area covered 1 inch deep (formula A)
$\dfrac{40 \times 12}{8} = 60$ sq ft

Area taken up by joists (formula B)
$\dfrac{60 \times .125}{4} = 5.6$ Sq ft
3

So, the area covered 1 inch deep by the 40 lb bag of loose fill, at a density of 8, excluding that taken up by joists, is 60 sq ft. The area covered 1-inch deep including joist area is 60 + 5.6 or 65.6 sq ft.

Formulas A and B may be used for bags of different weights.

The number of 40-lb bags of loose fill insulation for floors or ceilings may be read directly from Table 20–2A. The Table includes areas from 200 to 3,000 sq ft and densities per cu ft of 6, 8, and 10 lbs.

Table 20–2 gives the square foot area a 40-lb bag of loose fill will cover, including joists (or studs), for the densities and depths of insulation shown. Joists (or studs) are presumed to be 24 inches on center.

TABLE 20–2

Square Feet a 40-LB Bag of Loose Fill Will Cover, Including Studs or
Joists, for Densities And Depths of Insulation Shown.

Density-Lbs Per Cu Ft	Cu Ft of Material	Depth or Thickness of Loose Fill in Inches					
		1	2	3	4	5	6
6	6.67	85	43	28	21	17	14
8	5.00	64	32	21	16	13	11
10	4.00	51	26	17	13	10	9

Estimating labor. Pouring or packing in loose fill insulation is a relatively fast operation. Pouring in between joists takes less time than packing between studding. The physical conditions and thickness must be carefully considered.

TABLE 20–2A

Loose Fill Insulation for the Ceilings or Floors of Dwellings (mineral wool or fiberglass) Number of 40-lb Bags Required for the Depth and Density Shown (no openings have been deducted)

Ground Floor Area Sq Ft	Density 6 Lb/Cubic Foot Depth of Fill						Density 8 Lb/Cubic Foot Depth of Fill						Density 10 Lb/Cubic Foot Depth of Fill					
	1"	2"	3"	3½"	4"	6"	1"	2"	3"	3½"	4"	6"	1"	2"	3"	3½"	4"	6"
200	2.4	4.7	7.0	8.2	9.4	14.1	3.1	6.3	9.4	10.9	12.5	18.9	3.9	7.8	11.8	13.8	15.7	23.5
300	3.5	7.1	10.6	12.3	14.2	21.1	4.7	9.4	14.1	16.4	18.8	28.3	5.9	11.8	17.6	20.7	23.6	35.3
400	4.7	9.4	14.0	16.5	18.9	28.2	6.3	12.5	18.8	21.9	25.0	37.7	7.8	15.7	23.5	27.6	31.5	47.0
500	5.9	11.8	17.6	20.6	26.6	35.2	7.8	15.7	23.5	27.3	31.3	47.2	9.8	19.6	29.4	34.5	39.4	58.8
600	7.1	14.1	21.1	24.7	28.3	42.3	9.4	18.8	28.2	32.8	37.5	56.6	11.8	23.5	35.3	41.4	47.2	70.6
700	8.2	16.5	24.7	28.8	33.0	49.3	11.0	21.9	32.9	38.3	43.8	66.0	13.7	27.5	41.2	48.3	55.0	82.4
800	9.4	18.8	28.2	32.9	37.7	56.3	12.5	25.1	37.6	43.7	50.0	75.5	15.7	31.4	47.0	55.2	63.0	94.0
900	10.6	21.2	31.7	37.0	42.5	63.4	14.1	28.2	42.3	49.2	56.3	84.9	17.6	35.3	53.0	62.0	71.0	106
1000	11.8	23.5	35.2	41.2	47.2	70.4	15.7	31.3	47.0	54.6	62.5	94.3	19.6	39.2	58.8	69.0	78.7	118
1100	13.0	25.9	38.7	45.3	51.9	77.5	17.2	34.5	51.6	60.1	68.8	104	21.6	43.0	64.7	76.0	86.6	129
1200	14.1	28.2	42.3	49.4	56.6	84.5	18.8	37.6	56.3	65.6	75.0	113	23.5	47.0	70.6	83.0	94.5	141
1300	15.3	30.6	45.8	53.5	61.3	91.5	20.4	40.8	61.0	71.0	81.3	122	25.5	51.0	76.5	89.7	102	153
1400	16.5	33.0	49.3	57.6	66.0	98.6	22.0	43.9	65.7	76.5	87.5	132	27.5	55.0	82.4	96.6	110	165
1500	17.6	35.3	52.8	61.7	70.8	106	23.5	47.0	70.4	82.0	93.8	141	29.4	59.0	88.2	104	118	177
1600	18.8	37.6	56.3	65.8	75.5	113	25.1	50.2	75.1	87.4	100	151	31.4	63.0	94.0	110	126	188
1700	20.0	40.0	60.0	70.0	80.2	120	26.6	53.3	79.8	92.9	106	160	33.3	66.7	100	117	134	200
1800	21.2	42.4	63.4	74.1	84.9	127	28.2	56.4	84.5	98.4	113	170	35.3	70.6	106	124	142	212
1900	22.4	44.7	66.9	78.2	89.6	134	29.8	59.6	89.2	104	119	179	37.3	74.5	112	131	150	224
2000	23.5	47.0	70.4	82.3	94.3	141	31.3	62.7	93.9	109	125	189	39.2	78.4	118	138	158	235
2100	24.7	49.4	73.9	86.4	99.1	148	32.9	65.8	98.6	115	131	198	41.2	82.4	124	145	165	247
2200	25.9	51.8	77.5	90.5	104	155	34.5	69.0	103	120	138	208	43.0	86.3	129	152	173	259
2300	27.1	54.1	81.0	94.7	109	162	36.1	72.1	108	126	144	217	45.0	90.2	135	158	181	271
2400	28.2	56.5	84.5	98.8	113	169	37.6	75.2	113	131	150	226	47.0	94.0	141	166	189	282
2500	29.4	58.8	88.0	103	118	176	39.2	78.4	117	137	156	236	49.0	98.0	147	172	197	294
2600	30.6	61.2	91.5	107	123	183	40.8	81.5	122	142	163	245	51.0	102	153	179	205	306
2700	31.8	63.5	95.1	111	127	190	42.3	84.6	127	148	169	255	53.0	106	159	186	213	318
2800	33.0	65.9	98.6	115	132	197	43.9	87.8	132	153	175	264	55.0	110	165	193	221	329
2900	34.1	68.2	102	119	137	204	45.5	90.9	136	159	181	274	57.0	114	171	200	228	341
3000	35.3	70.6	106	124	142	211	47.0	94.0	141	164	187	283	59.0	118	177	207	236	353

Ordinary working conditions permit a worker to pour granule insulating wool 3 to 5 inches thick at the rate of 100 to 125 sq ft per hour. Packing between studding, should cover 40 to 60 sq ft per hour.

Granulated vermiculite, a dry and slippery material, pours easily and will cover floor areas at almost twice the rate of insulating wool. Vermiculite comes in 4 cu ft bags also. One bag will cover about 12 sq ft 4 inches thick. Density is not a concern.

TABLE 20–2B

Insulation Batts for the Ceilings of Dwellings (glass, mineral or rock wool)
Number of Batts Required

Ground Floor Area Sq Ft		Batt Size					
		15" By		19" By		23" By	
		24"	48"	24"	48"	24"	48"
200		76	38	60	30	50	25
300		114	57	90	45	75	38
400		152	76	120	60	100	50
500		190	95	150	75	125	63
600		228	114	180	90	150	75
700		266	133	210	105	175	86
800		304	152	240	120	200	100
900		342	171	270	135	225	113
1000		380	190	300	150	250	125
1100		418	209	330	165	275	137
1200		456	228	360	180	300	150
1300		494	247	390	195	325	163
1400		532	266	420	210	350	175
1500		570	285	450	225	375	188
1600		608	304	480	240	400	200
1700		646	323	510	255	425	212
1800		684	342	540	270	450	225
1900		722	361	570	285	475	238
2000		760	380	600	300	500	250
2100		798	399	630	315	525	263
2200		836	418	660	330	550	275
2300		874	437	690	345	575	288
2400		912	456	720	360	600	300
2500		950	475	750	375	625	313
2600		988	494	780	390	650	325
2700		1026	513	810	405	675	338
2800		1064	532	840	420	700	350
2900		1102	551	870	435	725	363
3000		1140	570	900	450	750	375

Unit costs. Assume loose fill rock wool insulation costs $5 a bag, and labor is $9 per hour. At a rate of 100 sq ft an hour a worker will use about 5 bags to pour to a depth of 4 inches and a density of 4 lbs. per cu ft. (Table 20–2 shows 21 sq ft per 40 lb. bag.)

Material 5 bags	@ $5.00 =	$25.00
Labor 1 hour	@ 9.00 =	9.00
	Unit cost per 100 sq ft =	$34.00
	Unit cost per sq ft =	$ .34

Flexible Insulation

In new construction, a popular type of insulation is wool batts and wool blanket or strip insulation. They are easily handled and come in convenient widths from 15 inches to go between framing 16 inches on center, and 23 inches for framing spaced 24 inches on center.

The material consists of mineral or glass wool and is made with or without paper. Most batts and blanket wool have a form of Kraft-faced paper with a stapling flange attached to a special breather-paper layer on the side opposite the facing. This allows for inset stapling while keeping the vapor-resistant facing toward the warm side. This provides some vapor resistance.

There is also an aluminum foil-faced type which reflects back radiant heat that comes through the walls and ceiling.

Batts and strip insulating wool are available in thicknesses of 2 to 6 inches. Batts come in lengths 6 to 8 feet.

Unfaced (no paper) batts are designed to hold in place between studs and joists until the finished surface is completed. Their main advantage is in eliminating gaps around pipes, vents and so forth.

Estimating material. The square foot area, including studs or joists, that batts will cover is readily determined. Multiply the length in feet of a 15 inch batt by 1 1/3; the length in feet of a 19 inch batt by 1 2/3; and the length in feet of a 23-inch batt by 2.

Table 20–3 shows the number of batts of different sizes needed to cover 100 sq ft including the area taken up by studs or joists. Also it shows the batts needed per sq ft of area.

Illustrative Examples:

```
A 15" batt 24" long covers 2' × 1⅓' = 2.67 sq ft
 " 15"  "  48" "     "      4' × 1⅓' = 5.33  "   "
A 19" batt 24" long covers 2' × 1⅔' = 3.33  "   "
 " 19"  "  48" "     "      4' × 1⅔' = 6.67  "   "
A 23" batt 24" long covers 2' × 2'  = 4.0   "   "
 " 23"  "  48" "     "      4' × 2'  = 8.0   "   "
```

TABLE 20–3

Number of Batts Needed to Cover an Area of 100 sq ft Including Studs and Joists

Batt Size In Inches	Square Feet Per Batt	Batts Per 100 Sq Ft	Batts Per Sq Ft
15 × 24	2.67	38	.38
15 × 48	5.33	19	.19
19 × 24	3.33	30	.30
19 × 48	6.67	15	.15
23 × 24	4.00	25	.25
23 × 48	8.00	13	.13

Estimating labor. The time to install 100 sq ft of batt or strip insulation between studding averages, on new work, 1 to 1 1/4 hours. On ceilings a little more time should be allowed.

The number of insulation batts for floors and ceilings may be read directly from Table 20–2B. The table includes areas from 200 to 3,000 sq ft and batt widths of 15", 19" and 23" by 24" and 48" long.

Unit costs. If the walls of a residence measure 100 lin ft in length and are 7'–6" high, what would it cost to apply 15" × 48" batts 3" thick?

Area to be covered including that taken up by the studs is 100' × 7.5' = 750 sq ft
Table 20–3 shows that it takes .19 batts per sq ft of area.
The number of batts needed is
.19 × 750 = 142.5 or 143 batts
If a carpenter applies batts at the rate of 100 sq ft per hour it would take
$$\frac{750}{100} = 7.5 \text{ hours}$$

Material (assume batts cost $ .80 each)	
143 batts @ $ .80	= $114.40
Labor (Assume wages $12.00 per hr)	
7.5 hours @ $12.00	= 90.00
Total cost for 750 sq ft	= $204.40
Cost per sq ft	= $ .27

Rigid Insulation

The term *rigid insulation* applies to types that are made in board-form from processed wood, sugar cane, ground cork and other vegetable products. Polystyrene and urethane, the plastic foam-type insulation material, are also made into board-like sheets. These are most often used to insulate the perimeter of concrete slabs on the ground.

Some familiar trade names for rigid insulating board are Celotex, Johns-Mansville Insulating Board, U.S. Gypsum Insulating Board, Beaverboard, and Gold Bond. These products are commonly referred to as *sheathing board.* They are available in sheets 4' × 6' to 12' long and range in thickness from 1/2 to 1 inch.

Installation methods and estimating quantities and costs are discussed in Chapter 10, "Rough Carpentry."

REFLECTIVE INSULATION

Materials high in reflective properties are used in the enclosed stud, joist and rafter spaces to retard the transfer of heat by radiation. The reflective material used must face an air space at least 3/4 inch deep for if the reflective surface comes in contact with other material, its reflective value is worthless.

The materials used for reflecting radiant heat include aluminum foil and paper products coated with a reflective oxide composition. Reflective insulation is equally effective facing the warm or cold side. Some types have aluminum foil on both sides. Some reflective insulation products include air spaces between surfaces and have nailing or stapling tabs. This type should be installed in the center of the studs to permit at least a 3/4-inch space on each side.

Estimating costs. Placing reflective insulation between studs takes more hours of labor than nailing directly on the face of the studs. The rate of application should be estimated at about 50–75 sq ft per hour depending on the physical working conditions and how much cutting has to be done. The aluminum foil type must be held by strips of lath nailed to the inside of the studs.

VAPOR BARRIERS

The insulating of buildings introduced a new problem that gave rise to applying *vapor barriers* in walls or floors. Warm moist air from inside is blocked from escaping through walls or floors. When the temperature inside the air space between studs or rafters is considerably lower than that within the building, the water vapor condenses and moisture collects in the wall space. If the wall or floor spaces are completely filled with insulating material, the problem can be greatly reduced. However, this is not usually the case.

The condensation will eventually cause peeling and blistering of outside paint and damp spots may show up on inside walls or ceilings.

Vapor barrier materials are used to prevent passage of the warm moist air, the most effective being aluminum foil, copper-coated papers, specially asphalt-treated felts and laminated Kraft papers. Also recommended where these specially treated papers cannot be used, is the application of two coats of aluminum paint underneath the interior wall and ceiling decorations.

Estimating costs

The cost of applying vapor barriers to studs should be estimated by computing the number of square feet in the area to be covered and allowing a labor rate of 250–300 square feet per hour. A unit cost may also be computed as follows using assumed material and labor costs.

1–500 sq ft roll of vapor barrier paper	$13.00
2 hours labor (500) @ $12	24.00
(250)	
	Cost per 500 sq ft $37.00
	Unit cost per sq ft $ .074

CHAPTER 21

Mechanical and Miscellaneous Trades

There are a number of building trades that are highly specialized as to methods of installation, and the materials and labor such as plumbing, heating, electrical work, ceramic tiling, floor coverings, structural steel, sheet metal and sprinkler systems. Losses that involve these trades are best estimated by persons engaged in the particular business, especially if the cost of repairs or replacement is other than a moderate amount. There are a few general contractors who become familiar with the methods of estimating those special trades through their frequent contact with subcontractors and by reviewing their details. They are competent to take off specifications for repairs and, by using prices and unit costs supplied by subcontractors, are able to approximate the damage with a fair degree of accuracy. However, if the amount of loss is sizable the general contractor will, in most cases, call in a subcontractor.

Whoever prepares the specifications and costs for these mechanical and miscellaneous trades should do so in detail, listing the materials and labor necessary.

The purpose of this chapter is to furnish the estimator with general information concerning the customary methods used to approximate repair costs and to check out estimates that are submitted. Many times a qualified subcontractor is remotely located or not available. For this reason an estimator should learn to develop a sound working knowledge of the trades in order to take off the details, identify materials, and discuss the probable cost of repairs with a qualified subcontractor when he has the opportunity.

PLUMBING

A plumber's work is generally separated into two categories; *rough* plumbing, called *roughing-in*, and *finish* plumbing which is the setting and installation of fixtures.

The term roughing-in involves labor and materials to place piping that will be concealed in walls and under floors. This includes installing and testing the water supply and the drainage piping. Figure 21–1 shows a schematic drawing of a typical plumbing system in a two-story house.

Water Supply Materials

Galvanized pipe, while still used to some extent, is rapidly being displaced by copper tubing because it is easier to install, takes less time and requires no threading or threaded fittings. Two types of copper tubing are available, *rigid* (hard-drawn) and *flexi-*

STACK VENT THRU ROOF

SHOWER

WATER CLOSET

AIR CHAMBER

LAVATORY

BATH TUB

DRUM TRAP

SOIL STACK

VENT

KITCHEN SINK

HOT WATER TANK

SHUT-OFF VALVE

VENT STACK

LAUNDRY TUBS

COLD WATER SUPPLY

CLEAN-OUT PLUG

"P" TRAP

BUILDING DRAIN

Figure 21-1

Plumbing arrangement in a two-story house with basement.

ble (soft-drawn). The latter can be installed with sweeping bends. For common household fixtures such as sinks, tubs, water-closets, showers and dishwashers the 1/2-inch pipe is minimum. Automatic washers should have 3/4-inch copper pipe.

Copper tubing joints and fittings are soldered (sweated) together using solder, flux and a small blowtorch.

Drainage Piping Materials

A drainage system in a building includes the piping that carries off sewage and other liquid waste to the sewer. Piping materials may be cast iron; galvanized wrought iron, copper, brass or plastic. Cast iron is usually used for underground drains or those buried under concrete floors.

Plastic Piping: Carefully selected and properly installed plastic pipe offers several advantages over conventional piping materials such as galvanized steel and copper pipe or tube. There are no perfect plumbing materials and all must be installed with knowledge of their physical properties and limitations.

Today's plastic pipe and fittings are often the most economical and are nearly immune to the attack of aggressive waters. At this time PE (polyethylene) pipe is used most commonly for underground service. Since it is furnished in long coils, it requires a minimum of fittings for long piping runs. For short runs, the friction loss in the insert fittings is a disadvantage. PVC (polyvinyl-chloride) pipe is available in nearly twice the pressure rating for the same cost as PE. PVC pipe is most often assembled with solvent-welded fittings. Heavy-wall PVC Schedule 80 pipe may be threaded. CPVC (chlorinated-polyvinyl-chloride) pipe is available for hot water service.

ABS (acrilonitrile-butadiene-styrene) pipe was once primarily used in potable water distribution in a size known as SWP (solvent-welded pipe). Today ABS is used in DWV (drainage-waste-vent) systems. PVC is also used in DWV systems.

To be sure of getting quality plastic pipe and fittings, make sure that the material is marked with the manufacturer's name or trademark, pipe size, the plastic material type or class code, pressure rating, standard to which the pipe is manufactured (usually an ASTM standard), and the seal of approval of an accredited testing laboratory (usually N.S.F.—the National Sanitation Foundation.)

Estimating Materials

In estimating materials, the lengths of pipe required are measured from center to center of the fittings. The nominal size of pipe is the *inside* dimension. A 3/4-inch pipe has an inside diameter of approximately 3/4 inches. Unlike combustible building materials, plumbing pipes and fixtures are seldom destroyed in a fire. They generally can be measured and identified, making it possible to write up specifications on the job in accordance with the conditions that existed at the time of loss.

Estimating Labor

Labor for cutting, threading, and installing piping is estimated according to the type of work; that is, connecting galvanized water pipe, copper pipe, waste or soil pipe, and also with the conditions under which the work is performed.

On new construction, as opposed to loss work, a contractor has workers installing pipe as the building is erected, and before the walls are plastered or floors laid. Carpenters do the cutting and notching of joists, girders and studding for the plumber.

Where a building is partially damaged, a plumber may be confronted with difficult physical working conditions, or may be connecting new piping to that existing in an old structure where corrosion of the existing pipe makes it troublesome to secure tight joints. These factors should be taken into consideration when estimating labor costs on losses. A detailed breakdown of labor hours for each operation is the best method for accuracy.

In some sections of the country, Chicago and Boston being good examples, it is common practice on new construction to estimate roughing labor at a percentage of the

TABLE 21–1

Approximate Hours Labor for Roughing by a Plumber
and Helper Together

Kind of Roughing	*Plumber and Helper Team—Hours
Water closet in dwelling	8 to 10
Lavatory in dwelling	6 to 8
Bath tub and shower in dwelling	8 to 10
Kitchen sink in dwelling	5 to 6
½ to 1-inch cast or wrought iron pipe per joint	¼ to ⅓
1¼ to 2-inch cast or wrought iron pipe per joint	⅓ to ½
4-Inch soil pipe per joint	¾ to 1
Complete roughing for one-story house	50 to 60

*Add plumber and helper wage rate and multiply by hours shown.

cost of the material. The percentage may vary from 80 to 100 percent depending upon whether it is an ordinary frame dwelling (for the former) or an apartment building (for the latter). These percentages are based on experience and cost records. For industrial and fire resistive structures the labor may be figured as high as 125 percent.

While it is not practical to set fixed hourly rates that will be useful in all situations, Table 21–1 shows the approximate hours for a plumber and a helper to make certain installations. The hours shown are a guide for new construction and will have to be adjusted if applied to specific loss repairs in accordance with the conditions in each situation.

PLUMBING FIXTURES

Plumbing fixture prices vary considerably depending on type, size, material, finish, construction and grade, all of which must be considered when estimating cost. Water closets and urinals are made of vitreous china, while bath tubs and most lavatories (sinks) for bathrooms are made of cast iron with porcelain enameled surface. Kitchen sinks are made of cast iron with porcelain enamel finish, or of stainless steel. The size, style, color and make of fixtures should be obtained before pricing.

Labor for setting fixtures after the roughing is complete is very often estimated according to the type of fixture, and the conditions under which the work is done. As with roughing, in some areas the labor to install fixtures is also based on a percentage of the cost of the fixture. Approximately 20 to 25 percent is customary.

No definite rates of labor will apply to each case, but for general estimating Table 21–2 will serve as guide. (See page 412.)

Illustrative Example

Assuming wages and prices shown, estimate the cost of roughing in and installing three bathroom fixtures. Wages are for a plumber and a helper.

New roughing materials for 3 fixtures complete	=	$ 200.00
Labor to rough in:		
Lavatory	6 hours @ $12.00 + $9.00 =	$ 126.00
Bath tub and shower	8 hours @ 12.00 + 9.00 =	168.00
Water closet	8 hours @ 12.00 + 9.00 =	168.00

Price of fixtures:

Lavatory		$ 100.00
Bath tub and shower		150.00
Water closet and seat		130.00

Labor setting fixtures:

Lavatory	3 hours @ $12.00 + $9.00 =	63.00
Bath tub and shower	4 hours @ 12.00 + 9.00 =	84.00
Water closet	3 hours @ 12.00 + 9.00 =	$ 63.00
		$1,252.00
	20% plumber's overhead and profit =	250.40
	Total	$1,502.40

TABLE 21-2

Approximate Hours Labor for a Plumber and Helper to Install
Plumbing Fixtures

Kind of Fixture	*Plumber and Helper Team — Hours
Water closet — ordinary grade	3 to 4
Lavatory — ordinary grade	3 to 4
Bath tub and shower — ordinary grade	4 to 6
Laundry tubs	4 to 6
Water heater — gas fired (45 gal)	4
Urinal — wall type	4
Urinal — floor type	6

*Add plumber and helper wage rate and multiply by hours shown.

DAMAGEABILITY OF PLUMBING

A loss that involves plumbing, particularly when a fire or freeze-up is the cause, should be inspected carefully by an experienced person; otherwise, the damage may be overestimated or underestimated. Often the system will have to be tested for leaks. Leaded joints should be examined. Fixtures in kitchens and bathrooms frequently have to be test-cleaned to determine whether heat and smoke have caused permanent injury. Fittings such as faucets, may or may not clean up, and in some cases where they are inexpensive, the cost of cleaning may exceed the cost to replace. Medicine cabinets that have been exposed to heat and smoke often require new mirrors or, if the finish is affected, they may need new chrome or baked enamel. The cost of replacing with new should be compared to the cost of refinishing. Towel racks, shower rods and shower heads fall into the same category.

The jackets on hot water heaters and other appliances installed by plumbers such as washers, dryers, and so forth, need careful inspection to determine whether they require refinishing or replacement. Motors, if submerged or wet, will have to be tested, and appliances with insulation that has been damaged by water should be checked carefully.

Many times after a fire, or when a basement has been flooded, the floor drains to the sewer have to be cleaned out.

The cost of final testing and inspecting of the plumbing after repairs should be included in the estimate.

HEATING

All but minor repairs to heating systems should be estimated by a heating contractor. If such a contractor is not available, the person making the general estimate can list the items that are damaged, identify materials, and discuss the probable cost of restoration with a competent subcontractor. Sketches and photographs will aid in the discussion, but if the loss is serious or more than a moderate amount of money is involved, the estimating should be turned over to someone in the trade. That person should be connected with a local reputable firm and preferably one that has had adequate experience in repairing damaged systems.

Where any doubt exists as to the extent of damage, or whether a fixture, control, pipe or appurtenance is injured, proper tests should be made.

Estimators should familiarize themselves with the installation and parts of warm-air, steam, and hot water heating systems. They should also know the various types of floor and space heaters.

Warm-Air Heating

Warm-air central heating includes the *gravity, forced,* and *radiant* systems. All three types employ a furnace, fired by coal, oil or gas, to supply the warm air.

The pipeless, gravity-type, warm-air system is one of the early models in which the unit is located in the basement near the center of the house. The cast iron furnace has a metal casing that is attached to a large grille in the floor above. Cool air flows down the outer edges of the grille and the warm air rises through its center.

In a piped, gravity-type, warm-air system the warm air from the furnace flows through round pipes to registers in floors and partitions. The air is moved by the difference in the weight of the lighter warm air, and the heavier cool air at the intake. Air to this intake located at the lower part of the furnace usually is drawn through return ducts leading from registers in the baseboard of the rooms.

Forced warm-air systems have a blower at the cool air entrance to the furnace. Ducts of galvanized steel or aluminum are rectangular in shape and connected to a main trunk. The air is filtered at the blower to prevent dust and lint from passing through the system. Advantages of the forced system are that the furnace need not be located near the center of the building (as in the gravity type), the ducts may be made smaller, and the flow of air can be controlled.

Warm-air radiant heating systems (radiant panel heating) employ ducts under the floor slab through which warm air is forced. The furnace and controls are similar to those used in the forced warm-air system.

Another type of warm-air system is the floor furnace which is suspended below the floor. A combustion chamber, fired by gas or oil, permits heat to pass over radiator sections and through a floor grille. Cool air flows down the outer edges of the grille by gravity to the bottom of the furnace pushing the warm air ahead of it. Expansion of the warm air causes the circulation.

Space heaters, utilizing warm air, are used where large areas, such as garages, factories, and supermarkets require heating. They are unit heaters suspended from the ceiling, and they operate by fans blowing air over steam coils, electric heating coils, or gas or oil heated radiator sections. There are many different sizes and styles of space heaters.

Steam and Hot Water Heating Systems

Steam and hot water heating systems are similar in many respects. Water is heated in a boiler which may be fired by coal, oil or gas. In a steam system the water is converted to steam and circulated through piping to radiators. The steam, after giving up its heat, condenses to water and is returned to the boiler. In a one-pipe system the water returns by gravity through the same pipes that supply the steam. In a two-pipe system the steam flows to the radiators and the water of condensation is returned by a separate system of piping connected to the opposite end of the radiators.

The hot water system provides for the circulation of water from the boiler rather than steam. The water is piped to radiators, passes through them and returns to the boiler through the return pipes. The water may circulate by gravity, in which case its movement depends on the difference between the weight of the hot water in the supply pipes and the cooler water in the returning pipes. The water may also be circulated by a pump located in the supply line near the boiler, and a one-pipe loop system.

Estimating

Steam and hot water heating systems are installed by a steamfitter and helper team. On large installations or heavy work two helpers may be used or more than one team. Much of the labor consists of pipe cutting, threading, fitting and hanging. Forced hot water systems are usually installed with copper tubing. The labor hours shown in Table 21–1, for plumbing, may be used as a guide in pipe cutting, threading and fitting. Measurements of pipe lengths are taken from center to center of the fittings. Piping for steam and hot water systems is black or galvanized wrought iron or steel. Weights and prices vary according to whether the piping is standard or extra heavy.

The specifications should include a full description of the work contemplated. Note the furnace (if involved) and its accessories by manufacturer, type, model, and any other identifying marks which will make it possible to price them properly. There are many kinds of burners and controls used in heating systems, and the price and cost of installation vary, so that adequate descriptions are very important. Insulation on piping, boilers, and duct work should be listed as to type and quantity.

Electric Heating

There are six principal methods of heating by electricity. Baseboard heaters, wall heaters with fans, electric heat pumps, electric furnace, ceiling cable, and wall panel heaters. There are also combinations of these systems. All electric heating systems specify particular thicknesses of insulation in walls, usually three inches; and in ceilings, usually six inches. The services of electrical contractors are recommended in estimating either repairs or replacement of electrical heating systems.

DAMAGEABILITY OF HEATING SYSTEMS

The furnaces, piping, and duct work of heating systems can withstand a considerable amount of heat during a fire without major injury. Where any question arises, tests should be made on the boilers, radiators, piping, and controls. This can be done by a competent subcontractor. Fireboxes, oil and gas burners that have been seriously wet down by water, or actually submerged, should be promptly dried out and checked for damage.

Puff-backs, or furnace box explosions, particularly in oil-fired systems, may do nothing more than blow the smoke pipe loose, crack or dislodge the asbestos covering on a boiler and deposit soot over the boiler room. If the explosion is severe, extensive damage is possible, not only to the firebox and boiler, but to the building structure. Many times a puff-back in an oil-fired furnace is preceded by leakage of oil into the firebox and in front of the furnace. A delay in ignition can ignite the oil and fire ensues. Damage may be confined to the exterior front of the boiler involving the burner, furnace and controls, or it may spread to the building itself.

Failure of a low-water cut-off may result in continued firing of a *dry* boiler. If not discovered in time, the heat can ruin the furnace, accessories, and controls.

Explosions in steam boilers are relatively infrequent due to safety devices to relieve pressure, but when they do occur the damage is usually more serious than that resulting from a firebox explosion.

STRUCTURAL STEEL

Steel beams, girders, columns and other structural members are priced per 100 pounds or per ton; therefore, each piece must be listed as to shape, size, and length in order to compute its weight. Structural steel handbooks show cross-sections for different steel shapes and list the weights per lineal foot for each type.

When measuring a steel member such as an angle iron, channel, H-beam, or I-beam, care should be taken to measure the exact width of the flanges, their thickness and, in beams and girders, the depth of the webs. Otherwise, they cannot be properly identified in the handbook.

The steel should also be listed as to its location, as different labor operations are involved. A fabricated truss, for example, should have all of its members itemized under the heading of "Truss." A girder or a column should be noted in the listing and its place in the structure identified.

It is customary to allow a percentage for rivets and details based on the weight of the steel. In built-up girders and columns, and in roof trusses, the allowance runs between 15 and 20 per cent. For plain girders and beams about 5 percent is adequate.

Small amounts of steel can be estimated by computing the weight and using a unit cost for material and labor based on a local quotation by someone familiar with the specific job. However, a steel contractor is the most qualified person to make an estimate where other than a moderate amount is involved.

DAMAGEABILITY OF STRUCTURAL STEEL

Unprotected steel members begin to fail between 1,000° and 1,500° F. Permanent injury is not always indicated by change of color, bending, or by the deforming of the shape of the steel. When it is economically sound to do so, structural steel members can be removed, straightened and replaced. This was demonstrated some years ago when structural steel framing was damaged by fire during construction of the Riverside Church in New York City. The members were removed, straightened and replaced satisfactorily.

CERAMIC TILE

Ceramic tiles are 3/8 inch thick. A popular size for floors is the 3/4 or 1 inch hexagonal, and for walls the 3" × 6" rectangular. Floor tile is laid over a concrete or cement base. The backing for wall tile may be cement board, W/R Gypsum Wallboard or portland cement plaster on wire lath. The base and backing for tile should be separately estimated.

The quantity of tile required is obtained from the square foot area to be covered to which 5 or 10 percent is added for breakage and waste in cutting. Cap, corner, cove and base tiles are individually priced for materials but the labor in setting these units is included in the over-all labor cost per square foot or for the job.

Tiles are set in portland cement mortar. A 1:3 mix is used which requires 1 sack of portland cement to 3 of sand.

A tile setter and a helper work together setting tile. The labor varies with the type and size of tile, the size and layout of the room, and the number of doors and windows. The small floor tile come attached to paper backing and can be set at about twice the rate as the individual tile. The labor for developing unit cost of an average bathroom may be figured as follows:

Kind of Tile	*Tile Setter and Helper*
3/4" or 1" paper backed floor tile	8 hrs per 100 sq ft
2" square, individual floor tile	16 hrs per 100 sq ft
4 1/4" × 4 1/4" rectangular glazed wall tile	16 hrs per 100 sq ft

Illustrative Example

Unit Costs for Ceramic Tile

Assuming the prices and wages used in the following examples, unit costs may be obtained.

Unit Cost for One-Inch Mosaic Hexagonal Floor Tile

100 sq ft 1" hexagonal tile	@ $ 1.00 =	$ 100.00
1 sack cement	@ 4.00 =	4.00
3 (100-lb) sacks sand	@ .30 =	.90
6 hours tile setter	@ 12.00 =	72.00
6 hours tile helper	@ 9.00 =	54.00
	Unit cost per 100 sq ft =	$ 230.90
	Unit cost per sq ft =	$ 2.31

Unit Cost for 4 ¼ " × 4 ¼ " Glazed Tile

100 sq ft 4 ¼ " × 4 ¼ " tile	@ $ 1.00 =	$ 100.00
1 sack cement	@ 4.00 =	4.00
3 (100-lb) sacks sand	@ .30 =	.90
8 hours tile setter	@ 12.00 =	96.00
8 hours tile helper	@ 9.00 =	72.00
	Unit cost per 100 sq ft =	$ 272.90
	Unit cost per sq ft =	$ 2.73

DAMAGEABILTY OF CERAMIC TILE

While ceramic tile can withstand considerable heat, it may be permanently injured by smoke stain. Glazed tile normally develops fine surface haircracks over a period of years, which are called "crazing." Frequently this is not noticed until hot smoke condenses on the surface. The condensate penetrates the cracks and deposits on the porous backing. This stain is impossible to remove, and the tile will have to be replaced, or some allowance must be made for the change in appearance.

Floor tile, which is not glazed, can be stained on the surface by excessive hot smoke. It can also be marred or scratched by broken glass, plaster and other debris being ground under foot.

In all cases the estimate of damage to stained or marred tile should be made after test-cleaning has been done.

Where sections of tile have been broken, cracked, or have come loose from the wall, repairs are made by cutting out and setting new tile, provided suitable matching is possible.

Water does not injure tile itself, but if the mortar used in the original setting is of poor grade, the tiles may come loose under conditions of submersion or excessive wetting from the back.

METAL AND PLASTIC WALL TILE

Metal and plastic wall tiles can be applied at approximately 100 square feet per 8 hour day. Resin or oil type adhesives are used. These tiles can be set over plaster and composition board walls. Assuming the prices and wage shown, the unit cost would be:

Unit Cost of Plastic Wall Tile

100 sq ft 4 ½ " × 4 ½ " plastic tile	@ $ .75 =	$ 75.00
50 lbs adhesive cement	@ .24 =	12.00
8 hours labor	@ 12.00 =	96.00
	Unit cost per 100 sq ft =	$ 183.00
	Unit cost per sq ft =	$ 1.83

The price of material for caps, corners, coves and bases is figured according to the individual job. The labor is included in the unit cost as shown for the over-all square foot area.

Metal tile can be estimated by using the same unit cost formula as shown for plastic tile and inserting local material prices and labor.

DAMAGEABILITY OF METAL AND PLASTIC TILE

Neither metal nor plastic wall tile can withstand excessive heat over a long period. The surfaces stain easily in the presence of hot smoke and they come loose from the adhesive backing. When they have been applied over plywood, plaster, or composition wall board that has been wet from the back side, a careful examination should be made to determine whether the adhesive bond has been injured.

FLOOR COVERINGS

Linoleum

Linoleum is a mixture of oxidized linseed oil, ground cork, other fillers and color pigments applied to a felt or burlap backing. It comes in rolls 6 feet wide and is available in three gauges; light (1/16-inch), standard (3/32-inches) and heavy (1/8-inch).

Linoleum should be laid over 15-pound saturated felt that has been pasted to a smooth surface. It is not recommended for concrete floors in basements or slabs on fill, or directly on wood floors.

Careful planning is required in measuring the quantity of linoleum needed. Remembering that it is sold in strips 6 feet wide, the direction of joints and adequate allowance for pattern-matching is important. Frequently, a pencil sketch of the room will aid in determining the number of square yards.

Linoleum paste covers a little over 12 square yards per gallon. There is a waterproof type, and also one that is not waterproof which costs about one third as much.

A unit cost for linoleum may be estimated as follows, based on an average labor rate of 4 square yards per hour.

Unit Cost for 1/8" Linoleum

100 sq yds 15-lb felt	@ $	.30 =	$ 30.00
100 sq yds heavy gauge linoleum	@	4.00 =	400.00
16 gals linoleum paste	@	5.00 =	80.00
25 hours labor	@	12.00 =	300.00
	Unit cost per 100 sq yds =		$ 810.00
	Unit cost per sq yd =		$ 8.10

Add cost of borders, designs, waxing and polishing as required.

FLOOR TILE

The five types of floor tile used in their order of cost are: asphalt, linoleum, rubber, vinyl and cork. Asphalt is the cheapest; cork the most expensive.

Unit costs for each type are estimated on the basis of 100 square feet of tile and lining felt, paste or adhesive, and the labor. For 6" × 6" tile approximately 4 hours are required per 100 square feet, 3 hours for 9" × 9" tile, and 2 hours for 12" × 12" tile. It takes about 50 percent more time to lay cork tile than asphalt, linoleum, rubber or vinyl.

A unit cost for vinyl tile would be developed as follows:

Unit Cost for Vinyl Tile

100 sq ft 15-lb saturated felt liner	@ $	.30 =	$ 3.00
100 sq ft 9″ × 9″ vinyl tile	@	1.50 =	150.00
2 gals paste	@	4.00 =	8.00
3 hours labor	@	12.00 =	36.00
		Unit cost per 100 sq ft =	$ 197.00
		Unit cost per sq ft =	$ 1.97

The price of any floor covering varies with the type, quality, pattern and color. It is not always easy to determine the quality of tile and gauge of linoleum when it is attached to the floor, and sometimes those inexperienced with floor tile have difficulty in distinguishing asphalt, linoleum, vinyl and rubber. If a large quantity is involved, and replacement has been decided upon, a sample cut from one corner should be taken to a floor covering shop or contractor for identification.

When old floor coverings have to be removed to lay new ones, the cost of taking up the old should be figured as a separate labor operation.

DAMAGEABILITY OF FLOOR COVERINGS

All floor coverings burn and most of them can be stained by severe hot smoke, and also by dyes from contents that have been wet. Broken glass, plaster and other debris ground into the surface by people walking over them during and after a loss may cause deep scratching and marring which cannot be removed.

Some of the adhesives used to lay floor coverings are water-soluble, and if the floors are seriously wet for any length of time, the covering comes loose. This condition will not always show up immediately, therefore, it is better to defer specifications for repair until enough time has elapsed to be certain. Usually several days is an adequate period to wait. It should not be assumed the floor covering will come loose merely because it has been exposed to water, or water is lying on the surface.

Wood floors under tiling or linoleum may expand and become corrugated if water reaches them, as in flooding, and the ridges will eventually show through the floor covering.

Linoleum or tile floors that are damaged in part may have to be completely replaced if they cannot be satisfactorily matched.

CHAPTER 22

Estimating Losses in Condominiums

Estimating losses under insurance policies that cover condominiums introduces a new concept as to property covered within a building and the interests insured. In a condominium, each occupant owns a *unit*, or *cube of space*. Each unit owner also owns an undivided share (percentage) of the common facilities and areas of the building complex including the land. An exception is when the land is leased.

TYPES OF CONDOMINIUMS

Residential Condominiums

Residential condominiums may consist of one-family homes utilizing the same land, recreation and parking areas. They may be rows or groups of 2 to 4-unit buildings, each having a private entrance but common recreational facilities and services. Another type of condominium is the high-rise apartment-like building with common recreation and other services. Residential condominiums are variously known as *Town Houses*; *Duplexes*; *Patio* or *Garden Apartments*; and *Plaza Houses*.

The FHA Model Condominium Statute defines *Building* as "one containing five or more apartments, or two or more buildings, each containing two or more apartments, with a total of five or more apartments for all such buildings and comprising a part of the property."

Conversion Condominiums

Many rental apartments and hotels have been converted into condominiums. A number of states have legislation that requires full disclosure before the owner may convert. The reason is that older buildings with the mechanical facilities (piping, wiring, heating, etc.) badly deteriorated, have been dressed up cosmetically at a small investment and a high profit upon conversion to a condominium. The legislation is designed to discourage this abuse.

Whenever loss and value are being estimated on conversion condominiums great care is necessary in determining the *quality* of repair and replacement.

Commercial Condominiums

This type of condominium is rapidly gaining in popularity for professional offices for lawyers, doctors and dentists where each owns a unit and all share in common facilities and services. There are also condominium office buildings and shopping centers.

Resort Condominiums

These condominiums, similar to the ones described under residential occupancy, are located in resort and seasonal locations such as along sea coasts, and in lake and ski country. Many are occupied only during resort seasons; some are rented in the off season.

LEGAL DOCUMENTS

State statutes set the standards for a property to qualify as a condominium. Each also requires a *master deed* (*declaration*), a *unit deed,* and that the *bylaws* be recorded at the registry of deeds. A fourth legal document, known as the *regulations,* is usually part of the bylaws.

Master Deed or Declaration

Some states refer to the master deed as the *charter, covenant,* or *deed restrictions.* It must be filed with and and approved by the proper authorities prior to committing the land to condominium use and before starting construction. The master deed or declaration specifies a number of things including:

1. Description of the land occupied.
2. Whether residential or commercial.
3. A description of the buildings.
4. A description of apartments or units.
5. Description of common areas and facilities.
6. Each unit owner's percentage interest in common areas and facilities.
7. Type of management organization.
8. Percentage of votes by unit owners necessary to decide whether to repair or sell the property in event of damage of all or part of the property.

Unit Deed

The unit deed for each unit describes the unit as to size, square feet and percentage of undivided interest in common areas and facilities. It is subject to easements in favor of other units, and to the provisions of the master deed and bylaws as to the floor plans of the condominium.

Bylaws

The bylaws provide the guidelines by which the unit owners will govern themselves. They deal with the establishment of a board of managers or directors, how many may serve and for what period, where meetings will be held, who can vote, and the type of organization, i.e., trust, corporation or association. Some bylaws describe methods for collecting assessments, rules for repair work, hiring personnel, etc.

Regulations

Regulations may be a part of the bylaws or attached to them. They stipulate rules of conduct of unit owners, such as keeping hallways, stairways, walkways and other common areas unobstructed; not cooking on the terraces; storage of bicycles, wagons, scooters, etc.; where children may play, and providing management with keys to units.

CUBE OF SPACE CONCEPT

The *cube of space* concept is that the unit owner has title to a cube of air composed of six sides enclosing the bare interior walls, floor and ceiling of the unit owned. Within this cube of space are the elements that are owned solely by the unit owner which convert the cube into dwelling quarters (in the case of a residential condominium). These elements include such items as non-load-bearing partitions, wall and ceiling decorations or paneling, floor coverings, plumbing fixtures, appliances, millwork and cabinetry. All of these may have been purchased when the building was erected, purchased from a previous owner, or installed by the present owner as improvements and betterments.

Load-bearing partitions and other structural items which reinforce or support floors, roofs, or ceilings above, are common elements even though they are within the unit owner's cube of space. Pipes, wires, conduits and similar utilities are also common elements and facilities, the units being conveyed subject to various easements in favor of the adjoining units.

Unit Boundaries

The master deed generally states the boundaries of each unit with respect to floors, ceilings, walls, doors and windows. For example, in one state it provides:

1. Floors: Upper surfaces of the subflooring.
2. Ceilings: The plane of the lower surface of the ceiling joists.
3. Interior Building Walls: The plane of the surface facing such unit of the wall studs.
4. Exterior Building Walls, Doors and Windows: As to walls, the plane of the interior surface of the wall studs; as to doors, the exterior surface thereof; as to windows, the exterior surface of the glass and the window frames.

(Not all master deeds provide the same description).

Common Areas and Facilities

The master deed describes these in detail.

Percentage of Unit Owners' Undivided Interest
in Common Areas

The master deed usually shows the undivided interest of the unit owners in common areas and facilities, and shows the percentage for each unit by number of unit.

INSURANCE FORMS

Authority for obtaining and maintaining property insurance on the common elements and facilities of the building is delegated to the management association by way of a Condominium Trust. The trust spells out the kinds of property insurance, perils to be insured against, and what property is to be covered. It requires the named insured to include the trust, trustees, unit owners and mortgagees.

Building Insurance

Policies covering the common elements and facilities are basically building forms that have been tailored to fit the condominium concept in which most of the structure is owned in common by the unit owners, while most of the real and personal property within the units is owned individually by the unit owners.

A summary of coverage under the Building policies may, in general, be stated as follows. (Not all policies are alike. Those covering the property involved apply.)

1. Loss to be adjusted with the named insured but payable to the designated Trustee.
2. The policy covers building(s), fixtures, machinery and equipment, a permanent part of the building(s) and pertaining to its service; yard fixtures, detachable building equipment, personal property in which unit owners have an undivided interest, used for the service or maintenance of the building(s), including floor coverings and out door furniture.
3. Personal property owned by the unit owners as tenants in common (jointly), and property of others in the care, custody, or control of the management association.
4. By endorsement, coverage is provided for fixtures, installations or additions *comprising a part of the building* within the unfinished interior surfaces of the perimeter walls, floors and ceilings of the individual condominium units, either initially installed, or replacements thereof, in accordance with the original condominium plans and specifications; or installed by or at the expense of the unit owners.
5. *Excluded* is property of any kind contained within the unfinished interior surfaces of the perimeter walls, walls, floors and ceilings of condominium units, except (1) common building elements and (2) pipes, wires, conduits and other utilities contained within the easements appurtenant to the common building elements and within such condominium units.

Other conditions, limitations, exclusions and extensions are to be found in these forms, such as co-insurance, deductibles, valuation and replacement cost.

Unit Owners' Insurance

The Homeowners Program, or policies patterned after it, provide the vehicle for covering the property of the unit owners. Some of these are actual cash value forms, others are subject to replacement cost coverage. The policies are the product of Insurance Services Office and other Bureaus, and individual insurers. While there are a number of Forms and Endorsements, and as with condominium building insurance Forms not all are alike, the following is, in general, a summary of the coverage under the unit owners' policies:

1. All property within the boundaries of the unit as originally installed by the developer, excluding any common elements.
2. Any variations in the original specifications installed by the developer or unit owner up to a limit of $_____.
3. Any additions to the original unit or to the unit as purchased from a previous owner.
4. Unit realty, walls and partitions inside the unit including structural items within the unit. Also included are fixtures, millwork, partitions, floor coverings, tile, plumbing fixtures, ranges and refrigerators. Also, utility pipes, wires and conduits which are not considered common elements.

Where there is an overlap of coverage between insurance covering common elements and facilities and the unit owner's insurance, the Other Insurance clause in the unit owner's policy makes it excess over any other insurance policy that covers the same property.

The MLB Condominium Forms waive subrogation against any unit owner. The Homeowners Condominium policies do not waive subrogation.

GUIDELINES FOR ESTIMATING CONDOMINIUM
BUILDING LOSSES

Examination of Legal Documents

Those charged with the responsibility for estimating losses and values in condominium buildings will be better prepared to do so by examining the following documents or otherwise being informed of their content.

1. Master deed (Declaration)
2. Unit deed
3. Bylaws
4. Regulations
5. State statutes pertaining to condominiums
6. City or county ordinance and laws

Examination of the first four documents listed is of less importance when the loss is relatively small and contained within units than when the loss involves both unit owner's premises and common elements and facilities. In the latter situation insurance covering the unit owner and that covering the common elements and facilities have to be examined.

When a loss is confined to common elements and facilities, all six of the legal documents listed above should be examined. State statutes override insurance policy provisions and those in the declaration, unit deed and bylaws.

Examination of the legal documents will usually disclose one or more of the following:

1. Date of purchase of the unit; from whom purchased and price paid. Purchased from the developer or from a previous owner.
2. How the master deed describes the horizontal and vertical boundaries of the units.
   a. To the surface of the wall studs, and to the floor joists above?
   b. To the unfinished floor below?
   c. To the unfinished surface of the perimeter walls and ceiling?
   d. To the inside or outside surface of the glass in the exterior walls?
3. Plans and specifications of the original fixtures, installations, or additions comprising a part of the building within the unit of the insured unit owner.
   a. Any replacements of above by the insured unit owner.
   b. Any fixtures, installations or additions comprising a part of the building made by or at the expense of the insured unit owner.
4. Any of the above covered under insurance policies carried by the management association.
5. How common elements and facilities are to be valued. By law—by trust agreement.
   a. Actual cash value with coinsurance.
   b. Agreed amount.

Examination of Insurance—Management Association

It is as important that the estimator examine and be familiar with insurance policies covering condominiums as it is to examine the legal documents. Policies held by the management association should be examined to establish whether:

1. the form is tailored for a condominium,
2. common elements and facilities are covered,
3. fixtures, installations and additions comprising a part of the building and located within an individual unit owner's premises are covered,
4. property within units, except common elements is excluded,
5. personal property of the management association is covered (office furniture, equipment, supplies, other),
6. any of the following is covered:
   a. trees, shrubs and plants,
   b. newly acquired buildings, additions and extensions,
   c. swimming pools, sauna baths, etc.
   d. piers, docks, retaining walls, walks, tennis courts, paved surfaces, etc.
   e. T.V. antennas
   f. storage and recreation buildings, cabanas, or storage sheds of unit owners.
7. loss is subject to replacement cost, whether a limited coverage or full replacement; what property is excluded from replacement cost,
8. there is co-insurance, or an Agreed Amount Endorsement, and
9. there are deductibles.

Examination of Insurance—Unit Owners

1. Is the unit owner's policy a condominium form or is it a Homeowners policy converted to a condominium policy by endorsement? Some other form?
2. Does the policy cover building items within the unit, originally there or installed by or at the expense of the unit owner, including
   a. wall and ceiling decorations, paneling, etc.,
   b. floor coverings,
   c. partitions—non-load bearing,
   d. cabinets, cupboards, plumbing, cooking and lighting fixtures, mill work, etc.?
3. Are there deductibles to be applied? How applied?

Other Insurance Forms and Endorsements—
Problem Areas

There are a number of forms and endorsements covering condominiums and insuring the interests of unit owners and management associations. Prior to preparing a detailed estimate of loss and value, as previously stated, the person making the estimate will be better prepared by examining the insurance policies to determine who is insured; what, when and where property is covered; what perils are insured against, and the amount of insurance applying. Fortified with this information, the estimator will be able to avoid many of the inherent problems that are built in to condominium loss adjusting.

For example, common elements within the units may, by policy terms, be covered under both the unit owner's and management association's insurance. Yet, the master deed may state that, for example, *all partitions* within the unit are owned by the unit owner. If it turned out that certain of the partitions were *load-bearing* (common elements), a question would arise as to which policy would be called upon to pay for their repair or replacement, in spite of the excess clause in the Homeowners and the primary insurance clause in the MLB Forms.

Other problems are created when insurance covering common elements is on a form not tailored to or designed for condominiums. Breakage of glass raises a problem where the master deed stipulates that glass breakage in perimeter walls, though a common element, is the responsibility of the unit owner.

Detailed Estimate—Step By Step

The following is an excellent example of a detailed fire damages estimate by a well-known construction company. It illustrates the importance of separating damage in the unit owner's apartment from the damage in the building proper which is commonly owned by all unit-owners and insured individually.

TELEPHONE 751-3816

GUNN & GUNN *Construction Company*

February 9, 1982   161 N. W. 52 STREET, MIAMI, FLORIDA 33127

American Insurance Company
111 S. W. 88th Street
Miami, Florida

ATT: Mr. Smith   Re: Miami Condominium
                                       000 S. W. 8th Street
                                       Miami, Florida

Dear Mr. Smith:

I have inspected the fire that occurred at the Miami Condominium. The
building is a 15 story high rise with 28 units per floor. The origin of
The fire was unit #1403 which suffered extensive damage. The surrounding
areas which were damaged were the 14th floor corridor, 13th floor corridor
and the exterior of the building. It was determined that the owner of
unit 1403's damage was limited to the painting of walls and doors inside
the unit, the floor covering inside the unit and the kitchen appliances.
All other damage inside unit 1403 and in the common areas will be considered
part of the building damage. The following repair specifications are divided between unit 1403 and the condominium building:

UNIT #1403, Living Room 16'2"x10'2"x8'

Paint drywall walls	2 coats	576 sf @	.22 sf	$126.72
Replace wall-to-wall carpet & pad		37 yds @	12.85 yd	475.45
Enamel applied base		72 lf @	.36 lf	25.92

Powder Room 5'x3'x8', 3'x2'x8'

Remove & prepare sheet vinyl floor		@		45.00
Replace sheet vinyl floor, small area cost		@		72.00
Paint plaster walls & ceiling		181 sf @	.22 sf	39.82
Enamel doorside & trim	2 coats	1 @		9.50
" applied wood base		15 lf @	.36 lf	5.40

Dining Room 10'x9'x8'

Paint drywall walls & soffit		277 sf @	.22 sf	60.94
Enamel base		29 lf @	.36 lf	10.44
Replace wall-to-wall carpet & pad		10 yds @	12.85 yd	128.50

Kitchen 9'6"x7'x8'

Remove & prepare sheet vinyl floor		67 sf @	.45 sf	30.15
Replace sheet vinyl floor covering		67 sf @	1.95 sf	130.65
Paint drywall walls		264 sf @	.22 sf	58.08
Enamel doorside & trim	2 coats	1 @		9.50
" wood base		8 lf @	.36 lf	2.88
Replace cook top with eye level oven & built-in hood, delivered		@		496.60
Replace dishwasher, delivered		@		289.88
" 18 c.f. refrigerator, over & under		@		791.68
" combination washer & dryer		@		369.89

Bedroom 14'6"x12'2"x8', Entry 3'x3'x8'

Paint plaster walls	2 coats	488 sf @	.22 sf	107.36
Enamel doorsides & trim	2 "	4 @	9.50 ea	38.00
" baseboard		61 lf @	.36 lf	21.96
Replace wall-to-wall carpet & pad		19 yds @	12.85 yd	244.15

page #2, Miami Condominium Fire Damage

Closet 5'3"x10'6"x8'
Paint plaster walls		252 sf @	.22 sf	55.44
Enamel doorsides & trim	2 coats	2 @	9.50 ea	19.00
" applied wood base		32 lf @	.36 lf	11.52
" shelves		21 lf @	.70 lf	14.70

Bathroom 8'x4'6"x8'
Paint walls & ceiling		236 sf @	.22 sf	51.92
Replace ceramic tile floor & base		@		265.00
Enamel doorside & trim	2 coats	1 @		9.50

	4017.55
Contractor's O. H. 10%	401.75
	4419.30
Contractor's Profit 10%	441.93
Owner's Total	$4861.23

CONDOMINIUM BUILDING DAMAGES Living Room 16'2"x20'2"x8'
Clean & seal concrete walls	288 sf @	.18 sf	$51.84
Scrape, clean & bond concrete ceiling	320 sf @	.38 sf	121.60
Acid clean concrete slab	320 sf @	.28 sf	89.60
Replace 1x2 p.t. wall furring	288 sf @	.45 sf	129.60
" drywall walls	576 sf @	.90 sf	518.40
Float, tape & finish drywall walls	576 sf @	.10 sf	57.60
Blow on acoustical plaster ceiling	320 sf @	1.45 sf	464.00
Replace 3153 single hung aluminum sash window with wood bucks & installation	@		218.00
Replace ceramic tile window sill	3 lf @	3.85 lf	11.55
" 6'0"x6'8" sliding glass doors & track installed, anodized heavy duty	@		486.56
Replace applied wood base	72 lf @	1.15 lf	82.80

Powder Room 5'x3'x8', 3'x2'x8'
Replace 3½" metal studs with top & bottom plates	15 lf @	8.50 lf	127.50
Replace ceiling framing	21 sf @	1.50 sf	31.50
" drywall walls & ceiling	181 sf @	.90 sf	162.90
Float, tape & sand drywall walls & clg.	181 sf @	.10 sf	18.10
Replace 30" formica vainty with formica top, installed	@		142.60
Replace 2'0"x6'8" h.c. flush door & jamb, butts & bathroom lock, installed	@		98.50
Replace wall mirror	@		56.98
" toilet paper holder	@		12.00
" towel bars	2 @	24.95 ea	49.90
Acid clean concrete floor	21 sf @	.28 sf	5.88
Clean & seal concrete ceiling	21 sf @	.18 sf	3.78
Replace applied wood base	15 lf @	1.15 lf	17.25

Dining Room 10'x9'x8'
Replace 3½" metal stud partitions	20 lf @	8.50 lf	170.00
" 1x2 p.t. furring on 1 wall	72 sf @	.45 sf	32.40
Clean & seal concrete block wall	72 sf @	.18 sf	12.96
Acid clean concrete floor slab	90 sf @	.28 sf	25.20
Scrape, clean & bond concrete ceiling slab	90 sf @	.38 sf	34.20
Replace metal soffit framing at a/c duct	@		183.60
" drywall walls & soffit	277 sf @	.90 sf	249.30
Float, tape & prepare walls & soffit	277 sf @	.10 sf	27.70
Replace applied wood base	29 lf @	1.15 lf	4.35
Blow on acoustical plaster ceiling	90 sf @	1.45 sf	130.50

Kitchen 9'6"x7'x8'
Replace drywall walls	264 sf @	.90 sf	237.60
Float, tape & prepare drywall walls	264 sf @	.10 sf	26.40
Clean & seal block walls & ceiling deck	199 sf @	.18 sf	35.82

continued next page

page #3, Miami Condominium Fire Damage

Kitchen continued:
```
Replace formica base cab'ts: 6'x3', 3'6"x3',
3'x3'                                            @              675.00
Replace formica wall cab'ts: 6'x2'6", 3'6"x30",
3'x1',3'x2'6"                                     @              837.00
Replace formica countertops        10 lf @ 15.95 lf    159.50
Carpenter & common labor for installation  @              167.00
Replace 2x2 tile ceiling & grid    67 sf @  1.45 sf     97.15
     "  3'0"x6'8" s.c. flush door & jamb,
butts, keylock & threshold, installed            @              244.40
Replace applied wood base           8 lf @  1.15 lf      9.20
Bedroom  14'x12'2"x8', Entry  3'x3'x8'
Clean & seal concrete block walls  208 sf @   .18 sf     37.44
Replace p.t. 1x2 wall furring      208 sf @   .45 sf     93.60
Scrape, clean & prepare concrete ceiling 178 sf @ .38 sf 67.64
Replace 3½" metal stud partitions   38 lf @  8.50 lf    323.00
     "  drywall walls             488 sf @   .90 sf    439.20
Float, tape & prepare drywall walls 488 sf @   .10 sf    48.80
Blow on acoustical plaster ceiling 177 sf @  1.45 sf    256.65
Replace applied wood base           61 lf @  1.15 lf     70.15
     "  2'6"x6'8" h.c. flush door & jamb,
butts, privacy lock, installed                   @              110.29
Replace 2'6"x6'8" h.c. flush door & jamb,
butts, passage lock, installed       2    @ 108.89 ea    217.78
Replace dbl. 3153 single hung windows with
mull bars, installed                 2    @ 128.00 ea    256.00
Replace ceramic tile sill            6 lf @  3.85 lf     23.10
Closet  5'3"x10'6"x8'
Clean & seal concrete block walls  128 sf @   .18 sf     23.04
Replace 1x2 p.t. furring           128 sf @   .45 sf     57.60
     "  drywall walls             252 sf @   .90 sf    226.80
Float, tape & prepare drywall walls 252 sf @   .10 sf    25.20
Replace applied wood base           32 lf @  1.15 lf     36.80
Scrape, clean & prepare concrete ceiling 56 sf @ .36 sf  20.16
Blow on acoustical plaster ceiling  56 sf @  1.45 sf     81.20
Replace shelves, brackets & rods    21 lf @  6.90 lf    144.90
Bathroom  8'x4'6"x8'
Clean, seal concrete block wall     64 sf @   .18 sf     11.52
Replace 3½" metal wall studs        17 lf @  8.50 lf    144.50
     "     ceiling framing          36 sf @  1.50 sf     54.00
     "  drywall walls & ceiling    236 sf @   .90 sf    212.40
Float, tape & prepare drywalls & ceiling 236 sf @ .10 sf 23.60
Replace 2'0"x6'8" h.c. flush door & jamb,
butts, privacy lock, installed                   @               98.50
Replace 2 towel bars, soap dish & paper holder   @               52.50
Replace formica vanity 4'6" with formica top     @              369.99
     "     wall mirror                            @               98.28
     "     tempered glass tub enclosure & rack   @              192.36
Replace built-in medicine cab't                  @               54.79
Replace 3/8" marble,cultured,full wainscot at tub @             780.00
Balcony  14'x7'
Acid clean concrete slab            98 sf @   .28 sf     27.44
Replace masonry handrail with solid concrete blocks
& open ornamental center blocks,stuccoed with cap            1050.00
West Exterior Wall
Replace 3153 single hung window & screen in apt. 1503        128.00
Labor & materials to clean & prepare exterior
wall areas, balconies, 14 floors                 @             2820.60
```

continued next page

page #4, Miami Condominium Fire Damage

West Exterior Wall continued:

Clean sliding glass doors & windows		@		1678.70
Remove loose & damaged stucco & apply bonding agent & restucco walls, balcony bottoms,window & door returns				5135.00

14th Floor Corridor 305'x5'6"x8', 45'6"x10'3"x8'

Replace drywall walls	1200 sf @	.90 sf	1080.00
Float, tape & prepare drywall walls	1200 sf @	.10 sf	120.00
Clean, seal ceiling deck a/ bonding agent	2144 sf @	.38 sf	814.72
Reblow acoustical plaster ceiling	2144 sf @	.89 sf	1908.16
Remove wallpaper,prepare drywall walls	182 rolls @	6.25 ea	1137.50
Materials to replace wallpaper walls	182 " @	12.50 ea	2275.00
Labor to hang wallpaper walls	182 " @	12.00 ea	2184.00
Replace applied wood baseboard	150 lf @	1.15 lf	172.50
Remove & prepare wallpaper mural wall	8 strips @	9.50 ea	76.00
Enamel applied wood base	666 lf @	.36 lf	239.76
Replace wallpaper mural,materials only	8 strips @	125.00 ea	1000.00
Labor to hang wallpaper	8 " @	25.75 ea	206.00
Replace bookmatch rosewood paneling	12 sheets@	74.98 ea	899.76
Labor installing paneling	@		395.28
Replace 3'0"x6'8" doors & jambs, trim, keylock, installed	12	@ 244.40 ea	2932.80
Replace raised molding on front apartment doors replaced	7	@ 49.78 ea	348.46
Replace wall-to-wall Alexander Smith woven back carpet & pad	307 yds @	26.70 yd	8196.90
Enamel multi-color front doors & trim	17	@ 24.50 ea	416.50
" doorsides & trim	42	@ 9.50 ea	399.00
Deodorize storage area	@		45.00
Replace wall-mounted mirror	@		59.95

12th Floor Corridor

Clean wall-to-wall carpet & pad	2144 sf @	.18 sf	385.92

Exterior Painting

Seal smoked area & paint 3 sections of stucco walls from penthouse to ground including balconies	@	2500.00

Sub Contractors & General Items

Electrical: replace damaged conduit, sub panel, wiring, outlets & switches. Hang new light fixtures. Install

kitchen appliances & test all circuits to panel	@		2100.00
Light fixture allowance in unit $1403	@		270.00
Temporary & emergency wiring done to date including fire alarm repair, house lights, temporary lights, 5 exit lights	@		1590.00
Hallway fixtures allowance, chandeliers	3	@ 150.00 ea	450.00
Wall mounted fixtures	2	@ 75.00 ea	150.00
Hall fixtures	13	@ 26.00 ea	338.00

Airconditioning: replace Florida heat pump EMV20 reverse cycle heat pump 1½ ton. Connect new equipment to existing condensor water piping. Replace all ductwork & grilles. Replace fire damper at the fresh

air intake:	@	1760.00
Clean & deodorize hallway air conditioner	@	375.00

Plumbing: remove fixrures & associated pipes. Furnish & install 1 new tub & trim, 2 water closets K3400EB, faucets, Moen Botique LKPSE, 3322 stainless steel sink, insinkerator disposal, refrigerator ice maker line, 40

gallon hot water heater	@	2522.50

page #5, Miami Condominium Fire Damage

Sub Contractors & General Items continued:

Labor for demolition		@	1500.00
Labor to hand carry materials to & debris from 14th Floor		@	1500.00
Labor & materials to cover corridor floors, elevator & lobby areas & cleanup daily		@	620.00
Scaffolding & shoring rental & erection		@	4963.06
Debris removal	12 loads @	65.00 ea	780.00
Engineer's report & inspections		@	475.00
Job superintendent		@	4000.00
Final job cleanup		@	620.00
Job permit fee		@	1260.00
			74,851.54
Contractor's O. H. 10%			7,485.15
			82,336.69
Contractor's Profit 10%			8,233.67
	Condominium Building Damage		$90,570.36

Owner's total $ 4,861.23
Condominium Associations's total 90,570.36
Grand total $95,431.59

This proposal is based on clear working areas and does not include moving, cleaning, storage or repairing contents items not mentioned in the above or correction of South Florida Building Code violations if existing.

Thank you for the opportunity of submitting this proposal.

Sincerely,

GUNN & GUNN CONSTRUCTION COMPANY

JOHN GUNN

JG/lg

KEY TO TABLES AND FORMULAS

Glossary of Building Terms

Adobe. An aluminous earth from which unfired bricks are made.

Aggregate. In mixing concrete, the stone, sand and gravel used.

Air-dried lumber. Lumber that has been stacked on end or piled in yards or sheds for any length of time.

Alkyd resins. Have generally replaced oils. They are treated vegetable oils in a form of glyptal resin, used principally for lacquer, paints, varnishes, and metal finishes.

Alligatoring. Coarse checking pattern characterized by a slipping of the new coating over the old coating to the extent that the old coating can be seen through the fissures.

Alternating current. An electric current which reverses its direction of flow at regular intervals.

American bond. A method of bonding brick in a wall whereby every fifth, sixth, or seventh course consists of headers; other courses being stretchers.

Ampere. The practical unit which indicates the rate of flow of electricity through a circuit.

Anchor bolts. Bolts to secure steel beams or columns or wooden sills to concrete or masonry floors, walls or piers.

Annealed wire. A soft pliable wire used extensively in the building trade for tie wires, especially for wiring concrete forms.

Anodize. An electrolytic oxide protective coating on aluminum.

Apron. A plain or molded piece of finish below the stool of a window, put on to cover the rough edge of the plaster. Also, the pitched portion of the garage floor in front of the door to carry off water.

Arcade. A range of arches, supported either on columns or on piers and detached or attached to the wall.

Arch. A curved or flat construction (usually masonry), for spanning an opening. An archway.

Armored cable. Rubber-insulated wires which are wrapped with a flexible steel covering; often called BX.

Asbestos. A variety of incombustible mineral fiber.

Ashlar. The facing of thin slabs of stone or terra cotta.

Asphalt. A bituminous material found in native form or as a residue from evaporated petroleum.

Astragal. A molding attached to one of a pair of swinging doors against which the other door strikes.

Attic ventilators. In home building, usually openings in gables or ventilators in the roof; also, mechanical devices to force ventilation by the use of power-driven fans. (Also see Louver.)

Awning type window. A type of window in which each light opens outward on its own hinges, which are placed at its upper edge.

Balloon framing. A type of framing with studs extending from sill to roof in a 1-1/2 or 2-story building.

Balusters. Small vertical spindles or members forming the main part of a railing for stairway or balcony, fastened between a bottom and top rail.

Balustrade. A row of balusters with the rails, generally used for stairs, porches, balconies, etc.

Bargeboard. The decorative board covering the projecting portion of a gable roof.

Base or baseboard. A board placed against the wall around a room next to the floor.

Batten. Narrow strips of wood or metal used to cover joints, or as decorative vertical members over plywood or wide boards.

Batter board. One of a pair of horizontal boards nailed to posts set at the corners of an excavation, used to indicate the desired level, also as a fastening for stretched strings to indicate outlines of foundation walls.

Bay. The wall space between two columns; the whole space between column centers.

Bay window. Any window projecting outward from the wall of a building and commencing from the ground.

Beam. A piece of timber, steel, or other material placed across an opening or from post to post to support a load.

Belt course—in masonry. A course of stones or brick projecting from the face of a brick or stone wall.

Belt course—in carpentry. A horizontal board across or around a building, usually made of a flat member and a molding.

Bevel siding. See page 248.

Blind-nailing. Nailing in such a way that the nailheads are not visible.

Blind stop. A rectangular molding, usually 3/4 to 1-3/8 inches or more, used in the assembly of a window frame.

Block bridging (solid bridging). In building construction, blocks nailed between joists.

Bloom. An efflorescence which sometimes appears on masonry walls, especially on a brick wall. Also, a defect on a varnished surface usually caused by a damp atmosphere.

Blue stain. A bluish or grayish discoloration of the sapwood caused by the growth of certain mold-like fungi on the surface and in the interior of the piece.

Board. Lumber less than two inches thick.

Board and batten. A type of siding composed of wide boards and narrow battens. The boards (generally twelve inches wide) are nailed to the sheathing so that there is one-half inch space between them. The battens (generally three inches wide) are nailed over the open spaces between the boards.

Board foot. See page 163.

Board measure. See page 163.

Bond—in masonry. The connection between bricks, stones, or other materials formed by laying them.

Bow window. A bay window, especially a bay with a curved ground plan.

Box column. A type of built-up hollow column used in porch construction; it is usually square in form.

Box cornice. See page 231.

Box gutter. A gutter built into a roof, consisting of a horizontal trough of wood construction lined with galvanized iron, tin, or copper to make it watertight; sometimes called concealed gutter.

Bracket. A projecting support for a shelf or other structure.

Breezeway. A covered passage between house and garage.

Brick veneer. A facing of brick laid against frame, cement block, or other dissimilar wall construction.

Bridging. See page 181.

Brown coat. See page 350.

Buck. Often used in reference to rough frame opening members. Door bucks used in reference to a door frame.

Building paper. Cheap, rosin-sized or asphalt-saturated paper.

Built-up roof. A roofing composed of three to five layers of rag felt or jute saturated with coal tar, pitch, or asphalt. The top may be finished with crushed slag or gravel.

Butt joint. The junction where the ends of two timbers or other members meet in a square-cut joint.

Buttering. In masonry, the process of spreading mortar on the edges of a brick before laying it.

Buttress. Masonry projecting from a wall to gain additional strength against the thrust of a roof or vault.

Buttress, flying. A detached buttress or pier of masonry at some distance from a wall and connected to the wall by an arch or part of an arch.

BX cable-(in electrical). See page 391.

Cantilever joists. Short joists used to support a projecting balcony or cornice where the overhang is parallel to the second-story joists.

Cap. The upper member of a column, pilaster, door cornice, molding, and the like.

Cased opening. Any opening finished with jambs and trim, but without doors.

Casein. The dried curd of a cow's milk; the principal ingredient in cheese. It is used extensively as an adhesive in making glue; also used in paint.

Casement frames and sash. Frames of wood or metal enclosing part or all of the sash, which may be opened by means of hinges affixed to the vertical edges.

Casing. The trimming around a door or window opening, either outside or inside, or the finished lumber around a post or beam, etc.

Caulking. Various types of elastic compounds introduced into cracks and openings around doors, windows, etc., to make waterproof and airtight.

Cement—Keene's. The whitest finish plaster obtainable that produces a wall of extreme durability.

Center-hung sash. A sash hung on its centers so that it swings on a horizontal axis.

Center stringer. The center stringer in a flight of stairs.

Ceramic tile. See page 416.

Chair rail. A wooden molding around the wall of a room at chairback height, to afford protection against damage when chairs are pushed back or rubbed against the wall.

Chamfer. To bevel the edge of anything originally right-angled; to round off or bevel an edge to fit on or into a connecting piece.

Checking. Fissures that appear with age in many exterior paint coatings.

Chimney hood. A covering for a chimney to make it more ornamental; also, to prevent rain water from entering the flues.

Chimney throat. That part of a chimney directly above the fireplace where the walls of the flue are brought close together as a means of increasing the draft.

Chord. In building construction, the bottom beam of any truss; in a Howe truss, where two parallel beams are placed one above the other, both beams are called chords.

Circuit breaker. See page 384.

Circuits. See page 391.

Cistern. An artificial reservoir or tank, often underground, for the storing of rain water collected from a roof.

Clapboards. (Similar to bevel siding.)

Coat. A thickness or covering of paint, plaster, or other work done at one time.

Coffer dam. A frame surrounding an excavation; or a frame placed in water, the water inside the frame being pumped out to build masonry piers.

Collar beam. A beam connecting pairs of opposite rafters above the attic floor.

Column. In architecture, a perpendicular supporting member, circular or rectangular in section, usually consisting of a base, shaft, and capital.

Comb. In carpentry, the ridge of a roof; a comb board. In masonry, a tool used to give a finish to the face of stone, a drag. In house painting, an instrument used for graining surfaces.

Concrete. A mass composed of cement, sand, and broken stone or gravel.

Concrete block. See page 136.

Conduit, (electrical). See page 390.

Coping. The highest and covering course of masonry in a wall.

Corner bead. A strip of formed galvanized iron, sometimes combined with a strip of metal lath, placed on corners before plastering to reinforce them.

Corner boards. See page 230.

Corner braces. Diagonal braces let into studs to reinforce corners of frame structures.

Cornerite. Metal-mesh lath cut into strips and bent to a right angle. Used in interior corners of walls and ceilings on lath to prevent cracks in plastering.

Cornice. A decorative element made up of molded members usually placed at or near the top of an exterior or interior wall.

Cornice return. That portion of the cornice that returns on the gable end of a house.

Counterflashing. A flashing usually used on chimneys at the roofline to cover shingle flashing and to prevent moisture entry.

Countersink. The cavity for the reception of a plate or the head of a screw or bolt, so that it will not project beyond the face of the work.

Course. A continued layer of bricks, stones, terra cotta, slate, shingles, etc.

Cove. The curved, concave portion of a cornice or ceiling as distinguished from the square parts or corners. Cove-molding, cove-ceiling, etc. refer to the concave portion of the molding, ceiling, etc.

Cove molding. A three-sided molding with concave face used wherever interior angles are to be covered.

Crawl space. A shallow space below the living quarters of a house.

Cricket. In architecture, a small false roof, or the elevation of part of a roof as a means of throwing off water from behind an obstacle; as a watershed built behind a chimney or other roof projection.

Crossover. In plumbing, a U-shaped fitting having the ends turned outward. It is used for passing the flow of one pipe past another when the two pipes lie in the same plane.

Crown molding. A molding used on a cornice or wherever a large angle is to be covered.

Cup. In lumber, a distortion of a board in which there is a deviation flatwise from a straight line across the width of a board.

Cupola. A small structure surmounting a roof.

Curtain wall. A non-bearing wall built between columns.

Cypress. The wood of an evergreen tree which grows in the southern part of the United States.

Dado. A rectangular groove in a board or plank. In interior decoration, a special type of wall treatment.

Dado joint. A joint formed by the intersection of two boards, usually at right angles, the end of one of which is notched into the side of the other for a distance of half the latter's thickness.

Damageability. Susceptibility to damage —the quality of being damageable.

Damper. An adjustable plate in the flue to regulate draft.

Deadening. Construction intended to prevent the passage of sound.

Deck paint. Paint with a high degree of resistance to mechanical wear, designed for use on such surfaces as porch floors.

Decking. Boards, plywood etc., applied to roofs. Roof sheathing.

Dimension lumber. Lumber as it comes from the saws, nominally two to four inches thick and from four to twelve inches wide.

Direct current. See page 384.

Door butt. Door hinge.

Doorjamb—interior. The surrounding case into which and out of which a door closes and opens. It consists of two upright pieces, called jambs, and a head, fitted together and rabbeted.

Door sill. Member at bottom of door opening that serves as a threshold.

Door stop. A device attached to door, floor or wall to prevent door from opening too far.

Dormer. An internal recess, the framing of which projects from a sloping roof to form a vertical wall suitable for windows.

Double-hung window. One with two sash that move up and down in tracks.

Douglas fir. A tree also known as Oregon pine.

Dovetailing. The method of joining boards or other material together; the joints tooth or dovetail shaped, fitting into each other.

Dowel. A pin used in the joining of two pieces of material.

Downspout. A pipe of metal or plastic for carrying rainwater from roof gutters.

Drainage. Relates to system of pitched roofs or floors, gutters, downspouts, conduits, etc. for the purpose of shedding water.

Dressed lumber. Lumber that has been machined and surfaced at the mill.

Drier (paint). Usually oil-soluble soaps of such metals as lead, manganese, or cobalt, which, in small proportions, hasten the oxidation and hardening (drying) of the drying oils in paints.

Drip cap. A molding placed on the exterior top side of a door or window frame to cause water to drip beyond the outside of the frame.

Drop siding. See page 294.

Dry-wall interior construction. See page 239.

Drywell. Hole in the ground lined with stone, used to catch waste or rainwater.

Dutch door. A door divided horizontally, so the lower section can be closed and fastened while the upper part remains open.

Dutchman. An odd piece inserted to fill an opening or to cover a defect in woodwork.

Eaves. The margin or lower part of a roof projecting over the wall.

Eaves board. A strip of wedge-shaped wood, used at the eaves of a roof to back up the first course of shingles, tiles, or slates.

Elevation. The drawing of the external walls of a building.

Expansion joint. Material used to separate blocks or units of concrete to prevent cracking due to expansion as a result of temperature changes.

Facia or fascia. A flat board, band, or face, used sometimes by itself but usually in combination with moldings, often located at the outer face of the cornice. Generally, the board of the cornice to which the gutter is attached or adjoins.

Filler (wood). A heavily pigmented preparation used for filling and leveling off the pores in open-pored woods.

Fir. A name commonly applied to a variety of trees yielding a soft wood which is used extensively for structural lumber, especially for interior finish and also for framing purposes.

Fire door. In building construction, a door made of fire-resisting or fireproof material.

Fire-resistive. Applies to materials for construction not combustible in the temperatures of ordinary fires and that will withstand such fires without serious impairment of their usefulness for at least one hour.

Fire stop. A solid, tight closure of a concealed space, placed to prevent the spread of fire and smoke through such a space. Usually consists of 2″ × 4″ cross-blocking between studs.

Flagstone (flagging or flags). Flat stones, from one to four inches thick, used for rustic walks, steps, floors, and the like.

Flashing. Piece of lead, tin, or sheet metal, either copper or galvanized iron, used around dormers, chimneys, or any rising projection such as window heads, cornices, and angles between different members or any place where there is danger of leakage from rain water or snow.

Flat paint. An interior paint that contains a high proportion of pigment, and dries to a flat or lusterless finish.

Floating. The equal spreading of plaster or cement on a surface by means of a board, called a float.

Flue. The space or passage in a chimney through which smoke, gas, or fumes ascend. Each passage is called a flue.

Flue lining. Fire clay or terra-cotta pipe, round or square, usually made in all of the ordinary flue sizes and in two-foot lengths, used for the inner lining of chimneys with the brick or masonry work around the outside.

Flush. Adjacent surfaces even, or in the same plane.

Flush door. A door with two flat surfaces—no panels.

Footing. The spreading course or courses at the base or bottom of a foundation wall, pier, or column.

Formica. A brand name for laminated plastic material used extensively in the building trades for wall covering, as a veneer for plywood panels, for kitchen cabinet tops.

Foundation. The supporting portion of a structure below the first-floor construction, or below grade, including the footings.

Frame construction. See page 161.

Framing. See *Carpentry, Rough.*

Framing—balloon. A system of framing a building in which all vertical structural elements of the bearing walls and partitions consist of single pieces extending from the top of the soleplate to the roofplate and to which all floor joists are fastened.

Framing—platform. A system of framing a building in which floor joists of each story rest on the top plates of the story below or on the foundation sill for the first story, and the bearing walls and partitions rest on the subfloor of each story.

Frieze. Any sculptured or ornamental band in a building. Also the horizontal member of a cornice set vertically against the wall.

Frostline. The depth of frost penetration in soil.

Fungi—wood. Microscopic plants that live in damp wood and cause mold, stain, and decay.

Fungicide. A chemical that is poisonous to fungi.

Furring. Strips of wood or metal applied to a wall or other surface to even it, to form an air space, or to give the wall an appearance of greater thickness.

Fuse. See page 384.

Fuse box. See page 393.

Gable. That portion of a wall contained between the slopes of a double-sloped roof.

Gable roof. See page 272.

Galvanizing. A coating of zinc applied to iron or steel to prevent rusting.

Gambrel roof. See page 274.

Gingerbread work. A gaudy type of ornamentation in architecture.

Girder. A large timber or steel member, either single or built-up, used to support floors or walls over an opening.

Grain. The direction, size, arrangement, appearance, or quality of the fibers in wood.

Grain—edge (vertical). Edge-grain lumber has been sawed parallel to the pith of the log and approximately at right angles to the growth rings.

Grain—flat. Flat-grained lumber has been sawed parallel to the pith of the log and approximately tangent to the growth rings; i.e., the rings form an angle of less than 45 degrees with the surface of the piece.

Grillage. A framework of heavy timbers or beams laid longitudinally and crossed by similar beams laid upon them, for sustaining walls to prevent irregular setting.

Groove. A long hollow channel cut by a tool into which a piece fits.

Ground floor. The floor of a building on a level, or nearly so, with the ground.

Grounds. Strips(s) of wood the same thickness as lath and plaster, attached to walls before the plastering is done. Used around windows, doors and other openings as a plaster stop and guide for plaster thickness.

Grout. Mortar made so thin by adding water that the mixture will run into joints or cavities of the mason-work and fill it up solidly.

Gunite. A construction material composed of cement, sand, or crushed slag and water mixed together and forced through a cement gun by pneumatic pressure.

Gutter. The channel for carrying off rain water from the roofs.

Gypsum. A mineral, hydrous sulphate of calcium. In the pure state, gypsum is colorless. When part of the water is removed by a slow heating process, the product becomes what is known as plaster of Paris.

Gypsum blocks. A type of building material usually grayish white in color; because of its friable texture, it is used only in nonload-bearing partition walls.

Gypsum board. See page 240.

Hardboard. A hard rigid board made from refined wood fibers into sheets 4′ × 8′ and larger and thicknesses 1/16″ to 3/4″.

Header. In building, a brick or stone laid with the end toward the face of the wall; one or more pieces of lumber used generally around openings to support free ends of floor joists,

studs, or rafters and transfer their load to other parallel joists, studs, or rafters.

Heartwood. The wood extending from the pith to the sapwood, the cells of which no longer participate in the life processes of the tree.

Heel of the rafter. The end or foot that rests on the wall plate.

Hemlock. An inexpensive wood resembling spruce in appearance and used for framing timber.

Herring-bone work. Bricks, tile, wood-flooring or other materials laid in a pattern of rows of short, slanted parallel lines in the direction of the slant alternating row by row.

Hickory. A hard, tough wood used for structural members which require bending.

Hip rafters. Rafters which form the hip of a roof as distinguished from the common rafters. A hip rafter extends diagonally from the corner of the plate to the ridge.

Hip roof. See page 273.

Housed joint. A joint made by cutting out a space in the end of a piece of wood to receive the tongue cut on another piece to which the first piece is to be attached; any fitted joint such as one made with a mortise and tenon.

Housed string. A stair string with horizontal and vertical grooves cut on the inside to receive the ends of the risers and treads.

Hydrostatics. The branch of physics having to do with the pressure and equilibrium of water and other liquids.

I-beam. A steel beam with a cross section resembling the letter "I."

Inlaid linoleum. See page 418.

Insulation—building. See Chapter 20.

Jack rafter. A short rafter of which there are three kinds. (1) those between the plate and a hip rafter; (2) those between the hip and valley rafters; (3) those between the valley rafters and the ridge board. Jack rafters are used especially in hip roofs.

Jalousie doors, jalousie windows. Doors and windows with movable glass louvers adjustable to slope upward to admit light and air yet exclude rain or snow.

Joist. One of a series of parallel beams used to support floor and ceiling loads, and supported in turn by larger beams, girders, or bearing walls.

Joist hangers. A steel or iron stirrup used to support the ends of joists which are to be flush with the girder.

Junction box. See page 387.

Kerfing. The process of cutting grooves or kerfs across a board so as to make it flexible for bending. Kerfs are cut down to about two-thirds of the thickness of the piece to be bent. An example is found in the bullnose of a stair riser which frequently is bent by the process of kerfing.

Keystone. The stone placed in the center of the top of an arch.

Kiln. A large oven or heated chamber for the purpose of baking, drying, or hardening, as a *kiln* for drying lumber, a *kiln* for baking brick, a lime *kiln* for burning lime.

Kiln-dried. A term applied to lumber which has been dried by artifically controlled heat and humidity to a satisfactory moisture content.

Kilowatt. See page 385.

Kilowatt hours. See page 385.

King-post truss. A truss framed with one tie member in the center.

Knee brace. In building construction, a member placed across the inside of an angle in a framework to add stiffness to the frame, especially at the angle between the roof and wall of the building.

Laced valley. In building, a valley formed in a tile roof by interlacing tile-and-a-half tiles across a valley board.

Lacing course. In masonry, a course of brickwork built into a stone wall for bonding and leveling purposes.

Lally column. A cylindrically shaped steel member, sometimes filled with concrete, used as a support for girders or other beams.

Laminate. In home contruction, the building-up with layers of wood, each layer being a lamination or ply; also, the construction of plywood.

Lath—gypsum. See page 347.

Lath—metal. See page 348.

Lath—wood. See page 347.

Lattice. An assemblage of wood or metal strips, rods, or bars made by crossing them to form a network.

Leader. See downspout.

Lean-to. A small building whose rafters pitch or lean against another building or against a wall.

Ledgerboard. The support for the second-floor joists of a balloon-frame house, or for similar uses; ribband.

Ledger strip. A strip of lumber nailed along the bottom of the side of a girder on which joists rest.

Light. Space in a window sash for a single pane of glass. Also, a pane of glass.

Linear measure. See page 410.

Linoleum. See page 418.

Lintel. A piece of wood, stone, or steel placed horizontally across the top of the door and window openings to support the walls immediately above the openings.

Live load. The moving load or variable weight to which a building is subjected, due to the weight of the people who occupy it; the furnishings and other movable objects as distinct from the dead load or weight of the structural members and other fixed loads.

Load-bearing walls. Any wall which bears its own weight as well as other weight.

Lookout. A short wood bracket or cantilever to support an overhanging portion of a roof or the like, usually concealed from view.

Louver. An opening with a series of horizontal slats so arranged as to permit ventilation but to exclude rain, sunlight, or vision. See *Attic ventilators*.

Lumber—dressed size. The dimensions of lumber after shrinking from the green dimension and after planing.

Lumber—matched. Lumber that is edge-dressed and shaped to make a close tongue-and-groove joint at the edges or ends when laid edge to edge or end to end.

Lumber—nominal size. As applied to timber or lumber, the rough-sawed commercial size by which it is known and sold in the market.

Lumber—rough. Lumber as it comes from the saw.

Lumber—shipping-dry. Lumber that is partially dried to prevent stain and mold in transit, and reduce freight costs.

Lumber—structural. Lumber that is two or more inches thick and four or more inches wide, intended for use where working stresses are required.

Lumber—surfaced. Lumber that is dressed by running it through a planer.

Main rafter. A roof member extending at right angles from the plate to the ridge. Same as common rafter.

Male plug. The two and sometimes three-prong connector used to establish an electrical circuit fitting into a receptacle.

Mansard. See page 275.

Maple. A tough, hard wood used in building construction for flooring and veneer.

Marquise (Marquee). The hood or canopy projecting over a carriage or other entrance to a building as a protection from the weather.

Masonite. See *Hardboard*.

Masonry. See Chapter 9.

Mastic. A term applied to a thick adhesive, consisting of a mixture of bituminous preparations such as asphalt and some foreign matter, usually fine sand; used for bedding and pointing window frames, bedding wood-block flooring and structural glass, and for repairing flat roofs.

Meeting rail. The bottom rail of the upper sash and the top rail of the lower sash of a double-hung window. Sometimes called the check rail.

Mezzanine. From the Italian word meaning "middle," a low story between two regular floors.

Millwork. Generally, all building materials made of finished wood and manufactured in millwork plants and planing mills are included under the term "millwork." It includes such items as inside and outside doors, window and door frames, blinds, porchwork, mantels, panelwork, stairways, moldings, and interior trim.

Mineral wool. Insulation—See page 400.

Miter. Two pieces with ends usually cut at 45 degrees to form a miter joint.

Molding:
Base—The molding on the top of a baseboard.
Bed—A molding used to cover the point between the plancier and frieze; also used as a base mold upon heavy work and sometimes as a member of a cornice.
Rake—The cornice upon the gable edge of a pitch roof, the members of which are made to fit those of the molding of the horizontal eaves.
Picture—A molding shaped to form a support for picture hooks often placed at some distance from the ceiling on the wall to form the lower edge of the frieze.

Moisture content of wood. Weight of water contained in the wood, usually expressed as a percentage of the weight of the oven-dry wood.

Mortise. A slot, a notch, a hole, or an opening cut into a board, plank, or timber, usually edgewise, to receive tenon (projecting piece) of another board, plank, or timber, shaped to fit, to form a joint.

Mullion. A slender bar or pier forming a division between panels or units of windows, screens, or similar frames. Often confused with muntin.

Muntin. The vertical member between two panels of the same piece of panel work. The vertical sash-bars separating the different panes of glass.

Neat cement. In masonry, a pure cement uncut by a sand admixture.

Neat plaster. A term applied to plaster made without sand.

Newel (newel post). Any post to which a railing or balustrade is fastened.

Nosing. The rounded edge of a board or step.

O.C.—(on center). The measurement of spacing for studs, rafters, joists, and the like in a building from center of one member to the center of the next member; i.e., 16"o.c.-24" o.c.

Offsets. When the face of a wall grows higher and thinner, the jogs are called offsets.

O.G.—Ogee. A molding with a profile in the form of a letter "S" having the outline of a reversed curve.

Open-string stairs. Stairs which are so constructed that the ends of the risers and treads are visible from the side.

Open wiring (knob and tube). Electric wires fastened to surfaces by the use of porcelain knobs; a circuit supported by insulators; wiring which is not concealed.

Outlookers. A cantilever to build the cornice on or to hang the cornice to.

Outriggers. A projecting beam used in connection with overhanging roofs. A support for rafters in cases where roofs extend two or more feet beyond the walls of a house.

Overhanging eaves. A type of roof in which the rafters and roofing extend two or more feet beyond the exterior face of a building.

Overhead door. A door which may either be mounted on a sliding track or pivoted canopy frame which moves upward to an overhead position when opened.

Panel. A large, thin board or sheet of lumber, plywood, or other material.

Panel box (service panel or panel board). A box in which electric switches and fuses for branch circuits are located.

Panel heating. A method of heating a building by using heating units or coils of pipes concealed in special panels, or built in the wall or ceiling plaster. Also called concealed heating.

Parapet wall. That portion of any exterior wall, party wall, or fire wall which extends above the roof line.

Particleboard. A rigid board made of wood chips and an adhesive (glue) formed under hot pressure into various thicknesses.

Parting stop or strip. A small wood piece used in the side and head jambs of double-hung windows to separate upper and lower sash.

Party wall. A wall built upon the dividing line between adjoining buildings for their common use.

Patio. An open area—usually paved—for outdoor living. It may be entirely or partially surrounded by parts of the house.

Pebble dash. In the building trade, a term used for finishing the exterior walls of a structure by dashing pebbles against the plaster or cement.

Perimeter heating. A system of heating in which ducts radiate from a central plenum chamber and release warm air through registers located along the outer walls.

Perlite. See *volcanic glass.*

Pier. The part of a wall between windows and doors; any detached mass of masonry to support an arch, girder, or column; a heavy column used to support weight.

Pilaster. A flat, square, or rectangular column attached to the wall and projecting from it to reinforce or strengthen the wall it attaches with or is a part of.

Pile. Timber or concrete shafts sunk into soft ground upon which foundations are built.

Pitch. See page 267.

Plank. Material two or three inches thick and more than four inches wide. Often used for scaffolding.

Plaster. See page 350.

Plate. (1) A horizontal structural member placed on a wall or supported on posts, studs, or corbels to carry the trusses of a roof or to carry the rafters directly. (2) A shoe or base member as of a partition or other frame. (3) A small, relatively flat member placed on or in a wall to support girders, rafters, etc.

Plate glass. See page 342.

Plumb. Exactly perpendicular; vertical.

Ply. A term to denote the number of thicknesses or layers of roofing felt, veneer in plywood, or layers in built-up materials, in any finished piece of such materials.

Plywood. A piece of wood made of three or more layers of veneer joined with glue and usually laid with the grain of adjoining plies at right angles. Almost always an odd number of plies are used to provide balanced construction.

Pointing. A term used in masonry for finishing of joints in a brick or stone wall.

Polystyrene. A thermoplastic polymer, an excellent insulator in board and in foam.

Portal. The arch over a door.

Portland cement. Hydraulic cement made by heating a mixture of limestone and clay in a kiln and pulverizing the resulting clinker.

Post. Generally, any vertical piece whose function is to sustain a vertical load.

Prefabricated modular units. Units of construction which are prefabricated on a measurement variation base of 4" or its multiples, and can be fitted together on the job with a minimum of adjustments. Modular units include complete window walls, kitchen units complete with installations as well as masonry, wall panels, and most of the other components of a house. Units are usually designed in such a way that they will fit functionally into a variety of house sizes and plan types.

Preservative (Wood). Any substance that, for a reasonable length of time, will prevent the action of wood-destroying fungi, borers of various kinds, and similar destructive life when the wood has been properly coated or impregnated with it.

Prestressed concrete. Prestressing is the imposition of preliminary interval stresses in a structure before working loads are applied, in such a way as to lead to a more favorable state of stress when the loads come into action.

Primer. The first coat of paint in a paint job that requires two or more coats.

Pulley stile. The member of a window frame which contains the pulleys and between which the edges of the sash slide.

Purlin. A piece of timber laid horizontally to support or strengthen the common rafters of a roof; also laid on roof trusses.

Quarry tile. In masonry, a name given to machine-made, unglazed tile.

Quarter round. A molding that presents a profile of a quarter circle.

Quonset hut. A type of hut or shelter semicircular in shape, made of corrugated metal.

Rabbet. A rectangular longitudinal groove cut in the corner of a board or other piece of material.

Radiant heating. A method of heating usually consisting of coils or pipes placed in the floor, wall or ceiling.

Radiate. To emit heat or light radially as the spokes of a wheel.

Rafter. One of a series of structural members of a roof designed to support roof loads. The rafters of a flat roof are sometimes called roof joists.

Rake. The trim members that run parallel to the roof slope and form the finish between wall and gable roof.

Raked joint. In brick masonry, a type of joint which has the mortar raked out to a specified depth while the mortar is still green.

Red oak. A coarse-grained wood, dark in color, used for interior trim in buildings.

Reflective insulation. Sheet material with one or both surfaces of comparatively low heat emissivity that, when used in building construction so that the surfaces face air spaces, reduces the radiation across the air space. This type of insulation also acts as a vapor barrier.

Reinforcing. Steel rods or metal fabric placed in concrete slabs, beams, or columns to increase their strength. (See page 128.)

Resin-emulsion paint. Paint, the vehicle (liquid part) of which consists of resin or varnish dispersed in fine droplets in water.

Resins. One of the many viscous substances of plant origin such as copal, rosin, and amber used in lacquers, varnishes, synthetic plastics, adhesives, etc.

Return. The continuation of a molding, projection, etc. in an opposite direction.

Reveal. The vertical sides of an opening between the front of the wall and the frame; used also to designate the return of a pilaster or pier, between the face of the pilaster or pier and the main wall face.

Ribbon. A narrow board let into the studding to add support to joists.

Ridge. The horizontal line at the junction of the top edges of two sloping roof surfaces. The rafters at both slopes are nailed at the ridge.

Ridge board. The board placed on edge at the ridge of the roof to support the upper ends of the rafters.

Ridge course. The last or top course of slates or tiles on a roof, cut to length as required.

Ridge roll. A strip of sheet metal, composition roofing, tiling, or wood used to cover and finish a roof ridge.

Rigid conduit. See page 390.

Rise and run. See page 267.

Rise of roof. See page 267.

Riser. Each of the vertical boards closing the spaces between the treads of stairways.

Rock wool. See mineral wool. Page 400.

Roll roofing. See page 295.

Roof sheathing (roof decking). See *Sheathing.*

Roof truss. See page 183.

Rough coat. The first coat of plastering applied to surface.

Rough hardware. All of the concealed hardware in a house or other building, such as bolts, nails, and spikes, which are used in the construction of the building.

Roughing-in. In building, a term applied to doing the first or rough work as roughing-in plumbing; installing pipes for water, sewage and drainage up to the plumbing fixtures.

Rough opening. An unfinished window or door opening; any unfinished opening in a building.

Rowlock. In masonry, a term applied to a course of bricks laid on edge. Also, the end of a brick showing on the face of a brick wall in a vertical position.

Rubber-emulsion paint. Paint, the vehicle of which consists of rubber or synthetic rubber, dispersed in fine droplets in water.

Rubble work. Masonry of rough, undressed stones.

Run. See page 267.

Sanitas. A wall covering somewhat similar to lightweight oilcloth, manufactured and sold under the trade name of Sanitas.

Sash. The framework which contains and holds the glass of a window.

Sash balance. A device, usually operated with a spring, designed to counterbalance window sash. Use of sash balances eliminates the need for sash weights, pulleys, and sash cord.

Saturated felt. (See *Building Paper*).

Scaffold or staging. A temporary structure or platform enabling workmen to reach high places.

Scotia. A hollow molding used as a part of a cornice and often under the nosing of a stair tread.

Scratch coat. The coat of plaster which is scratched to form a bond for the next coat.

Screed coat. In plastering, a coat of plaster laid level with the screeds.

Screeds. Narrow strips of plaster put on a wall as guides for the workmen. The strips usually are about 8 inches wide with a thickness of two coats of plaster, serving also as thickness guides when applying the remainder of the plastering; more often, a strip of wood to act as a guide for plaster or concrete work.

Scribing. Fitting woodwork to an irregular surface.

Scuttle hole. A framed opening with its cover through the roof or ceiling.

Sealer. A finishing material, either clear or pigmented, that is usually applied directly over uncoated wood for the purpose of sealing the surface.

Seasoning. Removing moisture from green wood.

Septic tank. In plumbing, a tank in which sewage is kept to effect disintegration of organic matter by natural bacterial action which dissolves most of the solids into liquids and gases within about 24 hours.

Service drop (electrical). See page 393.

Service entrance. The place where the service wires are run into a building.

Service panel. (electrical). The main distribution box in a building from which the many circuits distribute electrical service to various outlets.

Shake. A handsplit shingle, usually edge-grained.

Sheathing. In construction work, a term usually applied to surface materials or boards nailed to studding or roofing rafters, as a foundation for the covering of the outer surface of the side walls or roof of a house.

Sheathing paper. See *Building Paper.*

Shellac. A transparent coating made by dissolving lac, a resinous secretion of the lac bug (a scale insect that thrives in tropical countries, especially India), in alcohol.

Shingles. Roof covering of asphalt, asbestos, wood, tile, slate, or other material cut to stock lengths, widths, and thicknesses.

Shiplap. In carpentry, a term applied to lumber that is edge-dressed to make a close rabbeted or lapped joint.

Shore. A piece of timber usually placed in an inclined position to support a building, roof or wall temporarily while it is being repaired or altered.

Siding. See page 248.

Siding (bevel). See page 248.

Sill. The lowest member of the frame of a structure, resting on the foundation and supporting the uprights of the frame. The member forming the lower side of an opening, as a door sill, window sill, etc.

Sill cock. Water faucet on the outside of the house to which a hose may be connected.

Siphon trap. In plumbing, a trap fitted to water closets and sinks, having a double bend like an "S" , the lower bend containing the water seal which prevents the reflux of foul odors and gases.

Sizing. Working material to the desired size. A coating of glue, shellac, or other substance applied to a surface to prepare it for painting or other methods of finish.

Skirt. In architecture, the border or molded piece under a window stool commonly called apron; also, a baseboard. The same as skirting.

Skylight. A frame supporting glass sash, placed in a roof to light a passage or rooms below.

Slag. A vitreous mass as a byproduct of smelting.

Slag strip. In roofing, a strip of wood or metal nailed around the edges of a graveled roof to give the edge a finish and to prevent the gravel from rolling off the roof. Also called gravel strip.

Slate shingles. See page 293.

Sleeper. Timber laid on the ground to receive joists; or pieces of wood imbedded in concrete to fasten finished floors to.

Smoke shelf. A ledge in the front lower part of a chimney that prevents cold air from flowing down the chimney into the fireplace.

Soffit. The underside of the members of a building such as staircases, cornices, beams, and arches, relatively minor in area as compared with ceilings.

Soil pipe (soil stack). A vertical drain pipe conveying waste matter from a water closet to the drainage system of a building.

Span. The distance between structural supports such as walls, columns, piers, beams, girders, and trusses.

Specification. A description of the kind, quality and quantity of materials and workmanship that are to govern the fabrication and erection of a building or other construction.

Spread footing. A footing whose sides slope gradually outward from the foundation to the base.

Splitshakes. In carpentry, a type of shingle split by hand.

Square. A unit of measure—100 square feet.

Stain. A form of oil paint, very thin in consistency, intended for coloring wood without forming a coating of significant thickness or gloss.

Stair landing. A platform between flights of stairs or at the termination of a flight of stairs.

Starter strip. In roofing, the first strip of composition roofing material applied to a roof.

Stepped footings. If a house is built on sloping ground, the footings cannot all be at the same depth; hence, they are stepped.

Stile. The upright piece in framing or paneling. One of vertical members in a door or sash into which secondary members are fitted.

Stool (window). The flat, narrow shelf forming the top member of the interior trim at the bottom of a window. The base or support of a window as the shelf-like piece inside and extending across the lower part of the opening; sill.

Storm sash or storm window. An extra window usually placed on the outside of an existing window as additional protection against cold weather.

Stretcher. A brick or block of masonry laid lengthwise of a wall.

String, stringer. A timber or other support for cross members. In stairs, the support on which the stair treads rest; also, springboard.

Strut. In carpentry, any piece fixed between two other pieces to keep them apart, as a member which is designed to resist pressure or compressive stress endwise in a frame or structure.

Stucco. Most commonly refers to an outside plaster made with portland cement as its base.

Stud. One of a series of slender wood or metal structural members placed as supporting elements in walls and partitions. Usually 2" × 4" in standard dwelling construction. (Plural: studs or studding.)

Subfloor. Boards or sheet material laid on joists over which a finish floor is to be laid.

Sump. A hole or depression to collect water, usually in the basement floor. A pump on the sump automatically keeps it emptied.

Surface. To make plane and smooth.

Sweat joints. In plumbing, a type of joint made by the union of two pieces of copper pipe which are coated with solder containing tin. The pipes are pressed together and heat applied until the solder melts.

Tail beam. A relatively short beam or joist supported in a wall on one end and by a header on the other.

Templet. A form to lay out work; piece of timber or stone to distribute pressures over a larger area. When applied to steel, the templets are known as "bearing plates" or "slabs" as the case may be.

Termites. Insects that superficially resemble ants in size, general appearance, and habit of living in colonies; hence, they are frequently called "white ants." If unmolested, they eat out woodwork, leaving a shell of sound wood to conceal their activities, and damage may proceed so far as to cause collapse of parts of a structure before discovery.

Termite shield. A shield, usually of noncorrosive metal, placed in or on a foundation wall or other mass of masonry or around pipes to prevent passage of termites.

Terra cotta. Baked clay of a fine quality.

Terrazzo. A type of Venetian marble mosaic in which portland cement is used as a matrix. Though used in buildings for centuries, terrazzo is a modern floor finish used also for bases, borders, and wainscoting, as well as on stair treads, partitions, and other wall surfaces.

Thermostat. An electric instrument that controls heat and air-conditioning temperature.

Threshold. A strip of wood or metal beveled on each edge and used above the finished floor under outside doors. On interior doors sometimes called a carpet strip.

Timbers. Lumber at least five or more inches in dimension.

Toenailing. To drive a nail at a slant with the initial surface in order to permit it to penetrate into a second member.

Tongue. A projection of a board or other material to be inserted in a groove.

To the weather. In building, a term applied to the projecting of shingles or siding beyond the course above; that part of a shingle, siding, etc., exposed to the elements.

Transom. The bar or horizontal construction which divides a window. More commonly applied to the sash over the door.

Tread. The horizontal board in a stairway on which the foot is placed.

Trim. The finish materials in a building such as moldings applied around openings (window trim, door trim) or at the floor and ceilings of rooms (baseboard, cornice, picture molding).

Trimmer. The beam or joist into which a header is framed.

Truss. See page 183.

Undercoat. A coating applied prior to the finishing or top coats of a paint job. It may be the first of two or the second of three coats. In some usage of the word, it may become synonymous with priming coat.

Valley. The internal angle formed by the junction of two sloping sides of a roof.

Valley rafter. The rafter immediately under the valley and to which the jack rafters or purlins connect.

Vapor barrier. See page 407.

Veneer. Thin sheets of wood or other materials.

Vent. A pipe installed to provide a flow of air to or from a drainage system or to provide a circulation of air within such systems to protect trap seals from siphonage and back pressure.

Verge boards. The boards which serve as the eaves finish on the gable end of a building.

Vermiculite. A mineral closely related to mica with the faculty of expanding on heating to form lightweight material with insulation quality. Used as bulk insulation and also as aggregate in insulating and acoustical plaster and in insulating concrete floors.

Wainscoting. Matched boarding or panel work covering the lower portion of a wall.

Wallboard. Wood pulp, gypsum, or other materials made into large, rigid sheets that may be fastened to the frame of a building to provide a surface finish.

Warp. To bend, or turn from a straight line; a piece of lumber, when improperly seasoned, may become curved, twisted or turned from a straight flat form; a defect or permanent distortion of a timber from its true form, caused usually by exposure to heat or moisture.

Water-repellent. A liquid designed to penetrate into wood and to impart water-repellency to the wood.

Water table. A ledge or offset on or above a foundation wall for the purpose of shedding water. Level of water below ground level.

Weatherstrip. Narrow strips made of metal or other material so designed that when installed at doors or windows, they will retard the passage of air, water, moisture, or dust around the door or window sash.

Weep hole. In retaining walls and other similar structures, a small hole through which surplus water drains to the outside; hence, preventing damage to the wall by pressure of accumulated water back of or under the structure.

Western framing. In building construction, a method of framing where the studding of each story rests on a sort of sill, as opposed to the balloon type of framing; also called platform framing.

Wind. A term used to describe the surface of a board when twisted (winding) or when resting upon two diagonally opposite corners, if laid upon a perfectly flat surface.

Wiped joint. In plumbing, a joint formed between two pieces of lead pipe. Molten lead is poured upon the joint until the two pieces of pipe are of the right temperature. The joint is then wiped up by hand with a moleskin or cloth pad while the solder is in a plastic condition.

Wire-cut brick. Brick formed by forcing plastic clay through a rectangular opening designed for the purpose and shaping the clay into bars. Before burning, wires pressed through the plastic mass cut the bars into uniform brick lengths.

Wire glass. In building construction, a type of window glass in which wire with a coarse mesh is embedded to prevent shattering of glass in case it is broken; also, to protect a building against intruders.

Wire mesh reinforcing. See page 127.

Wooden brick. Piece of seasoned wood made the size of a brick and laid where it is necessary to provide a nailing in masonry walls.

Zonolite concrete. A form of concrete which acts as insulation; used as parts of floor slabs for houses without basements.

BIBLIOGRAPHY

RE: BUILDING CONSTRUCTION AND ESTIMATING

Adjustment of Property Losses
By Thomas-Reed, 1977
McGraw-Hill Book Company, Inc.
New York, N.Y.

Building Construction Estimating
By Alonzo Wass
Prentice-Hall, Inc.
Englewood Cliffs, N.J.

Building Construction Simplified
By Cooper and Badzinski
McGraw-Hill Book Company, Inc.
New York, N.Y.

Building Estimator's Reference Book 1980
By Frank R. Walker
Frank R. Walker, Publishers
Chicago, Ill.

Condominium Insurance Coverage
The late R.L. Lusk, former
Consultant to Property Loss Research Bureau,
Chicago, Ill.

Construction Estimates and Costs
By H.E. Pulver
McGraw-Hill Book Company, Inc.
New York, N.Y.

Construction Industry Production Manual
By Taylor Winslow
Craftsman Book Company of America, Publishers
Los Angeles, California

Electrical Construction Wiring
By Walter N. Alerich
American Technical Society, Publishers
Chicago, Ill.

Estimating For the Building Trades
  By Joseph Steinberg and Martin Stempel
  American Technical Society, Publishers
  Chicago, Ill.

Handbook of Tables and Formulas for Home Construction Estimating
  By Paul I. Thomas
  Prentice-Hall, Inc.
  Englewood Cliffs, N.J.

House Wiring Simplified
  By Floyd M. Mix
  The Goodheart-Willcox Company, Inc.
  Chicago, Ill.

How To Build and Buy Cabinets For the Modern Kitchen
  By Robert P. Stevenson
  Arco Publishing Company, Inc.
  New York, N.Y.

Materials of Construction
  By Professor Albert P. Mills
  John Wiley & Sons, Inc.
  New York, N.Y.

National Construction Estimator
  Craftsman Book Company
  Los Angeles, California

Practical Estimating For Painters and Decorators
  By William P. Crannel
  Frederick J. Drake & Company, Publishers
  Wilmette, Ill.

Practical Electric Wiring
  By H.P. Richter
  McGraw-Hill Book Company, Inc.
  New York, N.Y.

Wood Frame House Construction
  By L.O. Anderson, Engineer
  U.S. Department of Agriculture
  Agricultural Handbook No. 73, July 1970

American Plywood Association, 1119 A, Tacoma, Wa.
  Plywood Underlayment
  Plywood for Floors
  Plywood Siding

Asphalt Roofing Manufacturers Association, New York, N.Y.
  Manufacture, Selection and Application of Asphalt Roofing and Siding Products, 1974

American Insurance Association
    Blasting Claims—A Guide For Adjusters

Brick Institute of America, 1750 Meadow Road, McLean, Va.
    Technical Notes Nos. 10 and 10B

Painting and Decorating Contractors of America
    7223 Lee Highway, Falls Church, Va.
    Estimating Guide, 1974

Red Cedar Shingle and Handsplit Shake Bureau, Seattle, Wa.
    Certi-Split Manual of Handsplit Red Cedar Shingles

United States Gypsum, Dept. 122, 101 S. Wacker Drive, Chicago, Ill.
    Drywall Construction Handbook

Western Wood Products Association, Portland, Ore.
    Products Use Manual

Western Red Cedar Lumber Association, Portland, Ore.
    Bevel Siding

RE: LOSSES OF SUSPICIOUS ORIGIN

Anti-Action Arson Plan
    National Committee on Property Insurance,
    Boston, Massachusetts

Adjustment of Property Losses
    Thomas-Reed, 4th Edition, Gregg
    Division, McGraw-Hill Book Co.

Arson: A Multifaceted Problem
    Prepared for the Federal
    Emergency Management Agency, U.S.
    Fire Administration, by Public Technology,
    Inc. Washington, D.C. 1980

Arson Information Kit
    Alliance of American Insurers, 20 North
    Wacker Drive, Chicago, Ill. 60606

Arson Resource Directory
    Arson Resource Center, Office of
    Planning and Education, United
    States Fire Administration,
    Washington, D.C. 20472

Could it be Arson? How Can You Tell
    Property Claims Service,
    (PCS), American Insurance Association
    (AIA), Bulletin No. 23, Arson.

Fire—Is It Arson?
    By John S. Barracato, Aetna Life & Casualty.

Insurance Fraud
    ICPI Handbook For Insurance Personnel,
    Insurance Crime Prevention Institute,
    15 Franklin St., Westport, Conn. 06880
    James F. Ahern, Director.

Model Arson Penal Law
    Alliance of American Insurers, American
    Insurance Association, National
    Association of Independent Insurers.
    (AAI Fact Sheet #5)

Property Insurance Loss Register
    American Insurance Association

Model Arson Reporting-Immunity Law
    Alliance of American
    Insurers, Fact Sheet #6

Procedural Handling of the Arson or Fraud Fire Claim
    Newsletter
    No. 11, By Stephen T. Carmick,
    Attorney, Chicago, Ill.

Reclassifying Arson as a Part I Offense
    Alliance of American
    Insurance, Fact Sheet #2

The Role of An Attorney in the Investigation and Defense of an Arson Loss
    By John E. Lecomte, Attorney,
    Lecomte, Shea, Dangora and Barber,
    Boston, Mass.

Status Report No. 1, on Arson Reporting-Immunity Legislation
    By The Insurance Committee For Arson
    Control.

The Torch's Reward
    Illinois Advisory Committee on Arson
    Prevention. P.O. Box 614, Bloomington,
    IL 61701

Was It Arson?
    By R.G. Provencher, National Director Investigations
    Division, General Adjustment
    Bureau, Inc.

INDEX